SPACE GROUPS *FOR* SOLID STATE SCIENTISTS

Gerald Burns (1933 – 1991)

SPACE GROUPS *FOR* SOLID STATE SCIENTISTS

Third Edition

GERALD BURNS*
IBM Thomas J. Watson Research Center, Yorktown Heights, New York

A.M. GLAZER
Emeritus Fellow of Jesus College, Oxford and Emeritus Professor of Physics at Clarendon Laboratory, Oxford
Visiting Professor of Physics, University of Warwick

*deceased 1991

ELSEVIER

Amsterdam • Boston • Heidelberg • London
New York • Oxford • Paris • San Diego
San Francisco • Singapore • Sydney • Tokyo
Academic Press is an imprint of Elsevier

Academic Press is an imprint of Elsevier
225, Wyman Street, Waltham, MA 02451, USA
The Boulevard, Langford Lane, Kidlington, Oxford OX5 1GB, UK
Radarweg 29, PO Box 211, 1000 AE Amsterdam, The Netherlands

British Library Cataloguing in Publication Data
A catalogue record for this book is available from the British Library

Library of Congress Cataloging-in-Publication Data
Glazer, A. M. (Anthony Michael)
Space groups for solid state scientists/Michael Glazer, Gerald Burns. 3rd ed. p. cm.
Previous ed. by Gerald Burns. Includes bibliographical references and index.
ISBN 978-0-12-810061-5 (alk. paper)
1. Solid state physics. 2. Space groups. I. Burns, Gerald, 1932-II.
Title. QC176.B865 2013
530.4'1–dc23

2012022097

Printed and bound in China
13 14 15   10 9 8 7 6 5 4 3 2 1

CONTENTS

PREFACE

Hi Mike,

One of the things that I was looking forward to was being in Oxford in Sept–Oct and writing that book with you. Not only do I enjoy living in College, but enjoy working with you. And it is such a good learning process.

Well, I can't promise that I can make it. All of a sudden I've become aware that I am very sick, in fact, with all of the tests behind me, I'm headed to the National Cancer Institute in Bethesda, Md. tomorrow. Probably will be taken into a new immunotherapy treatment. From there, we have to hope.

It is not too bad on the body so I'm sort of in for about 5 days and out for a few. We shall see what happens.

Sorry for the bad news. But... What are you up to this summer?

My very best, Gerry

I received this email in July 1991 in which I learnt that my friend and co-author Gerry Burns would not be coming to Oxford: we had planned to write another book together and this news came as a total shock. Sadly, within a couple of months Gerry died. It has taken until now for me to feel able to consider a third revision of our book "Space Groups for Solid State Scientists". It was originally published in 1978, and used as its basis the 1952 Edition of the International Tables for Crystallography Volume I (*IT52*). The second edition, published in 1990, introduced the then International Tables Volume A (*ITA*) as a new addition to the space group Tables. Now that the *ITA* has become firmly established, with many important and new topics, it seems to me that a third edition based around the *ITA* would now be of appropriate interest.

The International Tables for Crystallography are published under the auspices of the International Union of Crystallography (IUCr) and are the culmination of over two centuries of scholarship by a very large number of people devoted to the understanding of symmetry in crystals. Beginning in the seventeenth and eighteenth centuries[1] with the work of such luminaries

[1] Good historical accounts of the development of crystal symmetry can be found in the book "Historical Atlas of Crystallography" by J. Lima-de-Faria (1990), Kluwer (IUCr) and by H. Kubbinga. *Acta Cryst.*, **A68**, 3 (2012).

as Johannes Kepler (1571–1630), Robert Boyle (1627–1691), Robert Hooke (1635–1703), Bishop Nicolas Steno (1638–1686), Maurice Capeller[2] (1685–1769), Jean-Baptiste Louis Romé de Lisle (1736–1790) and, most importantly at the time, the Abbé René-Just Haüy (1743–1822), it came to be realized that the defining feature of crystals is their high degree of symmetry. Of particular importance was the notion that within crystals there were repeating units of polyhedral blocks, Haüy's "molecules intégrantes". Today we call this repetition *translational symmetry* and these blocks "unit cells". Haüy also was probably the first person to identify the idea of a symmetry element. The nineteenth century saw an enormous increase in interest in crystals, particularly in France, where their molecular nature tended to be of prime interest, and in Germany, where the focus was more on the symmetry of crystals. Johann Friedrich Christian Hessel (1796–1872) derived the 32 crystal classes in 1830. The idea of a unit cell was devised by Gabriel Delafosse (1796–1878) in 1840 in order to describe the concept of repetition. The works of Moritz Ludwig Frankenheim (1801–1869) and of Auguste Bravais (1811–1863) were responsible for the realization that translational symmetry was describable in terms of point lattices and that there was only a finite number of unique lattice types within any particular spatial dimension: we call them Bravais lattices today. Sixty-five space groups were identified by Leonhard Sohncke (1842–1897) who defined a regular system of points by identical lines drawn from each lattice point to all the others (he originally found 66 space groups, but two of them were eventually found to be equivalent). Even Louis Pasteur (1822–1895) in France made an important contribution to the subject by his realization that optical rotation was connected with, what today, we call *molecular chirality*, and that such symmetry was essential for living organisms to exist, an idea that presaged the field of modern genetics 100 years later! Then, in 1891 in Germany, Artur Moritz Schoenflies (1853–1928), by considering the connection between point symmetries and lattices, published his theory of the 230 space groups and developed his own notation for them, this notation still being in common use today. It is fair to say that slightly before, Evgraf Stepanovich Fedorov (1853–1919), working in Russia, had developed the space groups separately. More or less contemporaneously, the amateur geologist William Barlow (1845–1934) in England also discovered the 230 space groups. He also suggested different ways of packing spheres in order to explain particular crystal forms and chemical constitution.

2 A Swiss physician, who was the first person to introduce the term 'crystallography'.

All this was of theoretical interest at the time, with little direct experimental evidence for the existence of translational symmetry in crystals. This had to wait for the discovery of X-rays in 1895 by Wilhelm Conrad Röntgen (1845–1923). Hundred years ago (1912) Max Theodor Felix von Laue (1879–1960), Paul Karl Moritz Knipping (1883–1935) and Walter Friedrich (1883–1968) demonstrated for the first time that crystals could diffract X-rays, thus pointing simultaneously to its wave nature and to the high degree of symmetry present within a crystal at the atomic level. It seems that others had tried this earlier but had failed to observe the effect. Laue and co-workers were in fact very lucky with their experiment, for they originally thought that the X-rays were monochromatic (had this been true it is doubtful that they would have observed any diffraction using a stationary crystal). In the same year father and son, William Henry Bragg (1862–1842), and, especially William Lawrence Bragg (1890–1971), realized that if the X-rays used by Laue were polychromatic then the observed diffraction patterns could be used to determine the actual atomic arrangements in crystals. Thus began the modern era of X-ray Crystallography which today is capable of solving the structures of the most complex materials, especially those of biological importance.

Now, an integral part of solving and understanding crystal structures is the means by which crystals can be classified according to their space group symmetries. While the original Schoenflies notation was adopted widely for spectroscopic and other types of research by chemists and physicists, another nomenclature was developed by Carl Hermann (1898–1961) and Charles Victor Mauguin (1878–1858), the Hermann–Mauguin notation, published for use by X-ray crystallographers in 1935, and revised in 1944. With the establishment of the IUCr in 1947[3], a new set of space group Tables was published in 1952 (and revised in 1965 and 1969) edited by Kathleen Lonsdale, née Yardley (1903–1971) and by Norman Fordyce McKerron Henry (1909–1983). These Tables remained the staple diet for crystallographers until 1983, when the new International Tables Volume A (*ITA*) were produced under the editorship of Theo Hahn. This has been followed by several new editions incorporating new concepts in 1987, 1992, 1995 and most recently in 2002. A new edition is expected to be published in 2013, edited by Mois Ilia Aroyo. In addition, two companion sets of Tables are currently available from the IUCr: International Tables for Crystallography Volume A1 (*ITA1*) on "Symmetry relations between space groups"

3 see http://www.iucr.org/iucr/history/early-history.

and Volume E (*ITE*) on "Subperiodic groups". These are available both as printed books and online (at http://it.iucr.org/resources/).

While all this has become standard knowledge to many in the crystallography community, it is true to say that few practicing crystallographers have complete familiarity with space groups and their notations. The situation is even worse for those in other relevant scientific disciplines, such as in chemistry, physics, materials science, mineralogy, biology, etc. These solid-state scientists need at some time to be able to handle crystal symmetry as part of their research. Topics such as electronic bands, phonon dispersion, Raman and IR spectroscopy, and just about any science to do with crystalline materials (and that is a lot!) cannot be properly treated without a good grounding in crystal symmetry. For many scientists, it is vital that they understand how a crystal structure is described by crystallographers if they are to make sense of scientific publications. The International Tables with its rich content is poorly understood or appreciated by solid-state scientists in general, and it is for this reason that Gerry Burns and I wrote the original first edition of this book.

The present version of Burns and Glazer has been revised, and re-organized, to bring it up to date with the latest material in the *ITA*. Since the earlier editions, there has been the introduction of a new symmetry operation, the double glide, given the symbol *e*, so that, for instance, the former space group *Cmma* (number 67 in the Tables) is now called *Cmme*. Most crystallographic computer programs have yet to catch up with this change. The *ITA* contains, in addition to the Tables themselves, a great deal of new information. In this book, I have added a section on Euclidean normalizers, which I hope readers will find useful as a way to compare crystal structural information. Some of the new concepts, such as crystal families, Bravais flocks, etc., have been included for reference. A major change is the inclusion of a whole chapter devoted to antisymmetry and its importance in describing magnetic structures. This coincides with the availability of a new set of Tables by Dan Litvin that can be freely downloaded from the internet and which records all the magnetic space groups in the style of the *ITA*. In addition, he has added space group operators in the form of Seitz operators specified from a common origin, as used in this book, and it is possible that this will eventually be included as a useful addition to a later edition of the *ITA*.

Finally, I would like to thank Dan Litvin for keeping me on the straight and narrow where magnetic symmetry is concerned, a topic that was new to me prior to this revision. I am indebted to Vinod Wadhawan for his critical and incisive reading of the manuscript for this book, and to Pamela Thomas for suggesting the inclusion of the structure of $KTiOPO_4$ as an example of a $Z' > 1$ inorganic structure. Thanks are due also to Maureen Julian for sending useful comments. Jens Kreisel and Dean Keeble have also helped me to make several corrections to the original manuscript. I am also grateful to Peter Strickland of the IUCr for his help in giving me access to the online editions of the International Tables and for permission to use a number of space group diagrams. Finally, I wish to express my indebtedness to Mohanapriyan Rajendran and his team who did such complex and accurate typesetting for this book.

Mike Glazer
Oxford & Warwick (2012)

CHAPTER 1

Point Symmetry Operations

Contents

The mathematical sciences particularly exhibit order, symmetry, and limitation; and these are the greatest forms of the beautiful.

Aristotle, 384–322 BC

WHAT IS SYMMETRY?

We all instinctively recognize symmetry when we see it, but in order for us to make use of it, we shall need to define it reasonably rigorously. We can define it simply by the following statement:

> *Symmetry is that property possessed by an object that, when transformed in some way (e.g. by rotation, inversion, repetition etc.), looks the same after as before the transformation. In other words symmetry is a demonstration of the invariance of an object to some sort of transformation.*

The fact that certain objects possess symmetry is a common expression. We might say a pencil 'has symmetry along its long axis' or a human being 'has symmetry through a bisecting plane'. In this chapter, we wish to make clear what is meant when we say an object possesses certain symmetry, and furthermore, we should like to develop a notation for these symmetry operations.

In this book, we shall be dealing mainly with symmetry in crystals, rather than in other everyday objects. Thus, before going on to deal with symmetry operations, we need to appreciate what is meant when we use the

Space Groups for Solid State Scientists
http://dx.doi.org/10.1016/B978-0-12-394400-9.00001-0

word **crystal**. On a macroscopic scale, we may define a crystal as a solid of uniform chemical composition which is formed with plane faces each making precise angles with one another. This is not a rigorous definition, and in fact, it can only be made so by considering the microscopic nature of crystals. The important aspect that makes crystals different from all other objects is that they are solids that consist of atoms or groups of atoms repeated regularly in three dimensions. This regular repetition of groups of atoms is a form of symmetry known as **translational symmetry**, a subject to be dealt with at great length in the chapters that follow. For the moment, however, we shall just deal with the problem of point symmetry.

The word 'object' in our definition usually refers to something material that we can see, but in fact it can refer to a more abstract concept such as time or even music. Something that needs to be borne in mind is that symmetry, unlike most other properties, is in fact discrete. You cannot have something existing in two symmetry types at the same time: it has to be one or the other. There is no such thing as fractional symmetry! To illustrate this, consider a spherical ball falling towards the ground, as in Fig. 1.1a.

Now, assuming that gravity is not actually distorting this ball, we all understand that this object is spherically symmetric. The symmetry of a sphere is such that it is invariant to any rotation or inversion operations: in other words, the ball looks the same from every point of view. Of course in reality, the ball is likely to have some blemishes or other surface markings and so we could then see if it were rotating. But we are assuming an ideal world here.[1]

Anyway, suppose that this ball now contacts the ground (Fig. 1.1b). The force of impact will then cause the spherical shape to become an oblate spheroid. This has a different symmetry from that of the original sphere: it has circular symmetry when viewed from above but elliptical symmetry when viewed from any other direction. At no time, can we say that the ball has spherical symmetry and at the same time oblate spheroidal symmetry: it is one or the other. This is an example of what is known in scientific terminology as **symmetry breaking**. So the ball starts with spherical symmetry and then at an instant of time the spherical symmetry is broken to place the object in a different symmetry state. You should be able to see from this example that, in going from the spherical to the oblate spheroidal state,

[1] Note however that an atom in its ground state is really perfectly spherically symmetric, provided it is free from external fields, and so it is impossible to tell if it is rotating. From the perspective of quantum theory it is *not* rotating!

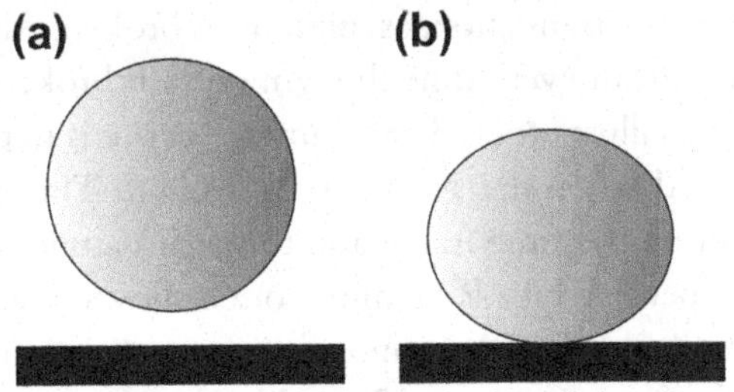

Figure 1.1 Change in shape and symmetry of a spherical ball as it falls onto a flat surface. An example of symmetry breaking.

some elements of the symmetry have been lost. The sphere is invariant to rotations about any axis we choose through its center, whereas the oblate spheroid is invariant to all rotations about the vertical axis. When we come on to study the idea of groups, we shall see that the breaking of the symmetry in this case has resulted in the formation of a subgroup of the original spherical group.

Why is this idea of symmetry breaking so important? Let us take an example from the musical world (Fig. 1.2). In the first line, we see two notes repeated over and over again, ad infinitum. This is an example of what we call translational symmetry, namely, continuous repetition of an object in space or time.

This line of 'music' is highly symmetric in the way it repeats over and over again, but if you hum it to yourself you will rapidly decide that this is really very boring indeed. It hardly rates the label of 'music'. The second line shows the same music but at certain places indicated by the stars there is

Figure 1.2 Three examples of 'music'.

a sudden change i.e. the translational symmetry is broken. Although this still is not great music, at least every time the symmetry is broken your attention is taken. So what this illustrates is that symmetry per se is in fact boring! But our interest *is* roused when the symmetry is broken. The third line shows symmetry breaking in the time frame and this again attracts attention away from the boring repetitive bits. Real music often shows a certain amount of symmetry (see Fig. 1.3), but the composer is always careful not to continue this for long, preferring instead to make sudden breaks that make the music more interesting to us.

This relationship between symmetry and symmetry breaking can be extended to many other areas. For instance, when a scientific discovery is made, we talk about a 'breakthrough': that is simply a way of saying that a sudden change to the norm has occurred. People sometimes talk about paradigm shifts. If a solid changes its atomic crystal structure when, say, the temperature is changed, we talk about a phase transition: the science of phase transitions is all about symmetry breaking. Indeed, symmetry breaking is associated with evolution and progress, whereas symmetry itself is associated with stagnation, an unchanging and dull universe. It has been said that 'symmetry is death' and that 'symmetry breaking is life'. In Japan, architecture is often built to be not quite symmetric in order to add interest to the eye. On the other hand, symmetry confers special stability to objects: for instance the 4-fold symmetry of the Eiffel Tower in Paris creates an extraordinarily robust building whose relative density is 1.2×10^{-3} times that of solid iron!

Figure 1.3 Die Kunst der Fuge, Contrapunctus XVIII by J.S. Bach.

Figure 1.4 The Oxford Museum of Natural History.

A nice example of symmetry breaking in architecture is afforded by the design of the famous Oxford Museum of Natural History (Fig. 1.4). At first sight, this building looks to be highly symmetric with the left-hand side related to the right side by reflection. However, if you look very closely at the windows on either side and at the small portals, you will see that they are not quite symmetrically balanced. The building does not have mirror symmetry. There is a rumour that this was the result of an argument between John Ruskin, who played a part in the design of the building and the University. His original idea was for an asymmetric structure but the powers to be in the University disagreed and forced him to accept a symmetric design. However, Ruskin got his revenge in the end, though in a rather subtle way!

Although you are now convinced that symmetry is boring, it is important to appreciate that only by studying it can concepts such as symmetry-breaking be understood. In order to do this, we first have to explain how symmetry can be described.

1.1. SYMMETRY OPERATIONS

In order to understand what is meant by a symmetry operation, consider a molecule such as benzene (C_6H_6), Fig. 1.5 (as a convention, in all figures a bold letter indicates a vector). For simplicity, only the six carbon atoms are shown. If this molecule is rotated by 60° about the **c**-axis

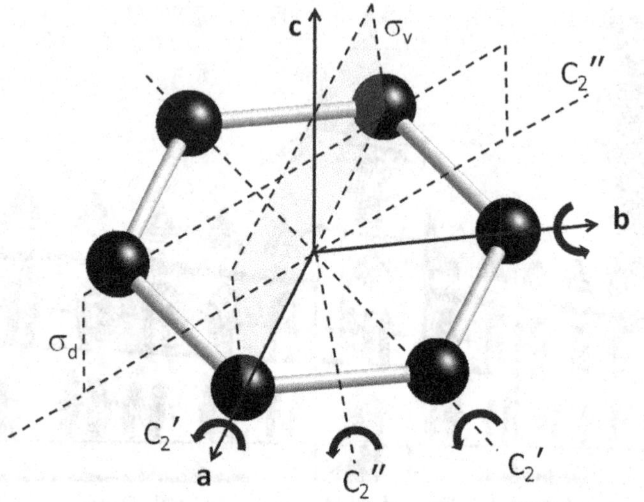

Figure 1.5 The carbon ring of a benzene molecule with some of the symmetry operations indicated.

perpendicular to the molecular plane, then it will appear exactly the same as before the rotation. We would then say that a rotation by 60° is a **symmetry operation** of this molecule. Hence, a symmetry operation for a molecule or crystal is defined as an operation that interchanges the positions of the various atoms and results in the molecule or crystal appearing exactly the same (being in a **symmetry-related position**) as before the operation. Furthermore, when the operation is carried out repeatedly the molecule or crystal must end up in some other symmetry-related position. It is worth noting that rotations by 120°, 180°, 240°, 300° and 360° are also symmetry operations in the benzene molecule. As we shall see later, there are other symmetry operations in this molecule. Some are rotations about other axes and some are operations other than rotations. The various types of point symmetry operations needed for crystals will now be discussed.

In order to be able to describe symmetry operations, it is necessary to adopt some sort of notation. In practice, there are two types of notation in common use. The Schoenflies notation, developed initially by the German mathematician Arthur Moritz Schoenflies towards the end of the eighteenth century, is mainly used in spectroscopy, whereas the so-called International Notation, developed by the French mineralogist Charles-Victor Mauguin and the German crystallographer Carl Hermann, is the optimal choice for describing crystals. In this book, we shall usually use both notations, with the International symbol first followed by the Schoenflies symbol in parentheses.

1.2. POINT SYMMETRY OPERATIONS

A **point symmetry operation** is a symmetry operation specified with respect to *at least* one point in space that does not move during the operation. For example, the rotations discussed in the last section are point symmetry operations. In later chapters, symmetry operations that involve translations will also be considered. Translations, although they may be symmetry operations as defined in the last section, cannot be considered to be taken with respect to any fixed point. When we talk about point groups (this term will be defined later), we shall consider a set of point symmetry operations all taken with respect to the same fixed point.

We should also like to describe the symmetry mathematically. In order to do this, we take three vectors, **a**, **b** and **c**, measured from a common origin, in such a way that **a** and **b** are not collinear and **c** is not coplanar with the **ab**-plane. Note that these three vectors act as **axes of reference** and need not be orthogonal. There are basically two ways to describe the effect of symmetry operations. We can define a symmetry operator which moves all points or position vectors of space with all vectors referred to a fixed set of axes. Alternatively, the symmetry operator can be made to move the axes of reference leaving all points in space, and hence position vectors, unmoved. The former type of operator is called an **active operator**, and the latter, which is its inverse, is called a **passive operator**. We shall use only active operators here. To illustrate what is meant, consider Fig. 1.6 where we show two points whose coordinates with respect to the axes of reference are (x, y, z) and (x', y', z'). Note that throughout this book, unless specifically indicated otherwise, we adopt a right-hand screw convention of axes with **a** down the page, **b** to the right, and **c** out of the page towards the reader. This is the same convention as used in the International Tables Volume A (*ITA*). In this figure, the axes of reference are right-handed but are drawn so that **c** points up and **b** to the right. The point in space denoted by the primed coordinates is obtained after operating with a point operator R on the point denoted by the unprimed coordinates. R is a symmetry operator, although what is said here is applicable to all operators. Thus we consider the axes **a**, **b** and **c** fixed in space and the point operators move the objects in

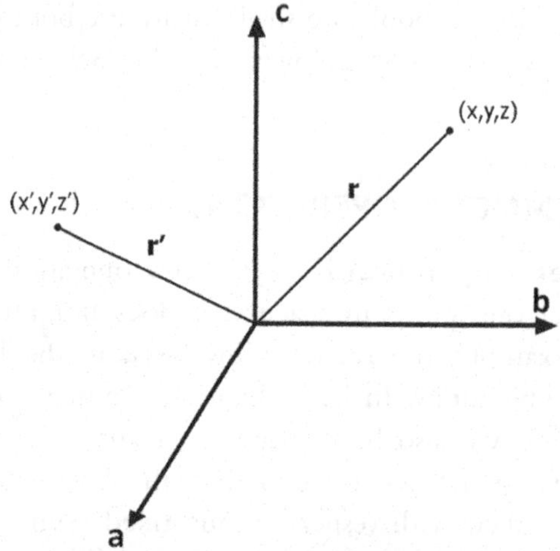

Figure 1.6 A point in space before and after a symmetry operation.

various ways. Then, after carrying out the point-symmetry operation, we can describe the new position in terms of the old by the matrix transformation:

$$\begin{bmatrix} x' \\ y' \\ z' \end{bmatrix} = \begin{bmatrix} a_{11} & a_{12} & a_{13} \\ a_{21} & a_{22} & a_{23} \\ a_{31} & a_{32} & a_{33} \end{bmatrix} \begin{bmatrix} x \\ y \\ z \end{bmatrix} \qquad [1.1]$$

or

$$\mathbf{r}' = R\mathbf{r} \qquad [1.2]$$

where Eqn [1.2] is the short-hand equation of Eqn [1.1] in which R stands for the matrix representing the point operation. This will become clearer when we write out some of the matrix operators.

1.2.1. Identity

The most important symmetry operation is the operation that can describe all objects; that is, the operation of doing nothing to the object. This trivial symmetry operation is given the symbol 1 in the International Notation or E in the Schoenflies notation. The matrix in Eqn [1.1], which describes this operation contains 1's along the diagonal and 0's off the diagonal, the so-called **unit** or **identity** matrix. This matrix, along with matrices for all the other symmetry operations, is given in Appendix 1 (see under heading, Origin).

1.2.2. Rotations

If rotation about an axis by 180° (π radians) is a symmetry operation for a particular object, then in the International (Schoenflies) notation we write it as $2(C_2)$. In general, if the symmetry operation is a rotation by an amount $2\pi/n$, where n is called the **order** of the rotation, the symbol is $n(C_n)$. This is sometimes called a **pure** or **proper rotation**. Here, only $n = 1$, 2, 3, 4 and 6 will be discussed, since for conventional crystals these are the only allowed rotations, as will be justified in the next chapter. Since rotations are taken about certain axes in the crystal, it is sometimes convenient to specify the orientation of a particular rotation axis. While there is no generally accepted way of doing this, we shall, when we think necessary, specify the orientation directly after the symbol for the rotation operator. Since the rotation axis is a line in a specific direction, we can describe it with respect to the **a**, **b** and **c** axes by the vector

$$\mathbf{S} = u\mathbf{a} + v\mathbf{b} + w\mathbf{c} \quad [1.3]$$

where the length of the vector **S** is adjusted to make u, v and w integers. The crystallographic convention for denoting this vector is to write it as $[uvw]$. Note that in some books, you may find that it is written in parentheses (incorrectly!) as (uvw); we write it with brackets, since to crystallographers parentheses signify the Miller indices of planes (Appendix 8) rather than directions. The rotation operation can then be written as $n[uvw]$ in the International Notation or as $C_n[uvw]$ in the Schoenflies notation. Angular brackets <> are used to specify all symmetry-related directions.

Let us consider a 2-fold rotation operation. Figure 1.7a shows a schematic diagram of two right hands with their palms up; these are related to each other by the symmetry operation $2(C_2)$. Of course in a crystal we do not really have 'hands' but collections of atoms. The hands in our diagrams serve only to illustrate in a convenient way the effects of the various symmetry operations. Thus, when this operation is applied about an axis perpendicular to the page, denoted symbolically by 2[001] (or C_2[001]), the hand pointing to the top of the page will rotate into the position of the hand pointing to the bottom of the page and vice versa. Notice that the hands both have their palms up and that both are right hands. One cannot be a left hand, unless both are left hands. We could think of this operation in a slightly different way. Just consider the hand that points to the top of the page and operate on it by a 2-fold (or C_2) operation. Then, the hand that points to the bottom of the page is generated and the composite picture has the **point symmetry** $2(C_2)$.

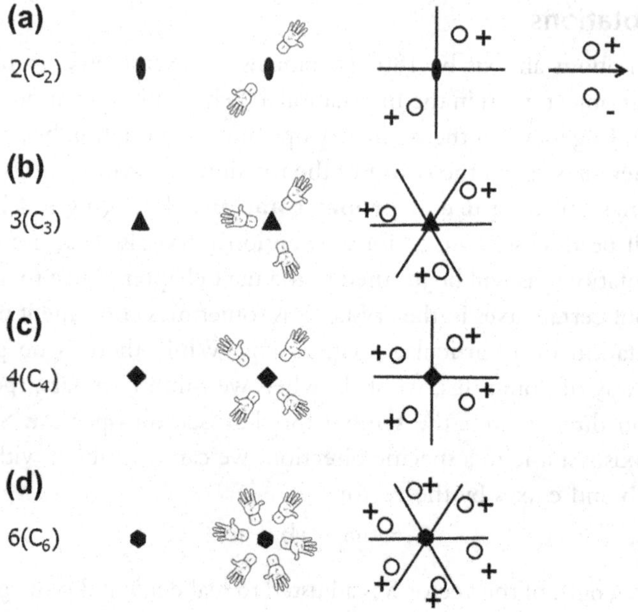

Figure 1.7 The point symmetry proper (pure) rotation operations and the symbols used to designate them. Hands are used to show the results of the operations. The symbols used to designate the symmetry operations are in the International (Schoenflies) notation.

Stereographic projections (or stereograms for short) are another way of representing these symmetry operations (see Appendix 5). Instead of the two hands, two circles with a + sign next to them are drawn. The circles are a convenient way of representing any general object such as a hand or a collection of atoms. The system of circles is used because it is compact and is the adopted symbol in the *ITA*. The '+' sign refers to the fact that each circle is above the plane of projection. A '−' sign would mean that the circle in question lies below this plane. Under a $2(C_2)$ operation, it is clear that these two circles interchange and that both remain at the same height above the plane of projection, so that the circles are in symmetry-related positions. The fact that only right hands are needed under this operation is denoted by the use of open circles. Notice that the 2-fold operation acts about a particular axis, [001] in this example, and that this is indicated on the diagram by a symbol similar to the shape of an American football. We also show, in the figure, two circles related by a 2-fold axis but observed perpendicular to the axis of rotation, [010] in this case. The line with the arrow is the *ITA* convention for a 2-fold rotation about an axis lying in the plane of projection. This and all of the other conventional diagrammatic symbols are listed in Appendix 6. As can be seen, now one of the circles is above the plane and one is below. The matrices describing the 2-fold symmetry operations about other principal directions are given in Appendix 1. Thus, for the symmetry operation 2[001] (or $C_2[001]$) operating on a general point at (x, y, z), we can write

$$\{2[001]\}(x, y, z) = \begin{bmatrix} -1 & 0 & 0 \\ 0 & -1 & 0 \\ 0 & 0 & 1 \end{bmatrix} \begin{bmatrix} x \\ y \\ z \end{bmatrix} = \begin{bmatrix} -x \\ -y \\ z \end{bmatrix} \quad [1.4]$$

to obtain a new point at ($-x$, $-y$, z) in agreement with the diagram.

The next symmetry operation is $3(C_3)$ denoted by a triangle. As can be seen in Fig. 1.7b, the right hand with palm up pointing to the top is rotated into the right hand with palm up pointing to the bottom left. At the same time the other hands are rotated into each other. Thus $3(C_3)$ is a symmetry operation for the three hands separated by 120° or $2\pi/3$ and equidistant from the origin as shown. Note that we adhere to a right-hand or **counter-clockwise** convention for a rotation (clockwise rotation is its inverse). The diagram to the right of the three hands shows the three circles above the plane of projection with $3(C_3)$ as a symmetry operation. It is important to realize that there is another symmetry operation closely related to $3(C_3)$. This is the operation 33 in the International notation (or C_3C_3 in the Schoenflies notation). This is normally written as 3^2 or C_3^2 where the squaring of a symbol has the usual meaning of operating twice. Thus, $n^2 = nn$, and this defines the product of n times n. Note that $3^2(C_3^2)$ is also a symmetry operation since it takes the right-hand palm up pointing to the top into the right-hand palm up pointing to the bottom right.

Again we point out a convention: if two operations AB are applied to an object, the operation on the right, B, is considered to be applied first and A second. Clearly, the order of operations is of no consequence when an operator is squared. The matrices representing the two symmetry operations $3(C_3)$ and $3^2(C_3^2)$ are given in Appendix 1 and can be applied directly to the coordinates of any general point.

The symmetry operations $4(C_4)$ and $6(C_6)$ are also shown in Fig. 1.7c and d. Note that $4^2(C_4^2) = 2(C_2)$, so that this is an operation that we have already discussed. This is a simple example of the fact that the existence of certain symmetry operations implies the existence of other symmetry operations. Thus if $4(C_4)$ is a symmetry operation for a particular crystal or molecule, then $2(C_2)$ is also a symmetry operation, as is $4^3(C_4^3)$. Similarly $3(C_3)$ implies that $3^2(C_3^2)$exists. From $6(C_6)$ we know that $6^m(C_6^m)$, where $m = 1, 2, 3, 4, 6$, are also symmetry operations. For $m = 2$, we have $3(C_3)$, for $m = 3$ we have $2(C_2)$, and for $m = 6$ we have $1(E)$, the identity operation.

Note that in all these pure rotations, $n(C_n)$, right hands are all that are required. At no time, do these operations require any left hands as well. Shortly, we shall see that other symmetry operations require both hands.

Figure 1.8 shows some examples of objects that have rotation symmetries, demonstrating that these basic concepts can be applied to physical objects in general.

Also note that the rotation symmetry operations are all described with respect to a line, namely, the **axis of rotation**. This axis is referred to by crystallographers as a **symmetry element**.[2] The rotation operation is accomplished by the movement of the object (hands or circles) the required amount about the axis of rotation. The hands or circles here have been placed at a **general position** – a position in space that is not situated directly on the axis. The effect of the symmetry operator is to produce a set of hands or circles that all lie on general positions but which are symmetry-related to one another, resulting in a set of **general equivalent positions**.

1.2.3. Inversion

The symmetry operation of **inversion**, sometimes called centrosymmetry, is more involved than the rotation operations because it changes right hands to left hands, and vice versa. By the operation of inversion, we mean that for every position in space given by the coordinates (x, y, z), we operate with the inversion operator $\bar{1}(i)$ to give the position $(-x, -y, -z)$. Note that the crystallographic convention is to put the minus signs above the coordinates: thus $(\bar{x}, \bar{y}, \bar{z})$ means $(-x, -y, -z)$, and is pronounced, bar x, bar y, bar z (or if you are American you probably say x bar, y bar, z bar). The operation can be written as

$$\{\bar{1}\}(x, y, z) = (-x, -y, -z) \qquad [1.5]$$

2 Actually, this term is more complex than appears at first sight. While it is obvious that a line or axis is the element relevant to a rotation, there are problems when trying to define formally the element relevant to a screw axis (see later). It is common to hear scientists incorrectly use the terms 'symmetry operation' and 'symmetry element' interchangeably.

Figure 1.8 Examples of rotation symmetry.

(see Appendix 1). In Fig. 1.10a, we see the result of $\overline{1}(i)$ operating on a right hand, palm up, above the plane of projection and pointing to the top of the page. The result is a left hand, palm down (knuckles up), below the plane and pointing to the bottom of the page. Thus, if two hands are placed as shown, $\overline{1}(i)$ is a symmetry operation linking them.

When circles are used to show the symmetry operations, the change in **chirality** or 'hand' is conventionally denoted by placing a comma in the center of the circle. The right hand is said to be enantiomorphically related to the left hand and the inversion operation is said to be an **enantiomorphic operation**. As we shall see later, a mirror image has this same property so that sometimes one says that two objects that are enantiomorphically related are mirror images of each other. One might equally well say one is a right-handed system and the other is a left-handed system.

You might ask if it is possible to change a left-handed object into a right-handed object by use of a pure rotation operation. In general the answer is no, unless one is prepared to carry out the rotation operation in a higher dimension, something with which mathematicians may be comfortable, but with which the rest of us would struggle! Fig. 1.9 illustrates this idea by demonstrating how an object in 'flatland', once brought into three

Figure 1.9 How to change chirality by a pure rotation through use of a higher dimensionality.

dimensions can be rotated to form its enantiomorph. Starting with a figure 7 (on the left) on a two-dimensional plane, this is rotated within three dimensions, and then projected back into flatland to produce its mirror image. Apart from this trick, however, it is safe to say that there is no way in which left and right-handed systems can be made equivalent using ordinary rotation operations. One point of definition: if two objects have the same chirality, they are said to be **congruent** to each other.

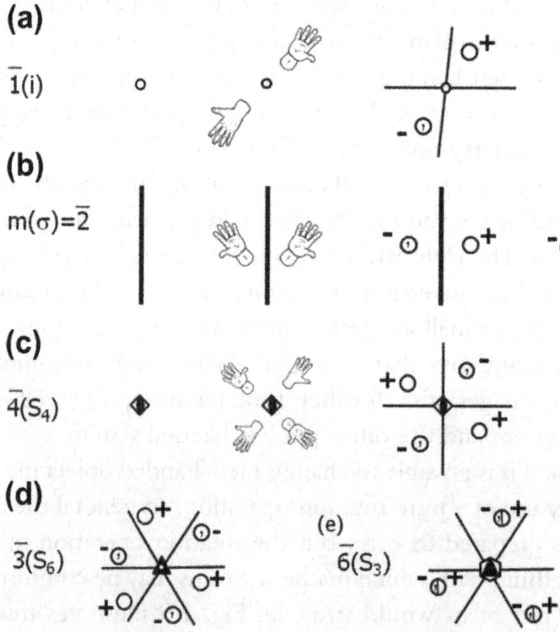

Figure 1.10 Inversion operations.

Note that, while rotation takes place about an axis, inversion takes place through a point, the **center of inversion** or **center of symmetry**. Examination of the benzene molecule in Fig. 1.5 reveals that there is a center of inversion situated at the molecular center. Such an object is said to be **centrosymmetric**.

Figure 1.11 shows an interesting demonstration of a center of symmetry, from as long ago as 1523 in a sketch by Holbein, through positioning of left and right hands in such a way that if a line is drawn from each finger tip on one hand to each finger tip on the other, all lines intersect at a point midway between the fingers. This point is a center of symmetry (to a good approximation).

1.2.4. Reflection Across a Plane

The operation of reflection across a plane, sometimes called a mirror reflection and denoted by the symbol m in the International notation (σ in the Schoenflies notation), is shown in Fig. 1.10b. The symmetry element for reflection is called a **mirror plane**. Given a point in space and a reflection operation, drop a perpendicular onto the mirror plane, and place a new

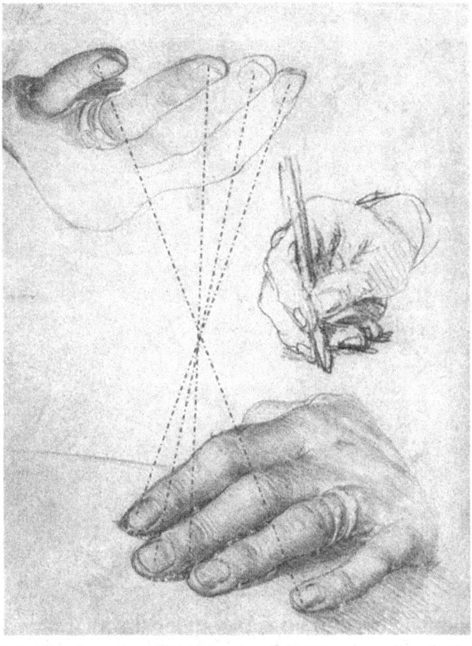

Figure 1.11 The hands of Erasmus of Rotterdam by Holbein, 1523.

point along this line an equal distance on the other side of the mirror; this is the mirror image of the first point. Thus, if the right hand is drawn palm up on the right side of the mirror, the mirror image is a left hand, also palm up, on the left side of the mirror as shown. For the coordinate system shown, we have

$$\{m[010]\}(x, y, z) = (x, -y, z) \quad [1.6]$$

We use the notation $m[010]$ to mean a reflection across a plane *perpendicular* to the [010] direction (i.e. the **b**-axis in this case). Figure 1.12 shows an example of a mirror reflection in practice (those of us of a certain age may recall this individual). In Fig. 1.10b, the two circles also show the $m(\sigma)$ operation; here we use a comma to indicate that one is the enantiomorphic (mirror) image of the other.

On the far right is another diagram of a mirror plane, only this time, the mirror is in the plane of projection. In this case one circle lies on top of the other. The convention used in the *ITA* to denote this is to divide the circle into two halves, one with and one without a comma.

The Schoenflies symbol for a mirror operation, σ, is usually given a subscript. For instance, if we define the **c**-axis to be the **principal axis,**

Figure 1.12 A mirror reflection, plus an apparent levitation! (*Harry Worth,* BBC comedy series).

then σ_h is a mirror reflection through a plane perpendicular to this axis (the 'h' means horizontal, assuming that the principal axis is vertical).

Clearly, we could just as well have put the comma in the circle on the right and have left it out of the circle on the left. Appendix 1 gives the matrices for some special orientations of a mirror plane.

Reflections through planes that contain the principal axis are either σ_v or σ_d (vertical or diagonal). σ_v contains the principal axis and the **a** or **b**-axis, while σ_d generally contains an axis that bisects the angle between the **a** and **b**-axes. The benzene molecule (Fig. 1.5) has these three mirror planes. The σ_h plane passes through all of the atoms in the molecular plane, σ_v passes through opposite pairs of atoms, and σ_d passes through the midpoints of opposite C–C bonds.

1.2.5. Rotation–Inversion (Improper Rotation)

This last type of symmetry operation is the most complicated to understand for two separate reasons. First, the approach taken in the International System is different from that taken by Schoenflies, and so the symmetry operations are given different names. In the International System, we talk about **rotation–inversion,** and in the Schoenflies system, **rotation–reflection.** Both types of operations are known as **improper rotations.** The second complication is that this symmetry operation is a **compound operation,** that is, it is the product of two other operations. In general, each of these other operations is not by itself a symmetry operation for a particular crystal or molecule but the product of the two is. We shall deal with the two approaches separately.

International approach: We shall continue to put the Schoenflies symbol in parentheses following the International symbol since there is a one-to-one correspondence, despite the fact that they are defined differently.

In the International approach, we first perform a rotation operation $n(C_n)$ and then follow it immediately with an inversion operation. Writing this as a product, we have $\overline{1}n(iC_n)$. The short-hand International symbol used instead is $\overline{n}$. First consider the operation $\overline{4}(S_4^3)$,[3] since it is probably the easiest to visualize (Fig. 1–10c). Start with a right hand, palm up, pointing to the bottom right and rotate it counter-clockwise through $2\pi/4$ and immediately

[3] It is important to distinguish between the operation and the point symmetry of a set of objects. Thus, in Fig. 1.10c and d, we write $\overline{4}(S_4)$, $\overline{3}(S_6)$ and $\overline{6}(S_3)$ to describe the point symmetry of the entire collection of objects shown. Notations such as $\overline{4}(S_4^3)$, $\overline{3}(S_6^5)$ and $\overline{6}(S_3^5)$ also refer to the particular operation linking one object to another.

invert it through the origin. The result is a left hand, palm down, pointing to the bottom left. To complete the diagram, reapply this operation to the hand just obtained. The result is a right hand, palm up, pointing to the top left. This shows that $\overline{4}^2 = 2$ (or $S_4^6 = S_4^2 = C_2$). Now apply the operation again, and the result is a left hand, palm down, pointing to the top right, corresponding to the operation $\overline{4}^3(S_4)$. Applying the operation yet again brings us back to the hand pointing to the top of the page in its original orientation, so that $\overline{4}^4 = 1$, the identity operation. $\overline{4}$ is therefore a symmetry operation. The conventional diagram with circles representing general positions, heights (+ and −) and commas is also shown. Thus for an arrangement of atoms with positions represented by these circles, we can state that $\overline{4}(S_4^3)$ and $\overline{4}^3(S_4)$ are also symmetry operations. Again notice how the symmetry operation $\overline{4}(S_4^3)$ implies the existence of a 2-fold rotation operation. Also note that for this system of hands or circles neither $4(C_4)$ nor $\overline{1}(i)$ by itself constitutes a symmetry operation in this case. For example, $4(C_4)$ operating on the first hand would result in a second hand, palm up, pointing to the top right. However in the diagram this hand in fact is palm down. Therefore, the compound operation $\overline{4}(S_4^3)$ is indeed a new type of symmetry operation whose component operations are not symmetry operations for this set of objects. This is an important point to grasp. The matrices for these various symmetry operations are given in Appendix 1.

Consider the operation $\overline{3}(S_6^5)$: starting with the circle at the top above the plane of projection and applying $\overline{1}3 = \overline{3}(S_6^5)$, we rotate through $2\pi/3$ to a position towards the bottom left and immediately invert through the center to obtain the circle towards the top right. This circle lies below the plane of projection and is enantiomorphically related to the first circle. After doing this six times, we come back to the circle that we started with. Therefore, if a crystal or molecule has this arrangement of atoms, $\overline{3}(S_6^5)$ is a symmetry operation for this crystal or molecule.

The $\overline{6}(S_3^5)$ symmetry operation can be carried out easily and is left for the reader to complete. When you do this you will notice that for some n in $\overline{6}^n$ new operations are obtained, while for other values of n, operations that we have already discussed are obtained.

It is clear that the operation $\overline{2}$ is simply a mirror operation, $\overline{2} = m(\sigma)$. You can now see why we choose to specify the orientation of a mirror plane by the direction of a line perpendicular to it. This line is parallel to the $\overline{2}$ axis, so that our use of a direction normal for the mirror plane is analogous to that used for axes of proper rotations. In this sense, a mirror reflection is really a type of rotoinversion operation.

Schoenflies approach: In this approach, the improper rotations are obtained in the following manner: first rotate by the appropriate amount and then immediately reflect across a plane perpendicular to the axis of rotation. The Schoenflies notation for this is $S_n = \sigma C_n$, but in order to make it clear that the mirror plane is perpendicular to the rotation axis we give the plane a subscript and write $S_n = \sigma_h C_n$. The '*h*' stands for a horizontal plane, naturally assuming that the axis of rotation is vertical and $S_n^m = (\sigma_h C_n)(\sigma_h C_n)(\sigma_h C_n)\ldots = (\sigma_h C_n)^m$ as always.

Now apply this approach to the $\overline{4}(S_4)$ diagram. Start from the hand pointing to the bottom right and apply S_4 (remember that when $\overline{4}$ was applied, the hand pointing to the bottom left was obtained). This time rotate by $2\pi/4$ and reflect. Now the hand pointing to the top right is obtained. Thus, we see that S_4 is equivalent to $\overline{4}^3$. To obtain the hand pointing to the bottom left, S_4 must be applied three times, i.e. $\overline{4}$ is equivalent to S_4^3. Thus, we obtain the hands or representative circles counter-clockwise by the operations

$$E, \quad S_4, \quad S_4^2 = C_2, \quad S_4^3, \quad S_4^4 = E$$

The Schoenflies approach obtains the position in a slightly more systematic order. For example, the positions in the $\overline{3}(S_6)$ diagram are obtained successively in a counter-clockwise order if one repeatedly applies S_6. In the International approach, the positions are obtained successively in a *clockwise* direction but the rotational parts of the symmetry operation must be applied counter-clockwise (i.e. our right-handed convention). Below, we list the corresponding symbols that are used to obtain the stereogram in the $\overline{3}(S_6)$ diagram in a counter-clockwise sense.

S_6	$S_6^2 = C_3$	$S_6^3 = i$	$S_6^3 = C_3^2$	S_6^5	$S_6^6 = E$
$\overline{3}^5$	$\overline{3}^4 = 3$	$\overline{3}^3 = \overline{1}$	$\overline{3}^2 = 3^2$	$\overline{3}$	$\overline{3}^6 = 1$

The list for $\overline{6}(S_3)$, again counter-clockwise, is

S_3	$S_3^2 = C_3^2$	$S_3^3 = \sigma_h$	$S_3^4 = C_3$	S_3^5	$S_3^6 = E$
$\overline{6}^5$	$\overline{6}^4 = 6^4$	$\overline{6}^3 = m$	$\overline{6}^2 = 6^2$	$\overline{6}$	$\overline{6}^6 = 1$

Before leaving the subject of symmetry operations, we should note that, from the definition of the term, the inverse of every symmetry operation must also be a symmetry operation. By the term **inverse**, we mean another operation such that the product of the two is the identity: if A, B, C, ... are symmetry operations and if $AC = E$, the identity, then A is the inverse of C, and vice versa, because one may prove that a left inverse is equal to a right inverse. We list the inverse operation of each symmetry operation in Schoenflies notation.

Symmetry Operation	Inverse	
S_n^m	C_n^{n-m}	
S_n^m	S_n^{n-m}	All m, n even
S_n^m	S_n^{2n-m}	m odd, n odd

Note also that $S_n^m = C_n^m$ for n odd and m even; E, i and σ are their own inverses. In addition notice how these 3 and 6-fold compound operations interchange depending whether the International or Schoenflies notation is used. For example S_3 is equivalent to $\bar{6}^5$. This is a potential source of confusion that will become even more important to understand when we deal with crystal systems.

In Appendix 1, the improper operations are listed side by side with their proper partners. You will notice that the determinant of any matrix of a symmetry operation is always ± 1, an important characteristic of orthogonal transformations. Moreover, the proper rotations all have determinant $+1$, whereas the improper rotations have determinant -1. It is therefore a very simple matter to determine whether a particular matrix operation is proper or improper.

1.3. HEXAGONAL COORDINATES

Figure 1.13 shows the effect of operating on a point with coordinates (x, y, z) using the operation $6^n(C_6^m)$ for $m = 1$ ... 6; the coordinates are specified with respect to the hexagonal axes, **a** and **b**, which are 120° to one another.

Using this diagram, we can see how the coordinates are obtained for these symmetry operations as well as for all the 3^m and 6^m operations. The figure shows that when the operation $6(C_6)$ is applied to the point B at (x, y, z), a new point B′ at ($x - y$, x, z) is obtained. Careful study of this diagram should make this clear. We see that the position of B′ measured parallel to the **a**-axis is equal to A′A″ − A′B′. Also A′A″ = AA′ = x, and by the geometry of equilateral triangles, A′B′ = AB, and hence equal to y. Therefore, the coordinate distance, if B′ is measured along **a**, is $x - y$.

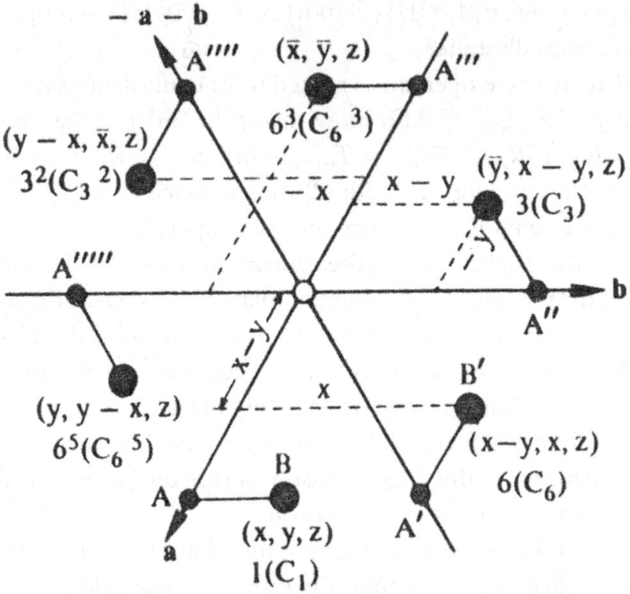

Figure 1.13 General equivalent positions with respect to hexagonal axes looking down the **c**-axis.

Similarly, the distance B′ measured parallel to **b** is equal to the distance of A measured parallel to **a**, i.e. it is equal to x. Hence B′ is at $(x - y, x, z)$. An alternative way of obtaining this result is to use the matrices given in Appendix 1. Thus, we have

$$\{6\}(x, y, z) = \begin{bmatrix} 1 & -1 & 0 \\ 1 & 0 & 0 \\ 0 & 0 & 1 \end{bmatrix} \begin{bmatrix} x \\ y \\ z \end{bmatrix} = \begin{bmatrix} x - y \\ x \\ z \end{bmatrix} \quad [1.7]$$

in agreement with the diagram. The reader should now be able to understand how the coordinates of all the other points related by $6^m(C_6^m)$ operations are obtained.

Problems

1. With reference to the list of improper operations given in Section 1.2.5, verify the correspondence between the two sets of symbols (International and Schoenflies).
2. Using the matrix multiplication method, find which operation is equivalent to the product $\{2[100]\}\{4[001]\}$, i.e. $\{C_2[100]\}\{C_4[001]\}$.

Also to the product {4[001]}{2[100]} or {C_4[001]}{C_2[100]}. What do the two results illustrate?

3. Show that an active operator is related to its equivalent passive operator by $R_{\text{active}} = (R_{\text{passive}})^{-1}$. Also show that if the order of the operations is reversed (i.e. if $R_{\text{active}}\ S_{\text{active}} = T_{\text{active}}$), then $S_{\text{passive}}\ R_{\text{passive}} = T_{\text{passive}}$.
4. Prove (1) $S_n^{n/2} = i$ for $n/2$ odd; (2) the existence of an S_n operation of even order implies the existence of a $C_{n/2}$ operation.
5. Verify that the determinants of the matrices in Appendix 1 are always ±1. Note that the product of two improper rotations is always a proper rotation.
6. What are the four symmetry operations that describe the H_2O molecule? What are the eight symmetry operations in an Egyptian pyramid?
7. On looking in a mirror, the right side of your body appears to become the left and your left the right. However, your head and feet are not interchanged. Does this suggest that a mirror only works in the horizontal direction and not in the vertical?
8. Recently, a plastic eight-sided disc was found in a Christmas cracker. On both faces, there was an arrow. When the disc was held as in (a) below and then rotated through 180° about the vertical to show the opposite face, as in (b), it was seen that the arrows pointed in the same direction

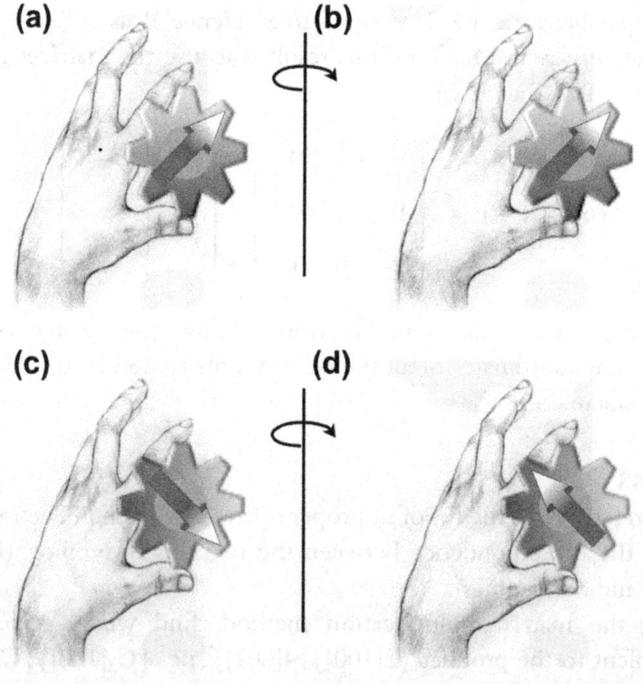

on opposite faces. However, when the disc was held as in (c), rotation through 180° about the vertical showed that now the arrow on the backside pointed in the opposite direction to the front side! Can you explain how this is possible?

CHAPTER 2

Crystal Systems

Contents

A casual glance at crystals may lead to the idea that they were pure sports of nature, but this is simply an elegant way of declaring one's ignorance. With a thoughtful examination of them, we discover laws of arrangement.

René-Just Haüy, Traité de Minéralogie, 1801

HAÜY'S LEGACY

Having explained the symmetry operations that are important in crystalline materials, we now ask the question: how do these symmetry operations allow us to classify and characterize crystals? Note that we really refer to 'fully ordered' crystals. In practice, however, there are thermal vibrations of the atoms and other forms of disorder, such as those found in metallic alloys, that make the local symmetry lower than that of the average structure. In this chapter, the basic symmetry operations are applied to a lattice and these impose certain restrictions on the axes and angles used to define it, leading to the so-called **seven crystal systems**. These form the coarsest level of classification of crystals. In the following chapters, we discuss the 14 Bravais lattices, the 32 point groups and the 230 space groups. Each of these classifications describes the crystal in increasing detail.

Space Groups for Solid State Scientists
http://dx.doi.org/10.1016/B978-0-12-394400-9.00002-2

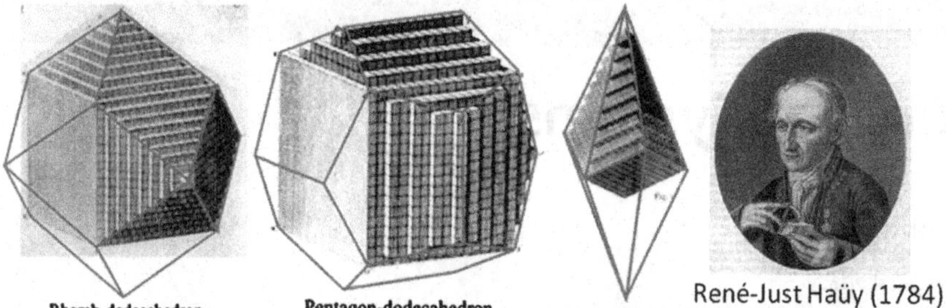

Figure 2.1 Models from Traité de Minéralogie by René-Just Haüy. Red lines added to emphasize external crystal habits.

We owe much of the classification of crystals to the pioneering work of René-Just Haüy. It is said that during a visit to a friend, he dropped a group of prismatic crystals of calcite and noticed that the fragments were recognizable as being of the same form as other crystals of calcite. Following this, he went to his laboratory and deliberately smashed other crystals of a scalenohedral crystal of calcite (dog-tooth spar) and another of rough rhombohedral habit, in each case obtaining similar rhombohedral fragments. This led him to the idea that crystals could be explained by small polyhedral units that were stacked together in a repetitive order (we now describe this repetition mathematically using the concept of a lattice). Figure 2.1 shows some of his models. At the time he could not know precisely what these units were, but we now think of them as what we call unit cells containing sets of atoms.

2.1. LATTICE

A **lattice** is an infinite array of points in space, in which each point has identical surroundings to all others. The simplest way of generating such an array is by invoking the property of **translational invariance**, which is the most fundamental feature of normal crystals. This property can be conveniently expressed by the **primitive translation vector**

$$\mathbf{t}_n = n_1\mathbf{a} + n_2\mathbf{b} + n_3\mathbf{c} \qquad [2.1]$$

where n_i is any integer (negative, zero or positive) and **a**, **b**, **c** are vectors chosen so that **a** and **b** are not collinear and **c** is not coplanar with the **ab**-plane: **a**, **b** and **c** start from the same origin and serve as **axes of reference**. The lattice points are given by the end points of the vector in Eqn [2.1] as points in the array. For simplicity, we have drawn a projection of a

3-dimensional lattice. If we choose a particular point as an origin (marked O) and consider the others with respect to it, then Eqn [2.1] gives us the vectors to these points.

For example, the point two units in the **a**-direction, three in the **b**-direction and five in the **c**-direction is given by $2\mathbf{a} + 3\mathbf{b} + 5\mathbf{c}$ from O. Thus by taking integral amounts of the **basis vectors**, **a**, **b** and **c**, and compounding them, an array of points is generated that display translational periodicity described by a new type of symmetry: **translational symmetry**. Note too that, no matter which lattice point is chosen as an origin, the array always looks the same when viewed from it.

Notice that we have not said anything about the angles between the vectors **a**, **b** and **c**. In fact, if there is no special relationship between the interaxial angles, the resulting lattice forms an oblique net of points, as in Fig. 2.2. On the other hand, if for example, the interaxial angles are 90°, that is the axes are orthogonal, either square or rectangular arrays are formed, depending on whether the vectors **a**, **b**, or **c** are equal in length. How restrictions on the axes and angles are obtained is the subject of this chapter.

2.2. UNIT CELL

By completing the parallelepiped formed by the vectors **a**, **b** and **c**, we enclose a volume in space, $\mathbf{a} \cdot (\mathbf{b} \times \mathbf{c})$, that, when repeated according to Eqn [2.1] fills all space and generates the lattice (Fig. 2.3a). Such a region of space is called a **unit cell**. If it contains only *one* lattice point, it is called a **primitive unit cell**. We should explain here what we mean when we say that a primitive unit cell *contains* one lattice point. If the origin of a unit cell is chosen so that it is on a lattice point, then in a three-dimensional net each lattice point can be thought of as being shared between eight primitive unit cells. Thus there are $8 \times 1/8$ lattice points associated with any one primitive

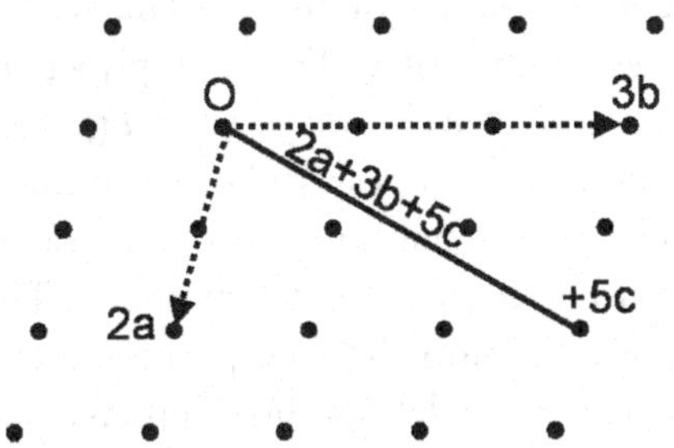

Figure 2.2 A lattice in projection.

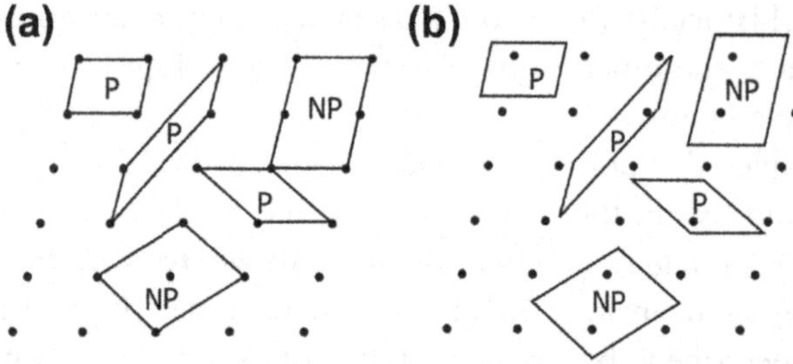

Figure 2.3 Some examples of primitive (P) and non-primitive (NP) unit cells.

unit cell. However, there is a much easier way of carrying out the counting by recognizing that in fact we can choose the origin of any unit cell wherever we wish. Figure 2.3b shows what happens if we displace the origin of each unit cell away from a lattice point: it is then a trivial matter to count the number of lattice points within the unit cell.

A primitive unit cell is not the only type of cell that is used to describe a lattice. A unit cell need not contain only one lattice point. If it contains more than one lattice point, it is called a **non-primitive** or **multiply primitive unit cell**. Figure 2.3 shows some examples of primitive and non-primitive unit cells, labelled P and NP, respectively.

Note that the volume (area in projection) of each primitive unit cell is the same irrespective of the choice of axes used. Clearly, there is an infinite number of primitive and non-primitive unit cells that can be chosen. However, some restrictions on our choice can be made if we observe certain conventions. Nevertheless, it is important to realize that no matter how we choose the axes, the lattice always remains the same; but the choice of a different unit cell may result in the lattice being given a different name or symbol.

2.3. CRYSTAL STRUCTURE

Before going on to consider lattices in more detail, we need to explain here what is meant by a **crystal structure**. This is simply a periodic arrangement of atoms or molecules. *Do not confuse the points in Fig. 2.2 with atoms: a lattice is an abstract mathematical concept.* A powerful electron microscope examining a crystal will not show a lattice, but it will show an arrangement of atoms or molecules set out according to translational symmetry. The role of the lattice then is to act as a template dictating how we position the atoms or molecules to form the crystal structure. It defines the symmetry operations that relate one atom/molecule to every other one. Thus, statements that you will see in many books such as the 'diamond lattice' or the 'copper lattice' are formally misleading, since they actually mean to say the 'diamond structure' or the 'copper structure'. Both crystals of diamond and copper actually have the same type of lattice but very different crystal structures. The confusion arises because many simple crystals, as described in undergraduate textbooks, often only consist of one atom per lattice point and so a drawing of the unit cell of the structure resembles a drawing of the unit cell of a lattice.

In some textbooks, you will see that the group of atoms is called the **basis** of the structure, not to be confused with the crystallographer's term for the set of axes defining the unit cell. This allows one to describe crystal structures in a simple way, by combining the concept of the lattice with the basis (we shall see later that the use of space group symmetry makes this process even more succinct). The formal relationship is through the mathematical process known as convolution.

2.3.1. Convolution

Consider two functions $g(\mathbf{r})$ and $h(\mathbf{r})$ spanning a space given by the vector $\mathbf{r}$. The convolution of these two functions $f(\mathbf{r})$ is given by

$$f(\mathbf{r}) = g(\mathbf{r}) * h(\mathbf{r}) = \int_{-\infty}^{+\infty} g(\mathbf{r}')h(\mathbf{r}-\mathbf{r}')\mathrm{d}\mathbf{r}' \qquad [2.2]$$

Suppose we have a collection of atoms (or a molecule) described by a function $B(\mathbf{r})$. In order to form a crystal structure, we need to repeat this function according to the translational symmetry of the lattice. The lattice function $L(\mathbf{r})$ is defined by

$$L(\mathbf{r}) = \sum_{uvw} \delta(\mathbf{r} - \mathbf{r}_{\mathrm{uvw}}) \qquad [2.3]$$

Here the vector $\mathbf{r}_{\mathbf{uvw}}$ points to each lattice point in the direction $[uvw]$, and then the crystal structure $C(\mathbf{r})$ is given by

$$C(\mathbf{r}) = L(\mathbf{r}) * B(\mathbf{r}_{\mathrm{uvw}}) \qquad [2.4]$$

Thus, for example, suppose we choose as $B(\mathbf{r})$ a molecule represented here by a spider (Fig. 2.4). Now convolute this with the lattice $L(\mathbf{r})$ and the result will be a 'crystal' structure. Of course, this arrangement of spiders is not a real crystal structure! But it does illustrate how the lattice acts as a template instructing how the basis is to be repeated.

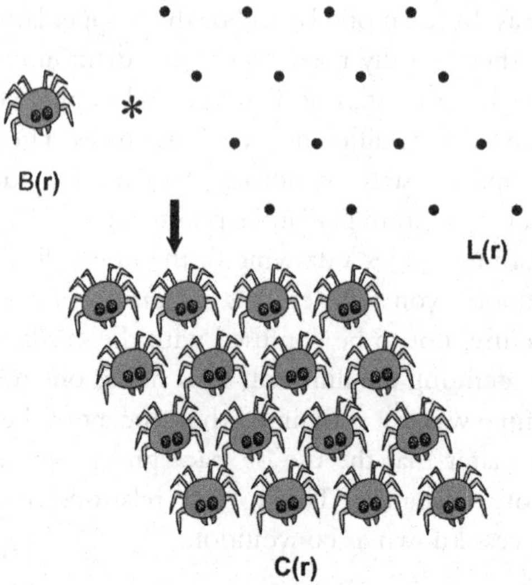

Figure 2.4 Building a crystal out of a spider and a lattice by convolution.

2.4. CRYSTAL SYSTEMS

Before deriving the crystal systems, it is worthwhile mentioning a convention. For our unit cell, we use the customary right-handed set of axes, labelled **a**, **b** and **c** with interaxial angles α, β and γ (Fig. 2.5).

In our treatment of the crystal systems, we shall start by considering the effect of the lowest symmetry operation on a unit cell and then, with the exception of 3 and 6-fold axes, work our way up to the highest symmetry (we shall leave the 3 and 6-fold operations till last since a number of complications arise that make their associated crystal systems

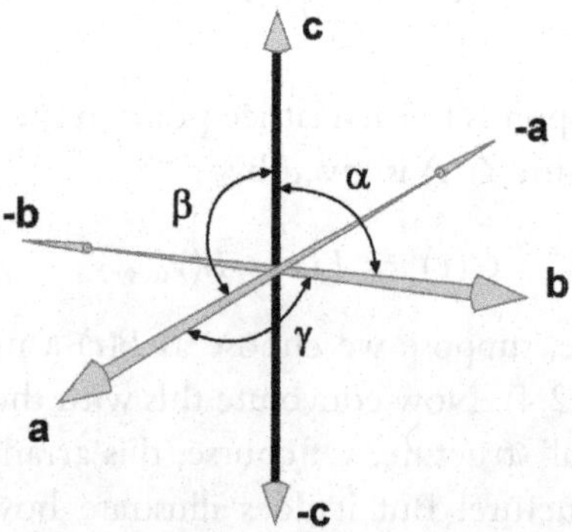

Figure 2.5 Conventional labels for axes and angles.

rather special). The seven crystal systems arise from applying proper and improper rotations to the unit-cell axes or translation vectors of the lattice. We can deal with this mathematically in general terms in the following way. Consider a symmetry operation R applied to a general position vector **r**, as we did in Chapter 1. This position vector is a vector from the origin of the unit cell, and can be expressed in terms of components along the **a**, **b** and **c** axes. These components are usually expressed in the form of coordinates measured as fractions of the unit-cell axial lengths. This means that a point in the unit cell specified by (x, y, z) lies at a distance from the origin with vector components $(x\mathbf{a}, y\mathbf{b}, z\mathbf{c})$. These fractional coordinates are often known as the **atomic position parameters** since we normally use them to represent atomic positions in a crystal structure. The position vector **r** joining the origin to the point (x, y, z) is given by

$$\mathbf{r} = x\mathbf{a} + y\mathbf{b} + z\mathbf{c} \qquad [2.5]$$

After applying the operation R, a new point at (x', y', z') is obtained that is related to the first point at (x, y, z) by

$$\begin{bmatrix} x' \\ y' \\ z' \end{bmatrix} = \begin{bmatrix} a_{11} & a_{12} & a_{13} \\ a_{21} & a_{22} & a_{23} \\ a_{31} & a_{32} & a_{33} \end{bmatrix} \begin{bmatrix} x \\ y \\ z \end{bmatrix} \qquad [2.6]$$

This is the same relationship given in Eqn [1.1], only here, because we are considering unit cells, the coordinates are fractional. The new point is specified by the vector $\mathbf{r}'$:

$$\mathbf{r}' = R\mathbf{r} = x'\mathbf{a} + y'\mathbf{b} + z'\mathbf{c} \qquad [2.7]$$

Since we take R as a symmetry operation, when comparing the vector components along the axes before and after applying this operation, we obtain the relationship between the unit-cell axes. We shall see that the symmetry operations of the lattice impose certain restrictions on the unit-cell geometry in that there will be relationships between the axial lengths and interaxial angles. In what follows, notice that the only rotations that are used in defining the crystal systems are the proper and improper rotations $n(C_n)$ and $\overline{n}(S_n)$ with $n = 1, 2, 3, 4$ and 6. The reason for this is that it is not possible with other values of n to construct unit cells that when joined together fill all space leaving no gaps (we do not consider here the possibility of using two or more differently shaped cells to fill space, as is done for the so-called quasi crystals – see Section 7.22). This can be demonstrated as follows.

Consider two lattice points A and A′ separated by a unit translation **t**, as in Fig. 2.6. A particular rotation operator R or its inverse R^{-1} acting at these points, respectively, will give rise to new points B and B′ by rotation of the vector AA′ through an angle α (the inverse of every symmetry operation is also a symmetry operation as discussed in Section 1.2). The condition for B and B′ to be lattice points is that the distance between them, **t**′, must be an integral number of basic translation units **t**. Therefore, we can write for integer m.

$$t' = mt \qquad [2.8]$$

and from the diagram

$$t' = -2t\cos\alpha + t \qquad [2.9]$$

Combining these two equations, we get

$$\cos\alpha = (1 - m)/2 \qquad [2.10]$$

Now if m is an integer, then so must $1 - m = M$. Furthermore, α must lie between 0 and 180° in order to obtain closure under the particular operation R, i.e. cos α lies between +1 and −1.

$$|\cos\alpha| \leq 1 \qquad [2.11]$$

Therefore

$$M \leq 2 \qquad [2.12]$$

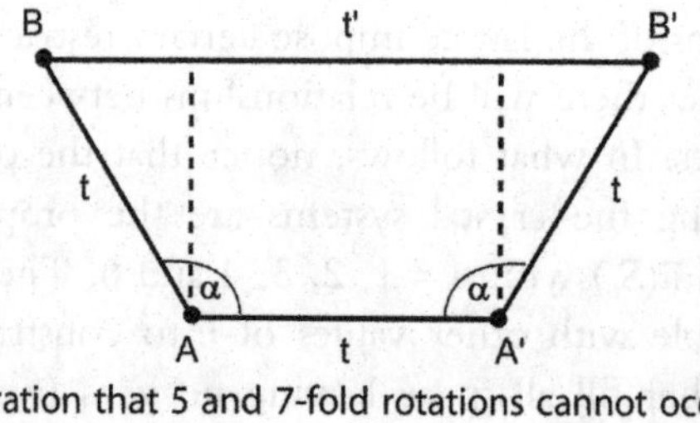

Figure 2.6 Demonstration that 5 and 7-fold rotations cannot occur in a conventional lattice.

and then

$$M = -2, -1, 0, 1 \text{ or } 2 \qquad [2.13]$$

This means that the only acceptable values of α are

$$\pi,\ 2\pi/3,\ \pi/2,\ \pi/3 \text{ or } 0$$

and hence the only allowed rotations, $2\pi/n$, are given by $n = 2, 3, 5, 6$ or 1, respectively. The same restrictions apply when considering improper rotations.

An alternative proof is as follows. Any proper rotation through an angle α about the **c**-axis, say, is given with respect to orthogonal Cartesian axes by the matrix

$$\begin{bmatrix} \cos\alpha & -\sin\beta & 0 \\ \sin\alpha & \cos\alpha & 0 \\ 0 & 0 & 1 \end{bmatrix}$$

It is well-known that the trace of such a matrix (sum of diagonal elements) must be equal to an integer provided that we are dealing with a symmetry operation (see for example Streitwolf, p. 60). Since $|\cos\alpha|$ can at most be equal to 1, the trace must lie between +3 and −1, i.e.

$$1 + 2\cos\alpha = 3,\ 2,\ 1,\ 0,\ -1$$

Therefore

$$\cos\alpha = 1, ½, 0, -½, -1$$

These five solutions give rotations with $n = 1, 6, 4, 3$ and 2, respectively, as expected.

We shall now develop the seven crystal systems one by one so that the effect of the progressive addition of higher symmetry can be seen. Note that in so doing we emphasize the fact that the crystal systems are defined in terms of the symmetry operations present and their effects on the unit-cell dimensions. Many books unfortunately tend to define the crystal systems in terms of the axial relationships instead, but this is incorrect, as it is entirely possible to find situations where the axes and angles are apparently related to one another accidentally. Only by looking then at the crystal structure does it become apparent that despite the axial relationships the underlying required symmetry is not present. *The crystal systems are defined according to the effect of symmetry operations which then place restrictions on the axes and angles, and not the other way round!*

2.4.1. Triclinic System

In this trivial case, the unit cell has no rotational symmetry other than $1(E)$ or $\overline{1}(i)$. Using the former we may write, according to Eqn [2.7],

$$\mathbf{r}' = \{1\}\mathbf{r} = \begin{bmatrix} 1 & 0 & 0 \\ 0 & 1 & 0 \\ 0 & 0 & 1 \end{bmatrix} \quad \mathbf{r} = x'\mathbf{a} + y'\mathbf{b} + z'\mathbf{c} \qquad [2.14]$$

where

$$\begin{aligned} x' &= 1x + 0y + 0z \\ y' &= 0x + 1y + 0z \\ z' &= 0x + 0y + 1z \end{aligned} \qquad [2.15]$$

The matrix is obtained from Appendix 1 for the $1(E)$ operation. Since we demand that this operation be a symmetry operation, we find that

$$\mathbf{r}' = x\mathbf{a} + y\mathbf{b} + z\mathbf{c} \qquad [2.16]$$

as expected for the identity operation. In this trivial example, the fractional coordinates are unchanged.

Proceeding in a similar way with the $\overline{1}(i)$ operation, we obtain

$$\mathbf{r}' = \{\overline{1}\}\mathbf{r} = -x\mathbf{a} - y\mathbf{b} - z\mathbf{c} \qquad [2.17]$$

In this case, all the signs have been reversed. Note that in both cases the coordinates x, y and z remain 'attached' to their axes **a**, **b** and **c**. This is simply another way of saying that there are no impositions placed on the axes by this symmetry; the axes do not bear any relationship to one another, and therefore no special restrictions are put on the unit-cell geometry. So the effect of symmetry operations $1(E)$ or $\overline{1}(i)$ defining the triclinic system on the metric relationships of the unit cell can be summarized as 'none'.

2.4.2. Monoclinic System

In this crystal system, the important symmetry operations are the 2-fold rotation $2(C_2)$ and/or the reflection $m(\sigma)$. Let the 2-fold axis be chosen to lie along **c**. This is the so-called **first setting** and is the convention often used by solid-state scientists (the **second setting**, where the 2-fold axis is chosen along **b**, is *usually* employed in crystallographic publications). Now, consulting Fig. 2.7, consider the restrictions that a 2-fold rotation imposes on a unit cell. Clearly, in order to relate **a** to –**a** by rotation through 180°, the **a**-axis must be perpendicular to the rotation axis, **c**, for if it were not, the 2-fold rotation acting on **a** would result in yet another axis, **a**′, at some angle from –**a**. Similarly, the **b**-axis must be perpendicular to the **c**-axis, but not necessarily to **a**. By using the matrices in Appendix 1, we can write that the effect of the 2-fold operation is

$$\mathbf{r}' = \{2[001]\}\mathbf{r} = -x\mathbf{a} - y\mathbf{b} + z\mathbf{c} \qquad [2.18]$$

and the effect of the $m(\sigma)$ operation (m is perpendicular to **c**) is

$$\mathbf{r}' = \{m[001]\}\mathbf{r} = x\mathbf{a} + y\mathbf{b} - z\mathbf{c} \qquad [2.19]$$

The differences in sign in Eqn [2.18] between the components along **c** and those along **a** and **b** imply perpendicularity as can be seen by taking the scalar product of these components before and after the transformation. Before the transformation for the **a** and **c**-axes, we have

$$x\mathbf{a}\cdot z\mathbf{c} \qquad [2.20]$$

After the transformation

$$x'\mathbf{a}\cdot z'\mathbf{c} = -x\mathbf{a}\cdot z\mathbf{c} \qquad [2.21]$$

where the right side of Eqn [2.21] is obtained by applying the result from Eqn [2.18] that $x' = -x$ and $z' = z$. Equations [2.20] and [2.21] must be equal by definition since we demand that the $2(C_2)$ operation be a symmetry operation, i.e. the crystal is unchanged. Thus, we obtain

$$xz|\mathbf{a}|\,|\mathbf{c}|\cos\beta = -xz|\mathbf{a}|\,|\mathbf{c}|\cos\beta$$

$$\cos\beta = -\cos\beta \text{ or } \beta = 90^\circ$$

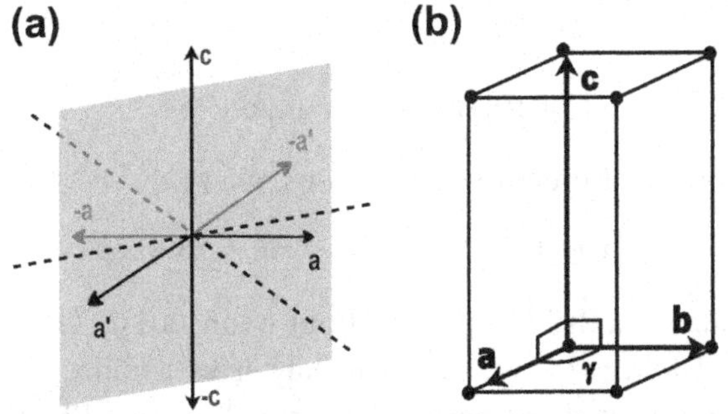

Figure 2.7 (a) The result of a 2-fold symmetry operation about the **c**-axis. (b) A monoclinic unit cell.

$\beta = 90°$ means that **a** and **c** are perpendicular. In the same way, **b** and **c** are found to be perpendicular ($\alpha = 90°$). The same scalar product for the **a** and **b**-directions yields an identity. The result is

$$x'y'|\mathbf{a}|\ |\mathbf{b}|\ \cos\gamma = xy\ |\mathbf{a}|\ |\mathbf{b}|\ \cos\gamma \qquad [2.23]$$

which tells us nothing. Hence γ is indeterminate. The fact that we have not interchanged the magnitudes of the axes under these operations means that no restrictions are placed on their lengths. Therefore, for the 'first setting', the monoclinic system has

$$\alpha = \beta = 90°$$

It should be noted that we normally choose the cell so that γ is greater than 90°; this is only a matter of convention and is not applied strictly, except in descriptions of crystal structures where confusion would otherwise occur. The use of a reflection, Eqn [2.19], instead of a 2-fold rotation, gives identical conditions for the axes and angles. Figure 2.7b shows a monoclinic unit cell.

Naturally, if one chooses the 'second setting' where the 2-fold axis is taken as the **b**-axis or the mirror plane is taken perpendicular to the **b**-axis, then one obtains

$$\alpha = \gamma = 90°$$

2.4.3. Orthorhombic System

In this system, we consider the effect of having two 2-fold or $\bar{2}$ axes ($\equiv$mirror planes). Suppose, that there are two 2-fold axes taken along **a** (or [100]) and **b** (or [010]). Then we can write

$$\{2[100]\}\mathbf{r} = x\mathbf{a} - y\mathbf{b} - z\mathbf{c} \qquad [2.24]$$

and

$$\{2[010]\}\mathbf{r} = -x\mathbf{a} + y\mathbf{b} - z\mathbf{c} \qquad [2.25]$$

Taking the product of these two operations, we find

$$\{2[100]\}\{2[010]\}\mathbf{r} = -x\mathbf{a} - y\mathbf{b} + z\mathbf{c} \qquad [2.26]$$

which is the same as a 2-fold rotation about **c** (or [001]). This means that if we have two 2-fold axes, we automatically have a third. Moreover, the change in signs tells us about perpendicularity. Equation [2.24] tells us that **a** is perpendicular to **b** and **c**, and Eqn [2.25] tells us that **b** is perpendicular to **a** and **c**, just as we found in the monoclinic system using Eqns [2.20] and [2.21]. Therefore, the three resulting axes are mutually perpendicular. Since the coordinates have not been interchanged with respect to the axes, there are no restrictions placed on the axial lengths. The two 2-fold axes yield a crystal system which is called orthorhombic and has the metric relationships

$$\alpha = \beta = \gamma = 90^\circ$$

We leave it as an exercise for the reader to show that appropriate combinations of mirror planes and 2-fold axes also lead to the same restrictions.

2.4.4. Tetragonal System

In this case, we consider the restrictions imposed by a 4-fold $4(C_4)$ or $\overline{4}(S_4^3)$ operation. By the same reasoning that we applied to the monoclinic system, we can see that if $4(C_4)$ is taken along the **c**-axis (the conventional choice), then **a** and **b** must be perpendicular to **c**. The 4-fold operation, moreover, means that +**a** must move to +**b**, +**b** to –**a**, –**a** to –**b** and –**b** to +**a**, in order that we do not produce a surplus of axes. This can be written mathematically as

$$\mathbf{r}' = \{4[001]\}\mathbf{r} = -y\mathbf{a} + x\mathbf{b} + z\mathbf{c} \qquad [2.27]$$

Similarly, we find that

$$\mathbf{r}' = \{4^3[001]\}\mathbf{r} = y\mathbf{a} - x\mathbf{b} + z\mathbf{c} \qquad [2.28]$$

with similar relations for $\overline{4}(S_4^3)$ and $\overline{4}^3(S_4)$. Again, the opposite signs show that **a**, **b** and **c** are perpendicular. Notice that there is now an interchange of x and y which means that **a** and **b** must have identical lengths. Therefore $4(C_4)$ or $\overline{4}(S_4^3)$ results in a new crystal system that we call the tetragonal system

$$\mathrm{a} = \mathrm{b} \quad \alpha = \beta = \gamma = 90^\circ$$

Of course, c can be greater or smaller than a = b.

2.4.5. Cubic System

This crystal system is familiar to solid-state physicists and chemists, and it is the system with the highest symmetry. However, despite its familiarity, we need to be careful about how we define it; it is not sufficient to use the criterion that all axes are equal and all angles 90°. As we continue to emphasize, the symmetry is the important thing in determining the crystal system. The symmetry conditions our choice of axes, not the other way round. What, then, are the important symmetry elements in the cubic system? Surprisingly perhaps, they are not the three mutually perpendicular 4-fold axes so readily observed in a cube, but rather, the four 3-fold axes corresponding to the body diagonals, <111>, of the cubic unit cell. (Recall that **angular brackets** signify the set of symmetry-equivalent directions. In this case, <111> means the set $[111]$, $[11\bar{1}]$, $[1\bar{1}1]$, $[1\bar{1}\bar{1}]$, $[\bar{1}1\bar{1}]$, $[\bar{1}\bar{1}1]$, $[\bar{1}11]$ and $[\bar{1}\bar{1}\bar{1}]$.) As we shall see later, it is possible to have a cubic crystal without any 4-fold axes of symmetry. It is also possible to prove either by group theory or by spherical trigonometry that if a crystal contains more than one 3-fold (C_3) axis, then it must contain four altogether, each one making an angle of 109° 28′ with any other.

We shall now show that the four 3-fold axes give rise to a cubic unit cell. Refer to Fig. 2.8. The 3-fold rotation about $[111]$ (number 1 on the diagram) acting on the vector **r** once and twice gives

$$\{3[111]\}\mathbf{r} = z\mathbf{a} + x\mathbf{b} + y\mathbf{c} \quad [2.29]$$

$$\{3^2[111]\}\mathbf{r} = y\mathbf{a} + z\mathbf{b} + x\mathbf{c} \quad [2.30]$$

Since the components have been freely interchanged, this implies that all axes are interchangeable and hence equal in length. The 3-fold rotation about $[11\bar{1}]$, marked (2), gives

$$\{3[11\bar{1}]\}\mathbf{r} = y\mathbf{a} - z\mathbf{b} - x\mathbf{c} \quad [2.31]$$

$$\{3^2[11\bar{1}]\}\mathbf{r} = -z\mathbf{a} + x\mathbf{b} - y\mathbf{c} \quad [2.32]$$

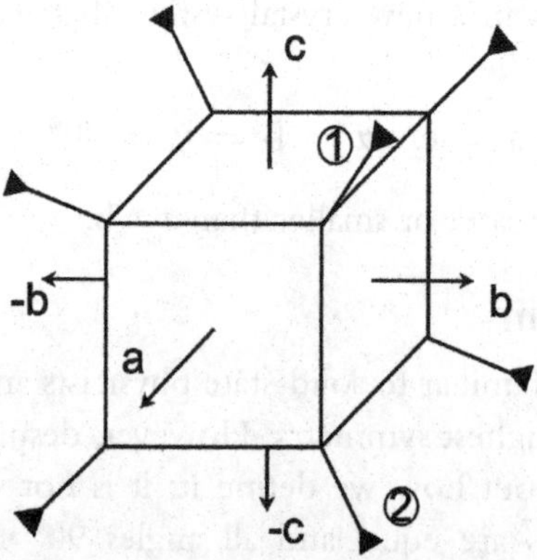

Figure 2.8 The four 3-fold symmetry axes of a cube.

with similar relations for the remaining 3-fold operations. We see then that in addition to the equality of the axes the signs themselves permute, implying orthogonality for all the axes. Therefore, our choice of four 3-fold axes gives a new crystal system, called the cubic crystal system, which has

$$a = b = c \quad \alpha = \beta = \gamma = 90^\circ$$

2.4.6. Trigonal and Hexagonal Systems

We have deliberately left the trigonal and hexagonal systems till last because they both present special problems that in some respects render them different from the other crystal systems. Considerable confusion about these systems is found in the literature; we hope that the following and the discussion in Section 3.2.6 on centering will clarify the problem. Perhaps, it is best to reread this section after reading Section 3.2.6.

Let us first deal with the **hexagonal** system. This crystal system can be defined as one determined by a single $6(C_6)$ or $\overline{6}(S_3^5)$ symmetry operation. Note that we immediately encounter a conceptual difficulty in that $\overline{6}$ is equivalent to an improper 3-fold rotation of the Schoenflies-type S_3^5, or alternatively to a proper 3-fold rotation with a perpendicular reflection ($3/m$ in International notation). Either way it can be confusing that the hexagonal system may be described both by 6-fold as well as 3-fold rotations in conjunction with other operations. The 6 or $\overline{6}$ symmetry leads, by similar considerations to those used in the other crystal systems, to the **a** and **b**-axes being at 120° to each other. This may be determined by using Fig. 1.13 or the matrices in Appendix 1

$$\mathbf{r}' = \{6[001]\}\mathbf{r} = \begin{bmatrix} 1 & -1 & 0 \\ 1 & 0 & 0 \\ 0 & 0 & 1 \end{bmatrix} \mathbf{r} = x'\mathbf{a} + y'\mathbf{b} + z'\mathbf{c} \qquad [2.33]$$

where

$$\begin{aligned} x' &= 1x - 1y + 0z \\ y' &= 1x + 0y + 0z \\ z' &= 0x + 0y + 1z \end{aligned} \qquad [2.34]$$

Therefore, we can write

$$\{6[001]\}\mathbf{r} = x(\mathbf{a}+\mathbf{b}) - y\mathbf{a} + z\mathbf{c} \qquad [2.35]$$

Similarly, we find that

$$\mathbf{r}'' = \{6^2[001]\}\mathbf{r} = x\mathbf{b} + y(-\mathbf{a}-\mathbf{b}) + z\mathbf{c} \qquad [2.36]$$

and so on.

The free interchange of coordinates x and y with respect to the **a** and **b** axes indicates that **a** and **b** must be equal in length. Furthermore, it is simple to show that these equations are consistent with the **a** and **b** axes being at 120° to one another. The scalar product of the components along the **a** and **b** axes before transformation can be related to the product after transformation. For $\{6[001]\}\mathbf{r}$, this is

$$x\mathbf{a} \cdot y\mathbf{b} = x(\mathbf{a}+\mathbf{b}) \cdot -y\mathbf{a} \qquad [2.37]$$

Therefore, we obtain

$$xy\,|\mathbf{a}|\,|\mathbf{b}|\cos\gamma = xy\left[-|\mathbf{a}|\,|\mathbf{b}|\cos\gamma - |\mathbf{a}|^2\right] \qquad [2.38]$$

and since $|\mathbf{a}| = |\mathbf{b}|$, this yields

$$\cos\gamma = -1/2 \qquad [2.39]$$

Thus the **a** and **b** axes are at 120° to each another. Similarly, we can show that **c** is perpendicular to **a** and **b**. The 6-fold symmetry therefore implies that

$$\mathrm{a} = \mathrm{b} \quad \alpha = \beta = 90^\circ \quad \gamma = 120^\circ$$

It should be realized that in addition to the **a** and **b** axes, there is another direction, –**a**–**b**, i.e. $[\bar{1}\,\bar{1}\,0]$, that is equivalent in magnitude to and 120° from **a** and **b** (see Fig. 2.9). Naturally, we could equally well choose this as an axis, so that there are four possible axes in this system. In fact, four axes are sometimes used, particularly when dealing with crystal morphology. However, we shall continue to use the three-axis notation here. Another point to note is that sometimes you will see in books that the unit cell of a hexagonal lattice is drawn as a hexagonal prism. This is misleading. A hexagonal prism is made up from three primitive unit cells. The conventional hexagonal unit cell, like all others used in defining crystal systems, is a parallelepiped (see Fig. 2.10a and b for example).

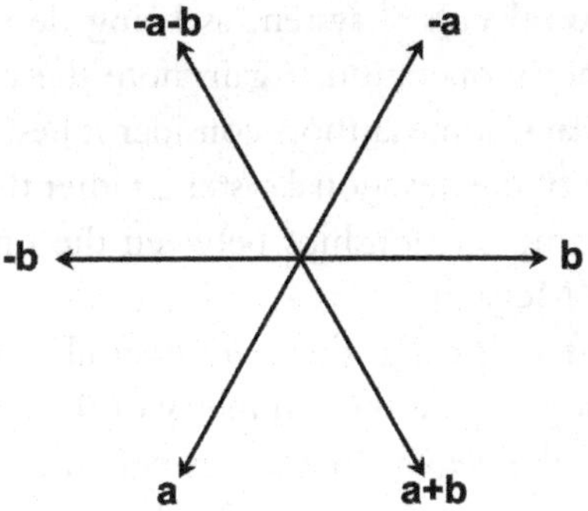

Figure 2.9 The hexagonal system of axes.

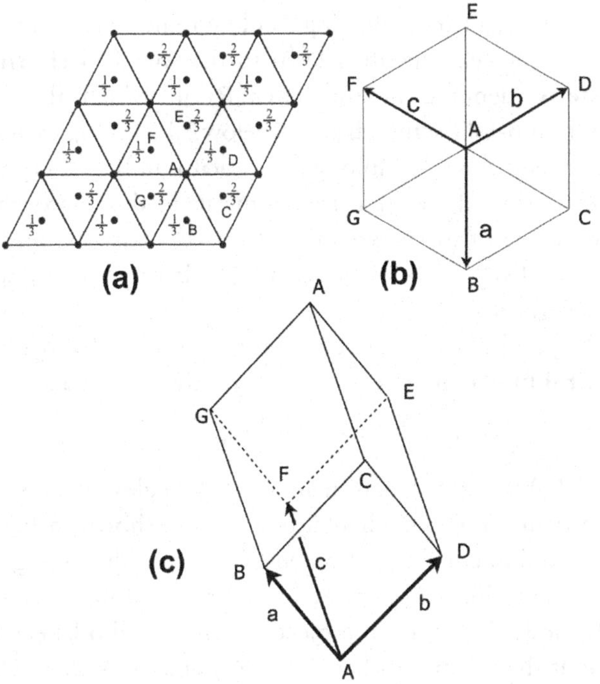

Figure 2.10 (a) The projection, down the **c**-axis, of a hexagonal lattice with additional points (centering) at (2/3, 1/3, 1/3) and (1/3, 2/3, 2/3). These points are labelled B and C, respectively, in the unit cell at the bottom right. (b) The rhombohedral unit cell looking down the hexagonal axis. (c) The rhombohedral unit cell in a 3-dimensional view. The capital letters A to G are used to indicate the relationship of the points in all three figures.

We define a **trigonal** crystal system as being determined by a single $3(C_3)$ or $\overline{3}(S_6^5)$ symmetry operation. Again note the conceptual difficulty with $\overline{3}$ and S_6^5. Nowadays, some authors consider it best to treat the trigonal system as a special case of the hexagonal system, rather than treat it separately since both have the same relationships between the unit-cell axes (see for example the book by Megaw).

The problem is the following: there are basically two ways of defining **crystal systems.** One is to use the symmetry of the *crystal*; the other is to use the symmetry of the *lattice*. In the former, the arrangement of the atoms or molecules is such as to display 3-fold symmetry, and hence it is reasonable to denote this as trigonal. In the latter case, a primitive hexagonal lattice (denoted by the symbol *P*) shows 6-fold symmetry, and this leads to the **hexagonal crystal system**. If you try to construct a primitive trigonal lattice you will find that this is identical to the hexagonal lattice. However, by adding extra lattice points, it is then possible to construct a lattice showing 3-fold symmetry (see Fig. 2.10a). By redefining the unit cell, this then leads to the so-called **rhombohedral crystal system** (denoted by the letter *R*) in which there is 3-fold symmetry but *no* 6-fold symmetry (see below). In this approach, there is no trigonal system, as such, although the total number of crystal systems remains seven. While there are certainly merits to this approach, we shall adhere to the usage of the present *ITA*, to have a separate hexagonal and a separate trigonal system, with the rhombohedral system as a special case of the trigonal system.

We shall therefore now discuss this special case by describing the **rhombohedral unit cell**. The conditions on the axes and angles are

$$a = b = c \qquad \alpha = \beta = \gamma$$

where the $3(C_3)$ or $\overline{3}(S_6^5)$ axis makes equal angles with **a**, **b** and **c**. Different views of the rhombohedral unit cell are shown in Fig. 2.10. To develop the rhombohedral crystal system, we start with a hexagonal lattice and center it; centering will discussed in more detail in the next chapter. Figure 2.10a shows the lattice in projection with **a** and **b** labelled and new points added at the positions (2/3, 1/3, 1/3) and (1/3, 2/3, 2/3); in the lower right unit cell, these points are labelled B and C, respectively. These points have been added so that the resultant total set of points (the original hexagonal lattice points plus these new points) has trigonal symmetry, and no longer has hexagonal symmetry. Once this is done, a rhombohedral unit cell can be chosen and this is indicated in Fig. 2.10b and c, where the capital letters, A to G, label the original positions of the points in Fig. 2.10a.

Centering in all the crystal systems is discussed completely in the next chapter and so we leave the details of centering the hexagonal lattice till 3.2.6. However, notice that the rhombohedral unit cell shown in Fig. 2.10c is primitive and is consistent with the symmetry of the trigonal crystal system but not with that of the hexagonal crystal system since it does not possess $6(C_6)$ or $\overline{6}(S_3^5)$ symmetry.

2.5. SUMMARY

In Appendix 2, the seven crystal systems are summarized in a convenient form. In most books, you will see that the crystal systems are defined in terms of axial and interaxial relationships. For example, the tetragonal system is sometimes defined by

$$a = b \neq c \quad \alpha = \beta = \gamma$$

This is only true if you are considering the lattice as a concept in isolation from a crystal structure. However, when thinking of a structure being described in terms of a particular lattice, the $\neq$ sign should be thought of as meaning 'not necessarily equal to'. Accidental equalities can occur in, say, orthorhombic, monoclinic or triclinic crystals, in which case inspection of the crystal structure (the arrangement of atoms) will show that there are not the required symmetry operations relating the atoms. It should be understood that the axial lengths and interaxial angles for the seven crystal systems are determined by the symmetry conditions, and not the other way round. Therefore, as a better way to define the crystal systems, we stress here only the equalities between axes and angles that are forced by symmetry. The results are also summarized below.

$1(E)$ or $\overline{1}(i)$	Triclinic	No conditions
$2(C_2)$ or $\overline{2}(\sigma)$	Monoclinic	$\alpha = \beta = 90°$ (first setting)
		$\alpha = \gamma = 90°$ (second setting)
Three 2-fold or $\overline{2}$ axes	Orthorhombic	$\alpha = \beta = \gamma = 90°$
$4(C_4)$ or $\overline{4}(S_4^3)$	Tetragonal	$a = b$; $\alpha = \beta = \gamma = 90°$
Four 3-fold or $\overline{3}$ axes	Cubic	$a = b = c$; $\alpha = \beta = \gamma = 90°$
$6(C_6)$ or $\overline{6}(S_3^5)$	Hexagonal	$a = b$; $\alpha = \beta = 90°$; $\gamma = 120°$
$3(C_3)$ or $\overline{3}(S_6^5)$	Trigonal	$a = b$; $\alpha = \beta = 90°$; $\gamma = 120°$
	Rhombohedral	$a = b = c$; $\alpha = \beta = \gamma$

Problems

1. **(a)** Show that the volume of a unit cell is given by

$$\mathrm{abc}\,(1-\cos^2\alpha-\cos^2\beta-\cos^2\gamma+2\cos\alpha\cos\beta\cos\gamma)^{1/2}$$

(b) Prove that in a given lattice all primitive unit cells are equal in volume irrespective of the choice of axes.

(c) Show that in the triclinic system the distance between two points at $\mathbf{r}_1$ and $\mathbf{r}_2$ is

$$\begin{bmatrix} (x_1-x_2)^2\mathrm{a}^2+(y_1-y_2)^2\mathrm{b}^2+(z_1-z_2)^2\mathrm{c}^2 \\ +2(x_1-x_2)(y_1-y_2)\mathrm{ab}\cos\gamma+2(y_1-y_2)(z_1-z_2)\mathrm{bc}\cos\alpha \\ +2(z_1-z_2)(x_1-x_2)\mathrm{ca}\cos\beta \end{bmatrix}^{1/2}$$

(d) Draw a cubic unit cell with lattice points on all corners. Add a lattice point to the center of each unit-cell face. Show, by shifting the origin of the unit cell, that there are four lattice points in such a cell.

2. Assuming the presence of a single $\overline{4}(S_4^3)$ symmetry operation prove, using the matrices in Appendix 1, that $\mathrm{a}=\mathrm{b}$ and that $\alpha=\beta=\gamma=90°$. Find the relationship between the axes and angles when a unit cell has a mirror plane $m(\sigma)$ containing a $2(C_2)$ axis.
3. Show that if there is a 3-fold axis along [111] and a 2-fold axis along [001] there are four 3-fold axes in total.
4. Sketch any object that has cubic symmetry but does not contain a $4(C_4)$ axis.
5. Show that by combining the four operations $1(E)$, $\overline{1}(i)$, $4[001](C_4[001])$ and $3[111](C_3[111])$ a total of 48 operations is obtained.

CHAPTER 3

Bravais Lattices

Contents

Nature is an endless combination and repetition of a very few laws. She hums the old well-known air through innumerable variations.

Ralph Waldo Emerson (1803–1882)

SYMMETRY AND LATTICES

The aim of this chapter is to show that there are only 14 distinct lattices that fill all space. These lattices are called the **14 space lattices**, or more often, the **14 Bravais lattices.** These were published by Auguste Bravais in 1845, although earlier Moritz Ludwig Frankenheim had come up with the same idea, but made the mistake of proposing 15 unique lattices. Note that in some textbooks you will find that Eqn [2.1] is used to *define* a Bravais lattice: this is misleading as it just defines a lattice, and does not directly describe the uniqueness of certain lattices.

As we have already seen in the previous chapter, there are seven crystal systems. It might be thought that by combining the seven crystal systems

Space Groups for Solid State Scientists
http://dx.doi.org/10.1016/B978-0-12-394400-9.00003-4

with the idea of a primitive lattice a total of seven distinct (Bravais) lattices would be obtained, one for each crystal system. However, it turns out that the trigonal and hexagonal lattices so constructed are equivalent, and therefore only six are, in fact, formed in this way. These lattices, to which we give the label *P*, define **primitive unit cells** in each case.

The other eight Bravais lattices arise by taking each of the six *P*-lattices and considering what happens when other lattice points are added at certain places called **centering**. The first question is that, after centering, is the new arrangement still a lattice? The second question is, does it form a *new* lattice? This results in eight centered lattices, seven of which are given a name (body-centered, all-face-centered, one-face-centered) and a new symbol (*I*, *F* and *A*, *B* or *C*). The eighth new lattice is a specially centered hexagonal lattice which, as we shall see in 3.2.6, can be regarded, after appropriate redefinition of axes of reference, as a primitive rhombohedral lattice.

For each crystal system, it is clear that the *I*, *F* or *C*-centered lattice has a unit cell that contains more than one lattice point since we have added lattice points in various centered positions. Unit cells with more than one lattice point are **multiply-primitive unit cells** as described in Section 2.2. It is convenient, and conventional, to use these multiply-primitive unit cells formed like this when dealing with Bravais lattices. These so-called **centered unit cells** are discussed in detail in Section 3.1. Using this convention, all the space lattices in a given crystal system are referred to the same axes. Thus all of these conventional, centered unit cells display exactly the same rotational symmetry as the corresponding primitive cell in a given crystal system. However, it is important to realize that for any of these new *I*, *F* or *C*-centered lattices, it is always possible to choose a smaller unit cell that has only one lattice point in it. This **primitive unit cell of** an *I*, *F* or *C*-centered lattice is a perfectly acceptable unit cell since by translation of its lattice vectors the entire space lattice will be reproduced. The disadvantage is that such a unit cell by itself does not obviously display the full rotational symmetry of the crystal system. Thus primitive cells of *I*, *F* or *C*-centered lattices are not conventionally used by crystallographers. On the other hand, solid-state scientists often use such a primitive unit cell for problems concerning electronic band theory and lattice vibrations or in any wave-vector counting problem. This is because a primitive unit cell has the full translational symmetry of the lattice and thus has the full translational symmetry of the Hamiltonian. These remarks will become clearer when these primitive cells are discussed in Section 3.3.

3.1. CENTERING OF LATTICES

As mentioned above, the assignment of axes of reference in relation to the rotational symmetry of the crystal systems defines six lattices which, by definition, are primitive. For these lattices, we need to determine if more points can be added in such a way that the lattice condition is still maintained. At the same time, addition of points must not alter the crystal system. These are the *two conditions* that are necessary to form a new Bravais lattice. For example, if we start with a cubic primitive lattice and add other lattice points in such a way that we still have a lattice, we also want to make sure that this new lattice still possesses cubic symmetry.

We discuss the addition of lattice points in general in this section and the specific results for each crystal system in the next section. Since the lattice condition must be maintained when these new points are added, the points must be added to highly symmetric positions of the primitive lattice. These types of positions are as follows: a single point at the body center of each unit cell; a point at the center of each independent face of the unit cell; a point at the center of one face of the unit cell (points at the centers of two independent faces of the unit cell do not permit a lattice to be formed) and the special centering positions in the trigonal system that give a rhombohedral lattice (see Section 2.4.6). We shall discuss each type of centering separately. Then in Section 3.2, we shall consider each crystal system separately and determine if within the crystal system the various kinds of centering maintain the lattice condition and form a new kind of lattice.

3.1.1. Body Centering (*I*)

For this kind of centering, an additional point must be placed at the end of the vector $\mathbf{a}/2 + \mathbf{b}/2 + \mathbf{c}/2$. The result is shown in Fig. 3.1b, which shows a point at the body center of the unit cell. The resulting lattice is given the symbol *I* (from the German Innenzentrierung). Notice that this unit cell

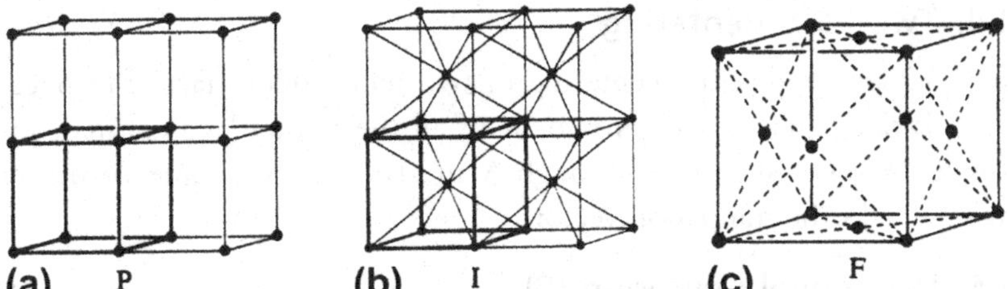

Figure 3.1 Two types of centering are shown. (a) Four primitive unit cells (not necessarily cubic). (b) Body centering – all four unit cells are shown. (c) All-face centering – only one unit cell is shown for clarity.

will now contain two lattice points, one at the origin of the unit cell (0, 0, 0), and one at the body center (½, ½, ½). The other lattice points in the diagram belong to neighbouring unit cells. Or alternatively, one may say that 1/8 of a lattice point is at each of the eight corners of the unit cell and one lattice point is at the body-center position, although there is no intrinsic difference between the lattice points at the corner or at the body center. (A better way of looking at this, as mentioned in Chapter 2, is to displace the origin of the unit cell away from a lattice point, remembering that the choice of origin is entirely arbitrary, in which case both lattice points are seen to be contained completely within the unit cell.)

3.1.2. Face Centering (*F*)

For this type of centering, three new points are added to the initial primitive unit cell. They are placed at the centers of each of the faces of the unit cell or at positions described by the end points of vectors $\mathbf{a}/2 + \mathbf{b}/2$, $\mathbf{a}/2 + \mathbf{c}/2$ and $\mathbf{b}/2 + \mathbf{c}/2$. The lattice so obtained is given the symbol *F*. The conventional unit cell shown in Fig. 3.1c contains four lattice points at (0, 0, 0), (½, ½, 0), (½, 0, ½) and (0, ½, ½). Alternatively, one may say that this unit cell contains 1/8 of a lattice point at each of the eight corners and 1/2 of a lattice point on each of the six faces.

3.1.3. One-Face Centering (Base Centering)

In this centering, only one face is centered, not all three. If the centering is on the **ab**-plane (at $\mathbf{a}/2 + \mathbf{b}/2$), the lattice which results is given the symbol *C*. Similarly, the lattice is given the symbol *A* if the centering is on the **bc**-plane and B, if the centering is on the **ac**-plane. In each case, there are two lattice points per unit cell. These are:

for *A* centering: (0,0,0) and (0, ½, ½)
for *B* centering: (0,0,0) and (½, 0, ½)
for *C* centering: (0,0,0) and (½, ½, 0)

3.1.4. Two-Face Centering

Figure 3.2 shows that the centering of two independent faces can never form a lattice, since the environment of all the points (as indicated, for example, by the dashed lines through A and B) is not the same no matter how one picks the translation vectors.

3.1.5. Hexagonal Centering (*R*)

We have already shown in the last chapter that a trigonal unit cell can be centered in a special way to give a rhombohedral cell. There are two

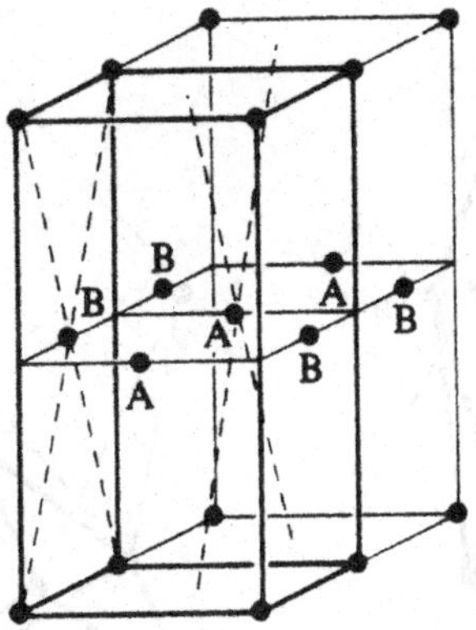

Figure 3.2 An impossible way to center a lattice.

possible rhombohedral centering positions, at ±(2/3, 1/3, 1/3) and ±(1/3, 2/3, 1/3). The resulting lattice is given the symbol *R*. It is easy to become confused by the rhombohedral lattice and it is best to consult Section 3.2.6. Part of this confusion results from the fact that the rhombohedral lattice can be referred either to rhombohedral axes, resulting in a rhombohedral unit cell with one lattice point, or to hexagonal axes with a unit cell that looks hexagonal but has three lattice points per cell to give a 3-fold axis of symmetry.

It is possible to body-center and face-center the rhombohedral lattice. However, such lattices are not new lattices since an *R*-lattice can still be formed, with different angles between the axes and with only one lattice point per cell. Examples of this are shown in Fig. 3.3.

3.2. THE 14 BRAVAIS LATTICES

In order to discuss the 14 Bravais lattices, we shall consider the seven crystal systems in turn and show which sort of *unique* space lattices can be formed in each case. As in the previous chapter, we start with the lowest symmetry, triclinic, and progressively go on to crystal systems of higher symmetry. As before, we shall leave the trigonal and hexagonal systems till last. Figure 3.4 shows the 14 Bravais lattices. Note that in the *ITA*, the Bravais lattices have special symbols given by a lower case letter (crystal system) followed by the lattice letter. These should be self-evident, although note that the letter '*a*' in the triclinic system stands for "anorthic" (Appendix 2).

3.2.1. Triclinic

In this system, there are no restrictions either on the magnitudes of the translation vectors, i.e. the lengths of the unit-cell axes, or on their

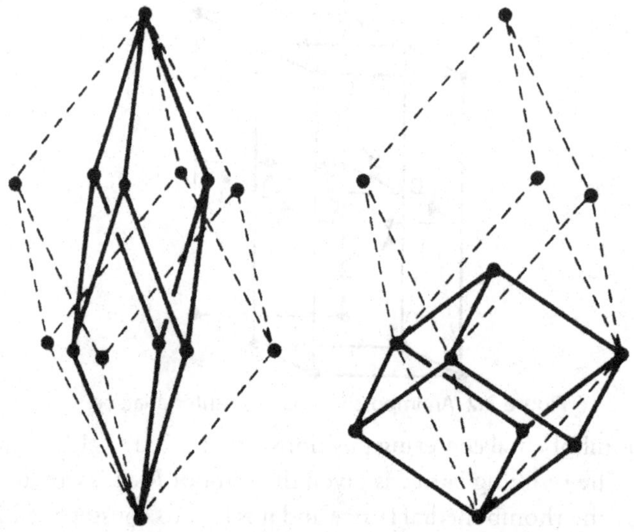

Figure 3.3 A rhombohedral unit cell (dashed lines) shown face-centered and body-centered. The full lines indicate primitive cells constructed from a hexagonal centered unit cell.

directions, i.e. the interaxial angles. Therefore, we can always take a triclinic lattice and center it and a lattice will indeed be formed compatible with the restrictions (none!) of the triclinic crystal system. However, there is nothing new about this lattice. A smaller primitive cell can be determined with the same complete arbitrariness of the cell edges and angles. Thus, for the triclinic crystal system, there is only one Bravais lattice, the primitive or *aP*-lattice.

3.2.2. Monoclinic

Figure 3.5 shows three ways to center a monoclinic unit cell. Here, we use the 'first setting' which takes the unique 2-fold axis as the **c**-axis. As can be seen in Fig. 3.5b, if you try to center the *C*-face (**ab**-plane) nothing new is obtained. The lattice can still be described as a *P*-lattice with different values of a and γ but the fundamental monoclinic conditions, **c** perpendicular to both **a** and **b**, with γ and all axial lengths unrelated (and described by a single 2-fold axis or mirror plane of symmetry), are retained. Thus, we say $mP \equiv mC$ for the monoclinic crystal system.

However, a new lattice is obtained if the *B*-face is centered as shown in Fig. 3.5c. This is because it is not possible to maintain the fundamental monoclinic conditions and yet describe this lattice as an *mP*-lattice. An

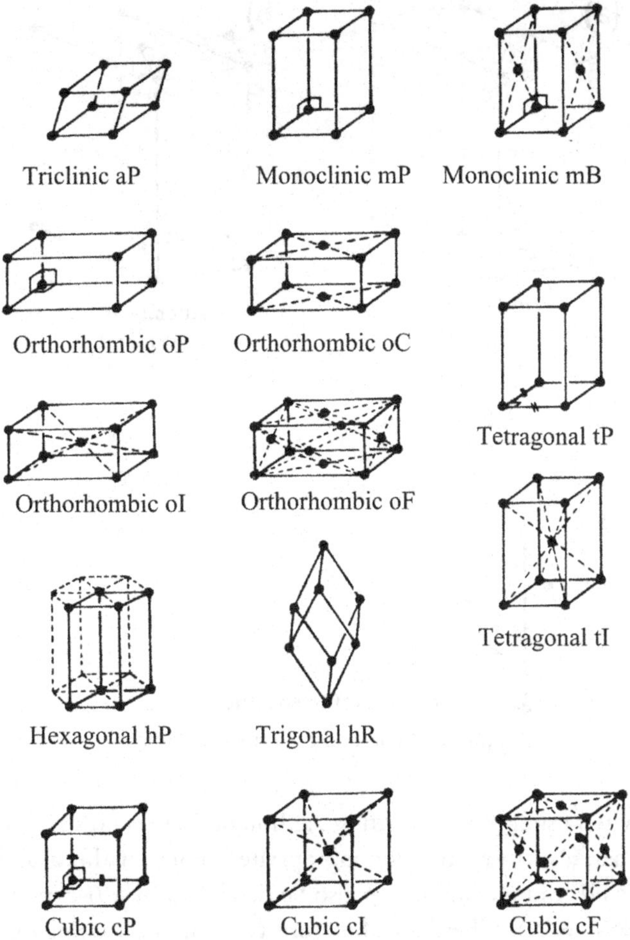

Figure 3.4 The conventional unit cell for each of the 14 Bravais lattices. (The hexagonal unit cell is outlined in black. The dashed prism shown only serves to indicate the angles of the unit cell.)

attempt at this is shown by the dashed cell, where with regard to this primitive cell the 2-fold symmetry appears to be lost. Of course, we know that the lattice still possesses a 2-fold axis. The multiply-primitive unit cell outlined with a full line in Fig. 3.5c does display the 2-fold axis and so this is the conventionally chosen unit cell. A lattice centered in this way is called an *mB*-lattice and it is also shown in Fig. 3.4, where all the 14 Bravais lattices are shown. In the same way, the **bc**-plane can be centered to give rise to an *mA*-lattice. It is usual, if one is dealing with a crystal not previously described, to choose the axes to give an *mB*-lattice rather than the equivalent *mA*-lattice

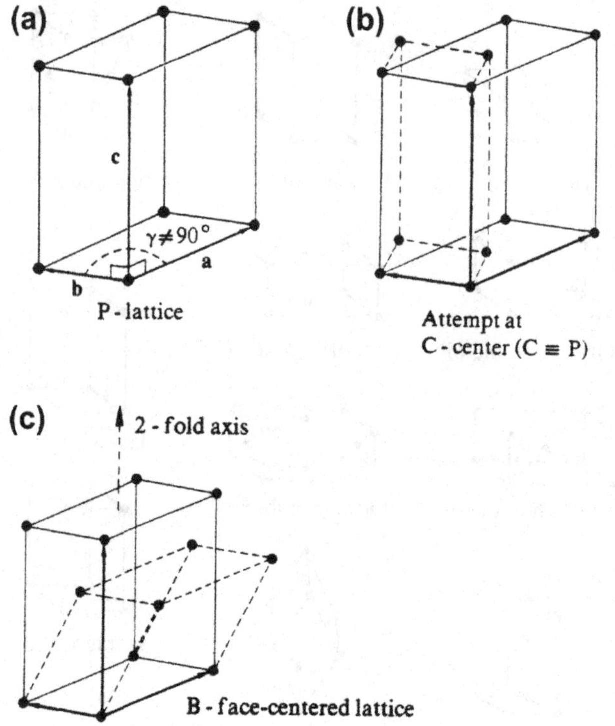

Figure 3.5 Centering a monoclinic lattice.

($B \equiv A$) when using the 'first setting', although there is no strict rule about this. One can also show that by appropriate choice of the **a** and **b**-axes, monoclinic *mF* and *mI* lattices can also be described by a *B*-centered lattice ($B \equiv F \equiv I \equiv A$). There are therefore two unique monoclinic Bravais lattices, *mP* and *mB* (Fig. 3.4).

The 'second setting', preferred by crystallographers, takes the unique 2-fold axis as the **b**-axis. In this case, the above discussion is still appropriate except for the interchange of axes. Thus, in this setting, the *mP* and *mC* lattices are unique monoclinic Bravais lattices and $C \equiv F \equiv I \equiv A$, *C* being the usual choice.

3.2.3. Orthorhombic

We may consider a primitive orthorhombic lattice to arise from a primitive monoclinic lattice with the added restriction that the third angle must also be 90°. Then all the unit-cell translation vectors are 90° to one another but with unrelated lengths. In the orthorhombic cell, any one face may be centered, but if we try to construct a primitive unit cell from it, in just the same way as we did for the monoclinic *C*-centered cell in Fig. 3.5b, we would find that the axes so obtained are not orthogonal. Thus we have an *oC* lattice which can also be described as an *oA* or *oB* lattice by interchange of the orthogonal axes. In this crystal system, the all-face-centered *oF* lattice and the body-centered *oI* lattice are also distinct from the *oP* or *oC* lattices. All unit cells have three mutually perpendicular 2-fold axes (or mirror planes) of symmetry, as required in the orthorhombic system. Thus there are four unique Bravais lattices, *oP*, *oI*, *oF* and *oC*. All four are illustrated in Fig. 3.4. Again we note that *C*, *A* and *B*-face centering form identical lattices but with a redefinition of the labels for the axes.

3.2.4. Tetragonal

We already know in general that a lattice is not obtained if two faces, the *A* and *B* faces say, are centered (Section 3.1.4). Furthermore, the tetragonal requirement that 4 or $\bar{4}$-fold symmetry should be present is not maintained if only one of these two faces is centered. Thus, for one-face centering, only the *C*-face centering need be considered. Centering this face does result in a lattice, but, as can be seen in Fig. 3.6, the lattice is the same as a *tP* lattice rotated by 45° about the **c**-axis. Thus, in the tetragonal crystal system, $tP \equiv tC$. Since the primitive cell is the smaller of the two cells (one lattice point per cell compared with two), the primitive cell is usually chosen.

Now consider the effect of body centering. As in the orthorhombic lattice, body centering of a tetragonal lattice still gives us a lattice. The surroundings of every point are still identical, with each point having eight nearest neighbours at the same distance and in the same directions. The 4-fold symmetry is maintained. Thus the *tI* lattice is a new lattice in the tetragonal crystal system.

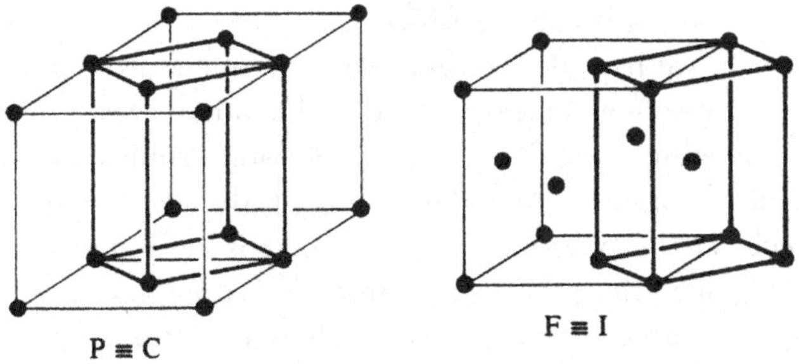

Figure 3.6 Centering a tetragonal lattice.

Face centering also gives a lattice in the tetragonal crystal system. However, just as a *tP* lattice can be obtained from a *tC* lattice by new tetragonal axes rotated 45° to the old ones, a *tI* lattice can be obtained from the *tF* lattice. This is shown in Fig. 3.6. Thus $tF \equiv tI$ in the tetragonal system, and since the *I*-cell is smaller than the *F*-cell, the *I*-cell is usually preferred.

Therefore, we see that for the tetragonal crystal system there are only two distinct Bravais lattices, *tP* and *tI* (and $tC \equiv tP$ and $tF \equiv tI$).

3.2.5. Cubic

Figure 3.1b shows a primitive cubic lattice that has been centered by placing a lattice point at the body-center position. It is clear that a new lattice is formed. Each point is surrounded by eight other nearest-neighbour points, all in the same relative positions no matter from which point one observes, and the four 3-fold axes, necessary to define the cubic system, are preserved. Therefore, the cubic crystal system can have a *cI* lattice, sometimes called the **bcc (body-centered cubic)** lattice.

Figure 3.1c shows the result when extra points are added at the face centers of the primitive cubic unit cell. Close observation shows that a lattice is again formed. Every lattice point is surrounded by 12 nearest-neighbour points. Once more, it is obvious that the four 3-fold axes remain. This *cF* lattice is sometimes called the **fcc (face-centered cubic)** lattice.

It is clear that the cubic crystal system cannot have a base-centered lattice because centering only one face would destroy the four 3-fold axes of symmetry. Thus, we conclude that there are three cubic Bravais lattices, *cP*, *cI*, and *cF*, which are shown in Fig. 3.4.

3.2.6. Hexagonal, Trigonal (and Rhombohedral)

As we have already seen in Section 2.4.6, when dealing with these systems, complications crop up that can cause confusion. In this section, we discuss in detail the centering of hexagonal lattices.

In Fig. 3.7a, we have drawn four primitive unit cells with hexagonal axes. Let us try to center it as we did the other crystal systems. In Fig. 3.7b, we show how this grouping of four primitive cells can be centered. First, consider the solid points at the base center of each primitive cell (at the position $\mathbf{a}/2 + \mathbf{b}/2$). It is clear that a hexagonal lattice is not formed since the 6 or $\overline{6}$ symmetry is not retained. In fact, the crystal system is now orthorhombic as shown by the dashed lines in the figure. If, on the other hand, the solid points are placed at the

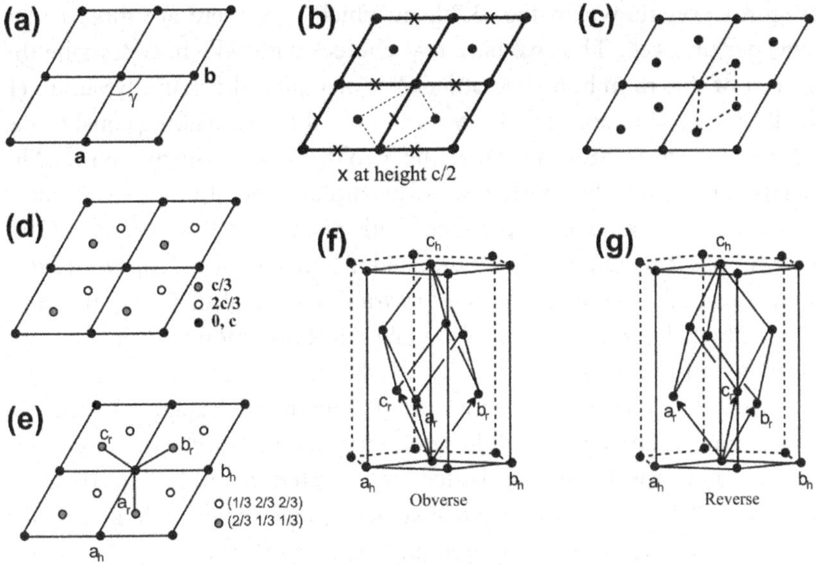

Figure 3.7 Hexagonal, trigonal and rhombohedral lattices.

body-centered position (at $\mathbf{a}/2 + \mathbf{b}/2 + \mathbf{c}/2$), a hexagonal lattice is still not formed. Now consider face centering. Centering on the side faces is represented by the X's in Fig. 3.7b. Once again, a hexagonal lattice is not formed. Figure 3.7c shows the same four primitive cells but centered at positions (1/3, 2/3, 0) and (2/3, 1/3, 0). A hexagonal lattice is now formed, since the 6-fold symmetry elements are maintained. However, this lattice is really still primitive, as can be seen from the dashed lines in Fig. 3.7c; these form a hexagonal primitive unit cell but in a different orientation and with different axial lengths from the original cell, thus enclosing a volume one-third of the original unit cell.

On the other hand, by centering the primitive lattice at the two positions (1/3, 2/3, 2/3) and (2/3, 1/3, 1/3), which is equivalent to ±(1/3, 2/3, 2/3), we do produce a new lattice (Fig. 3.7d), where the heights along the **c**-axis are indicated by the key. The surroundings of every point are identical and so a lattice is formed. Notice, however, that the lattice no longer has 6 or $\overline{6}$ symmetry but $\overline{3}$ instead. As explained in Chapter 2, it is now possible to define a new cell in the shape of a rhombohedron. Its edges, shown in Fig. 3.7 and Fig. 2.10, are the basic translation vectors of a new lattice, the rhombohedral lattice *hR*.

It is evident from the figure that the rhombohedral unit cell is primitive. We have also seen that this Bravais lattice can always be referred to the hexagonal axes shown in Fig. 3.7d, in which case there are three lattice points per unit cell. Thus we have two choices with which to describe the unit cell of this rhombohedral lattice. We may pick the primitive unit cell which has $a = b = c$ and $\alpha = \beta = \gamma$ with the 3-fold axis making equal angles with the three axes, or we may take the 3-fold axis as a principal axis **c**. The latter is conventionally referred to as the rhombohedral unit cell specified with **hexagonal axes of reference**, with $a = b$, $\alpha = \beta = 90°$, $\gamma = 120°$. The symbol *hR* is used for the rhombohedral lattice *no matter whether specified on hexagonal or rhombohedral axes of reference.* As we shall see in the next chapter, the Schoenflies symbol also makes no distinction between the two sets of axes of reference.

Note another point of confusion: we centered a hexagonal lattice to produce a new lattice with 3-fold symmetry. Such a lattice could be called trigonal. The rhombohedral lattice constructed from it can then be considered to belong to the trigonal system. Nevertheless, when specified with respect to the original centered unit cell, we still talk about 'hexagonal axes of reference' rather than 'trigonal axes of reference'. The reason for this is that as far as the primitive lattice (Fig. 3.7a) is concerned, there is no distinction between hexagonal and trigonal (see Section 2.4.6); only when we consider the centering of the unit cell, can we make any distinction. Although the primitive rhombohedral unit cell has the advantage of containing only one lattice point, it is frequently simpler to consider the unit cell with hexagonal axes of reference because hexagonal coordinates are easier to deal with than rhombohedral coordinates, especially when one is trying to visualize the crystal structure.

A further complication arises because the rhombohedral axes can be orientated in two ways relative to the set of hexagonal axes. Instead of choosing our centering at $\pm(2/3, 1/3, 1/3)$, we could have chosen $\pm(1/3, 2/3, 1/3)$. The rhombohedron, thus constructed, is turned through 180° with respect to the previous one. The two settings are given the names **obverse** for the first and **reverse** for the second. These are shown in Fig. 3.7f and g with respect to a fixed set of hexagonal axes. It is generally considered best, in order to avoid confusion, to keep to the obverse setting wherever possible, and this is the usage normally (but not always!) adopted for published crystal structures.

It is worthwhile writing the relationships connecting the rhombohedral and hexagonal coordinate systems. The axes are defined in Fig. 3.7e and f, where the subscript r refers to rhombohedral and h to hexagonal. To convert from hexagonal to rhombohedral axes

$$\begin{bmatrix} a_r \\ b_r \\ c_r \end{bmatrix} = \begin{bmatrix} 2/3 & 1/3 & 1/3 \\ -1/3 & 1/3 & 1/3 \\ -1/3 & -2/3 & 1/3 \end{bmatrix} \begin{bmatrix} a_h \\ b_h \\ c_h \end{bmatrix} \quad [3.1]$$

Thus,

$$\mathbf{a}_r = 2/3\mathbf{a}_h + 1/3\mathbf{b}_h + 1/3\mathbf{c}_h \quad [3.2]$$

On taking the scalar product of $\mathbf{a}_r$ with itself, we find

$$\mathbf{a}_r \cdot \mathbf{a}_r = a_r^2 = 4/9a_h^2 + 1/9b_h^2 + 1/9c_h^2 + 4/9a_h b_h \cos 120° \quad [3.3]$$

which simplifies to

$$a_r = 1/3\,(3a_h^2 + c_h^2)^{1/2} = (a_h/3)\,(3 + c_h^2/a_h^2)^{1/2} \quad [3.4]$$

The rhombohedral angle can be found by proceeding in a similar way. Taking the scalar product of $\mathbf{a}_r$ and $\mathbf{b}_r$ we get:

$$\begin{aligned} \mathbf{a}_r \cdot \mathbf{b}_r &= [2/3\mathbf{a}_h + 1/3\mathbf{b}_h + 1/3\mathbf{c}_h] \cdot [-(1/3)\mathbf{a}_h + (1/3)\mathbf{b}_h + (1/3)\mathbf{c}_h] \\ &= -2/9a_h^2 + 1/9b_h^2 + 1/9c_h^2 + 1/9a_h\, b_h \cos 120° \end{aligned} \quad [3.5]$$

But

$$\begin{aligned} \mathbf{a}_r \cdot \mathbf{b}_r &= a_r^2 \cos \gamma_r \\ \cos \alpha_r &= \cos \beta_r = \cos \gamma_r \end{aligned} \quad [3.6]$$

Therefore,

$$\cos\alpha = \left[1/3(c_h/a_h)^2 - 1/2\right] / \left[1/3(c_h/a_h)^2 + 1\right] \quad [3.7]$$

To transform from rhombohedral to hexagonal axes of reference, we have

$$\begin{bmatrix} a_h \\ b_h \\ c_h \end{bmatrix} = \begin{bmatrix} 1 & -1 & 0 \\ 0 & 1 & -1 \\ 1 & 1 & 1 \end{bmatrix} \begin{bmatrix} a_r \\ b_r \\ c_r \end{bmatrix} \quad [3.8]$$

This transformation matrix is the inverse of the preceding one, yielding

$$\begin{aligned} c_h/a_h &= 3[1 - 4/3\sin^2(\alpha_r/2)]^{1/2}/[2 \sin (\alpha_r/2)] \\ a_h &= 2a_r \sin (\alpha_r/2) \end{aligned} \quad [3.9]$$

Note that if we wish to transform *coordinates* from one orientation to the other, the transformation matrices are the *transposed* inverses of the preceding ones. That is, if the matrix **A** converts unit-cell axes from orientation 1 to orientation 2, and matrix **B** converts coordinates from orientation 1 to orientation 2, $\mathbf{B} = (\mathbf{A}^{-1})^{\mathrm{T}}$, where the superscript T means the transpose. This can be proved simply as follows. Matrix **B** converts coordinates (x_o, y_o, z_o) to (x_n, y_n, z_n) and matrix **A** converts unit cell axes $\mathbf{a}_o$, $\mathbf{b}_o$, $\mathbf{c}_o$ to $\mathbf{a}_n$, $\mathbf{b}_n$, $\mathbf{c}_n$. That is

$$\begin{bmatrix} \mathrm{a}_n \\ \mathrm{b}_n \\ \mathrm{c}_n \end{bmatrix} = \mathbf{A} \begin{bmatrix} \mathrm{a}_o \\ \mathrm{b}_o \\ \mathrm{c}_o \end{bmatrix} \quad \text{and} \quad \begin{bmatrix} x_n \\ y_n \\ z_n \end{bmatrix} = \mathbf{B} \begin{bmatrix} x_o \\ y_o \\ z_o \end{bmatrix} \qquad [3.10]$$

Now, the vector $x_o\mathbf{a}_o + y_o\mathbf{b}_o + z_o\mathbf{c}_o$ must be invariant under the transformation (passive operation) of the unit-cell axes, and so

$$x_n\mathbf{a}_n + y_n\mathbf{b}_n + z_n\mathbf{c}_n = x_o\mathbf{a}_o + y_o\mathbf{b}_o + z_o\mathbf{c}_o \qquad [3.11]$$

This is

$$[\mathrm{a}_n\mathrm{b}_n\mathrm{c}_n] \begin{bmatrix} x_n \\ y_n \\ z_n \end{bmatrix} = [\mathrm{a}_o\mathrm{b}_o\mathrm{c}_o] \begin{bmatrix} x_o \\ y_o \\ z_o \end{bmatrix} = [\mathrm{a}_o\mathrm{b}_o\mathrm{c}_o]\mathbf{B}^{-1} \begin{bmatrix} x_n \\ y_n \\ z_n \end{bmatrix} \qquad [3.12]$$

Therefore,

$$[\mathrm{a}_n\mathrm{b}_n\mathrm{c}_n] = [a_o b_o c_o]\mathbf{B}^{-1}$$

To form column vectors as in Eqn [3.10],

$$\begin{bmatrix} \mathrm{a}_n \\ \mathrm{b}_n \\ \mathrm{c}_n \end{bmatrix} = (\mathbf{B}^{-1})^{\mathrm{T}} \begin{bmatrix} \mathrm{a}_o \\ \mathrm{b}_o \\ \mathrm{c}_o \end{bmatrix} \qquad [3.13]$$

But this matrix $(\mathbf{B}^{-1})^{\mathrm{T}}$ is clearly the same as **A** in Eqn [3.10].

In this case, in order to transform hexagonal coordinates to rhombohedral coordinates, we use

$$\begin{bmatrix} x_r \\ y_r \\ z_r \end{bmatrix} = \begin{bmatrix} 1 & 0 & 1 \\ -1 & 1 & 1 \\ 0 & -1 & 1 \end{bmatrix} \begin{bmatrix} x_h \\ y_h \\ z_h \end{bmatrix} \qquad [3.14]$$

and for conversion in the opposite direction:

$$\begin{bmatrix} x_h \\ y_h \\ z_h \end{bmatrix} = \begin{bmatrix} 2/3 & -1/3 & -1/3 \\ 1/3 & 1/3 & -2/3 \\ 1/3 & 1/3 & 1/3 \end{bmatrix} \begin{bmatrix} x_r \\ y_r \\ z_r \end{bmatrix} \qquad [3.15]$$

3.3. PRIMITIVE CELLS OF THE 14 BRAVAIS LATTICES

For the *I*, *F* or *C* Bravais lattices, there are 2, 4 and 2 lattice points, respectively, in the conventional unit cells shown in Fig. 3.4. As mentioned in Section 3.1, for each of these lattices, it is possible to pick unit cells that are smaller in volume by factors of 2, 4 and 2, respectively, so that each of these new unit cells will contain just one lattice point. Thus, for each of the centered unit cells, it is always possible to find a primitive cell. As we shall see, these primitive cells by themselves (when isolated from the lattice) do not display the full symmetry of the particular system. They are unit cells, nevertheless, since when translated parallel to themselves by their translation vectors they fill all space and produce the original lattice. The examples will make this clear.

Figure 3.8a shows one example of a primitive cell constructed from a *cF*-lattice (note that an infinite number of ways of doing this is possible in practice). The basic translation vectors of the primitive cell are given in terms of the translation vectors of the conventional *cF* Bravais lattice. Similar constructions are shown for the *cI* and *tC* Bravais lattices in Fig. 3.8b and c. Now, if one considers the primitive cell in Fig. 3.8a in isolation, it is a rhombohedron in shape. This implies that there is a single 3-fold axis in such a cell and so it does not appear to display the full symmetry expected in the cubic lattice from which it is derived, namely the presence of four 3-fold axes. Thus, the isolated primitive cell of a cubic *F* lattice does not appear to have cubic symmetry. Nevertheless, it is still a unit cell constructed from a cubic lattice. However, from a visual point of view, it is easier to think of the crystal structure when working with a unit cell that has the full symmetry of the lattice. Therefore, one can appreciate why the conventional unit cells of the Bravais lattices are usually used, rather than primitive unit cells in every crystal system.

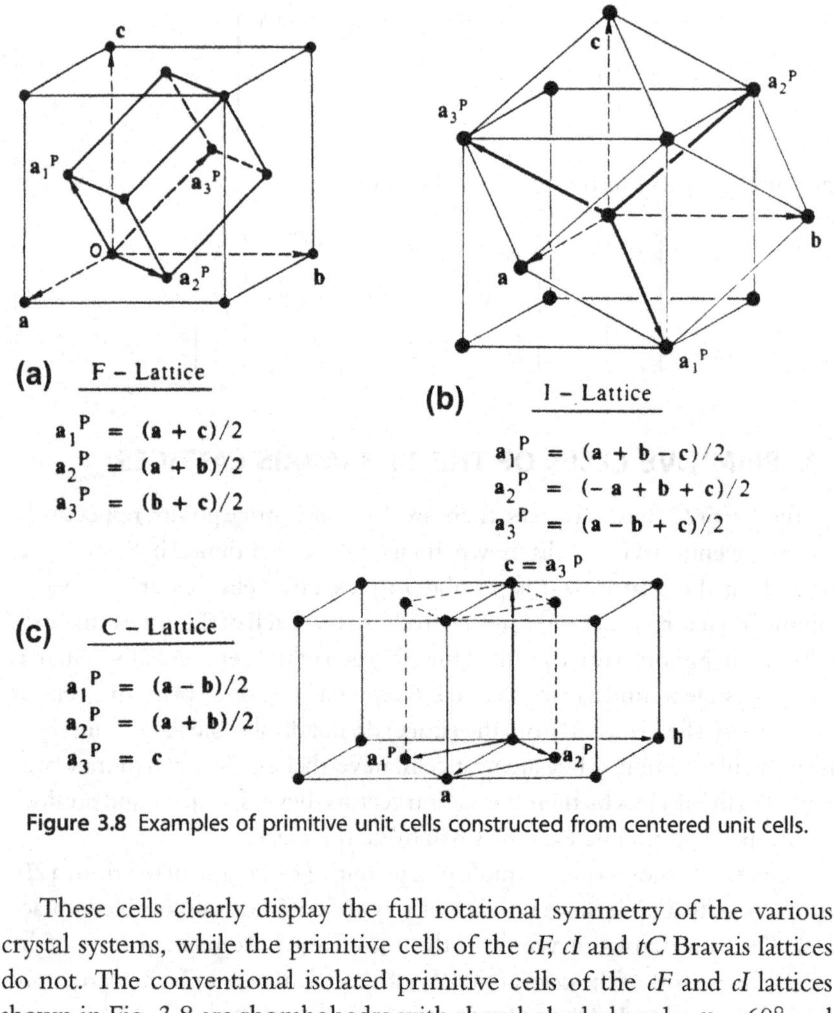

Figure 3.8 Examples of primitive unit cells constructed from centered unit cells.

These cells clearly display the full rotational symmetry of the various crystal systems, while the primitive cells of the *cF*, *cI* and *tC* Bravais lattices do not. The conventional isolated primitive cells of the *cF* and *cI* lattices shown in Fig. 3.8 are rhombohedra with rhombohedral angles $\alpha = 60°$ and $109°28'$, respectively.

It is worth repeating here that solid-state physicists and chemists sometimes prefer to work with the primitive cell, even though this means working with a cell of apparent lower symmetry. This is because the primitive cell contains one lattice point per cell: thus, it is the *smallest cell that describes the full translational invariance of the Hamiltonian.* For counting problems such as in the evaluation of normal modes of vibration or electronic states in band theory, working with the primitive cell will always give the correct number of states. One can use the ordinary Bravais cell for these problems, but care must be taken to allow for the effect of centering and divide the resulting number of modes or states by the number of lattice points in the cell.

Remember that the choice of primitive cell is not unique, as was shown in Fig. 2.3. However, the primitive cells shown in Fig. 3.8 are the conventional primitive cells used by most solid-state scientists.

3.4. THE WIGNER–SEITZ UNIT CELL

We take this opportunity to emphasize that the conventional centered unit cells of the 14 Bravais lattices shown in Fig. 3.4 or their corresponding primitive cells shown in Fig. 3.8 are not the last word on unit cells.

Occasionally, some special unit cell is chosen to emphasize certain special aspects of the crystal structure. For example, there might be a change of structure at some temperature (a phase transition) from a very simple high-temperature structure to a much more complicated low-temperature structure. The unit cell chosen in the simple structure could be relatively complicated (contain many lattice points centered at various positions), but might be picked to show how the low-temperature structure follows in a natural way from the high-temperature structure. Thus, the orientation of the axes in the high-temperature structure might be chosen so that the same orientation of axes above and below the phase transition is maintained. This choice would make visualization of the phase transition easy, although it may make the choice of the unit cell in the high-temperature structure unusual.

Besides this type of special choice of unit cell, there is another primitive unit cell of more general use, particularly for the description of electronic bands or phonon dispersion. This is known as the **Wigner–Seitz cell** (sometimes called the **proximity cell**, **Dirichlet domain** or **Voronoi cell**). This cell is obtained by starting at any lattice point, that we take as the origin, and drawing vectors to all neighbouring lattice points. Planes are then constructed *perpendicular* to and passing through the midpoints of these vectors. The Wigner–Seitz cell is the smallest volume cell about the origin bounded by these planes.

Figure 3.9 shows a step-by-step construction of the Wigner–Seitz cell for a *cI* lattice. Note that the Wigner–Seitz cell contains just one lattice point and displays the full rotational symmetry of the crystal system, as do the conventional unit cells of the 14 Bravais lattices. Of course, the Wigner–Seitz cell does not necessarily form a parallelepiped but it is, nevertheless, a perfectly acceptable unit cell. Figures 3.10a and b show the Wigner–Seitz cells for *cI* and *cF* lattices, where the full cubic symmetry displaying four 3-fold axes is again obvious. Wigner–Seitz cells of a *tI* Bravais lattice, for

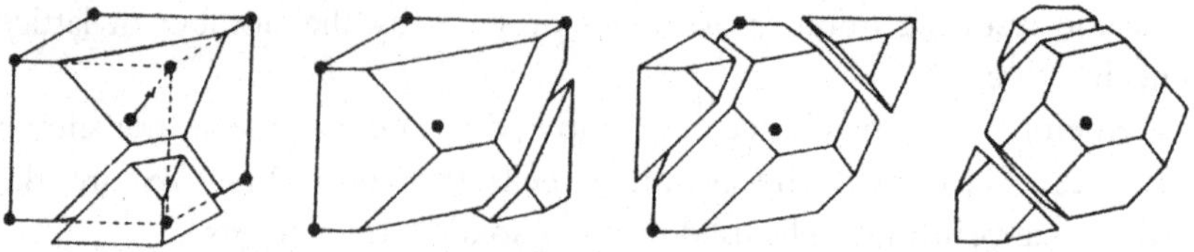

Figure 3.9 Construction of a Wigner-Seitz cell for a body-centered cubic cI lattice.

small and large c/a ratios, are shown in Fig. 3.10c and d; the similarity to the *cI* Wigner–Seitz cell is readily apparent for $c < a$. Figure 3.10e shows the stacking of the Wigner–Seitz cells for a *cI* lattice to form a complete lattice. The stacking is obtained by translations of the type $(\pm n_1/2, \pm n_2/2, \pm n_3/2)$ type, where the n_i are integers.

For each of the 14 Bravais lattices, there is at least one Wigner–Seitz cell. If different ratios of the crystal axes are considered, it can be shown that there are 24 distinct topological forms of Wigner–Seitz cells (see Appendix 3). For example, it can be seen that in a tetragonal *tI* lattice there are two topologies, one for small *c/a* (Fig. 3.10c) and one for large *c/a* (Fig. 3.10d).

The Wigner–Seitz cell construction is often used in reciprocal space, or **k**-space, to define the **first Brillouin zone**. Furthermore, to those readers who

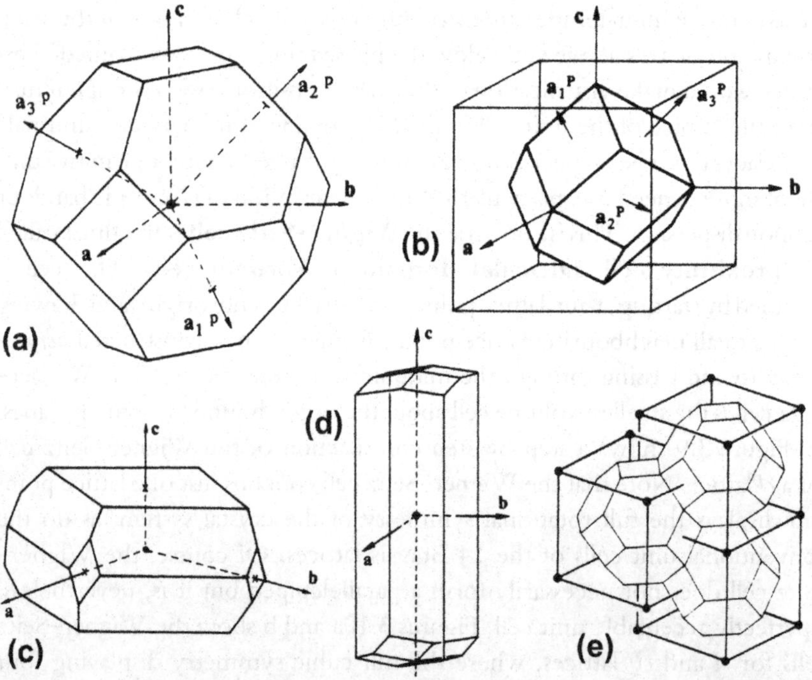

Figure 3.10 (a), (b), (c), and (d) show the Wigner-Seitz cell of a *cI* lattice, a *cF* lattice and a *tI* lattice with small and large c/a ratios, respectively. (e) The stacking of Wigner-Seitz cells of a *cI* lattice.

are familiar with Brillouin zones, we remind them that a *cI* lattice in real space (direct space) has a reciprocal lattice that is a *cF* lattice. Thus the first Brillouin zone of a *cI* lattice looks like the Wigner–Seitz cell of a direct space *cF* lattice and vice versa.

The Brillouin zone is of great importance in electronic band theory where it describes the eigenvectors of the allowed electron states, and in lattice dynamics where it is used in representing the phonon dispersion relations. *Incidentally, it is not often realized that the* ***first Brillouin zone*** *can be defined simply as a unit cell in reciprocal space* (Appendix 8). The Wigner–Seitz construction is normally used for this because of its property of displaying the full lattice symmetry while at the same time enclosing the smallest volume of **k**-states. However, especially for low-symmetry systems, e.g. triclinic and monoclinic, this construction is clumsy and unnecessary. In such cases, the normal parallelepiped unit cell may have advantages since it is simpler to visualize and thus less prone to error (see, for instance, the book by Bradley and Cracknell).

3.5. TWO-DIMENSIONAL LATTICES

It is a simple matter to derive the 2-dimensional lattices by taking the 14 space lattices and projecting them onto a plane. Clearly, all primitive lattices remain primitive; in two dimensions they are given the symbol *p* (for 2-dimensions, small letters are used). *C*-centered lattices when projected onto the **ab** plane produce a centered 2-dimensional lattice, given the symbol *c*. *A* and *B*-centered lattices, when projected onto the **ab** plane, give halved unit cells which can be redefined as primitive; all-face-centered lattices do both. Note that in rhombohedral centering, projecting onto a plane perpendicular to the **c**-axis produces a primitive lattice with **a** and **b** axes redefined.

The 2-dimensional lattices can adopt four possible shapes, oblique (where there is no relationship between the **a** and **b**-axes), rectangular, square, or hexagonal. Proceeding in the same way as for the 14 three-dimensional space lattices, we find that there are five 2-dimensional space lattices in four crystal systems. The 2-dimensional space lattices are

oblique	*mp*
rectangular	*op*
rectangular	*oc*
square	*tp*
hexagonal	*hp*

All other combinations are equivalent to these five.

Problems

1. (a) Show that every lattice point is at a center of inversion and that midway between the points there is another center of inversion. (b) If there is a 2-fold rotation axis at every lattice point, by drawing a suitable diagram, show that halfway between neighbouring lattice points that have a 2-fold axis, another 2-fold axis is generated. Do the same for mirror planes. (Hint: Your diagrams can be checked by consulting the *ITA* for space groups $P\bar{1}$, $P2$ and Pm.)
2. Consider *cP*, *cI* and *cF* lattices with the lengths of the conventional unit-cell axes in each case given by **a**. Make a table for the three lattices showing the number of nearest neighbours for each lattice point, the distance to these neighbours, the number of second neighbours and the distance to these second neighbours.
3. Again consider the *cP*, *cI* and *cF* lattices. Show that the maximum amount of volume that may be filled with hard spheres centered on the lattice points is 0.52, 0.68 and 0.74, respectively, of the unit-cell volumes.
4. Work out the matrix relating reverse rhombohedral axes to hexagonal axes. A point in the hexagonal cell has coordinates $(\bar{y}, x - y, z)$. What are the coordinates of this point with respect to the reverse rhombohedral cell?
5. (a) For the Wigner–Seitz cell of a bcc lattice, show that the volume of the largest sphere that can be contained inside of it is $2.721a^3$. (b) Show that the ratio of the value just calculated to that found in an fcc cell is $(3/2)^{3/2}$.

CHAPTER 4

Crystallographic Point Groups

Contents

Je ne te parlerai que cristaux.

Louis Pasteur, in a letter to Charles Chappuis from Strasbourg (July, 1850)

INTRODUCTION TO GROUPS

In Chapter 1 we discussed symmetry operations and their notations. Here, we shall consider what happens when several of these symmetry operations are combined. For example, what is the result when a rotation operation is combined with other rotations perpendicular to the first, or with reflections? These combinations of symmetry operations enable us to describe the symmetry of 3-dimensional objects.

Here, we shall discuss how the **32 crystallographic point groups** can be determined. The word 'crystallographic' implies that we only allow rotations through $2\pi/n$, where $n = 1, 2, 3, 4$ and 6. As discussed in Chapter 2, other rotations are not compatible with filling all space when periodic translation is considered; we only want to consider point groups that are allowed in normal crystals. This restriction on allowed values of n results in 32 point groups, and so they are sometimes simply called the **32 point groups**. The word 'point' means that we demand that all the symmetry

http://dx.doi.org/10.1016/B978-0-12-394400-9.00004-6

operations act about one point in common, which remains fixed throughout the application of the operations. This point is called the origin. We can take the word 'group' to mean a collection or a set of symmetry operations. The set of symmetry operations for every point group (crystallographic or not) forms a group in the mathematical sense. The concept of point groups is important and useful for isolated molecules as well as for crystalline solids; the only difference is that for molecules there is no restriction on the allowed rotations.

While we do not require group theoretical concepts to develop the 32 point groups, it is useful to appreciate the intimate connection between these point groups and mathematical groups. A set of operations forms a **group** if and only if the following four conditions are met:

(i) The product of any two operations is also a member of the set (closure). $PQ = R$

(ii) The set includes an identity $1(E)$. One can show that a right identity is also a left identity. $P1 = 1P = P$

(iii) For each operation R there is an inverse R^{-1}, such that $RR^{-1} = 1(E)$ (e.g. $C_n^m\ C_n^{n-m} = E$, $\sigma\ \sigma = E$, $i\ i = E$, etc.).

(iv) The multiplication of operations is associative (this is axiomatic for symmetry operations). $(PQ)R = P(QR)$

The number of elements h of the group is defined as the **order of the group**.

One other group concept should be noted. Consider two groups A and B with elements a_i ($i = 1$ to n) and b_j ($j = 1$ to m), respectively. Furthermore, these two groups have only the identity element in common and all the elements $\{a_i\}$ commute with all the elements $\{b_j\}$, that is $a_i b_j = b_j a_i$ for all i and j. The **product** of these two groups is defined as the set of nm elements

$$\{a_1 b_1, a_1 b_2, \ldots a_1 b_m, a_2 b_1, \ldots, a_2 b_m, \ldots a_n b_m\}$$

or

$$\{a_i b_j\}, \quad i = 1 \text{ to } n \quad \text{and} \quad j = 1 \text{ to } m$$

One can show that this set of nm elements still forms a group and this group is of order $h = nm$.

We should note that the set of symmetry operations forms a group in the mathematical sense as a consequence of our definition of a symmetry operation in Chapter 1. The set has closure because the product of any two symmetry operations always produces a third operation of the set. In addition, there is an identity and an inverse for each operation. Finally, it is clear that associativity applies to symmetry operations.

There are several reasons why we study point groups. First, it allows us to classify different crystals. Historically, the study of crystals began with the study of their external faces. By characterizing them with directions taken normal to each face and drawn through one point, it was found that any crystal could be classified as belonging to one of **32 crystal classes** (that there are 32 was not actually established until around 1830). Each of these crystal classes corresponds to one of the 32 point groups. Second, finite objects that can be described by one of the 32 point groups can be repeated throughout space according to the lattice in order to derive some of the space groups, as discussed in Chapters 5 and 6. Or, to put it another way, the space groups may be built up from objects with different point symmetry by the addition of translation. Third, the symmetry operations possessed by an object, molecule, or crystal can be described simply by means of a symbol. From this symbol, especially the International symbol, a solid state scientist can determine all the symmetry operations. Fourth, every space group has one of the 32 crystallographic point groups associated with it (the point group of the space group, usually just called the point group of the crystal). This is important in many areas of solid state science. For example, most of the macroscopic symmetry aspects of the physical properties of solids are related to the point group, as given by the so-called Neumann principle, a subject discussed in Section 5.3.

4.1. DEVELOPMENT OF CRYSTALLOGRAPHIC POINT GROUPS

There are several ways to develop the 32 crystallographic point groups. One method is to find all the point groups that can be made from proper rotations only. There are 11 such **pure rotational crystallographic point groups**. After this, a further 11 **centrosymmetric point groups** can be obtained by adding inversion as a symmetry operation to each of these. From the centrosymmetric point groups one can find 10 distinct non-centrosymmetric subgroups that are different from the 11 pure rotational point groups. This gives us the 32 crystallographic point groups. A second procedure is to start with the five point groups $1(C_1)$, $2(C_2)$, $3(C_3)$, $4(C_4)$ and $6(C_6)$, the so-called cyclic groups, and add symmetry operations to each of them. For example, by adding 2-fold rotations perpendicular to the cyclic axis of symmetry, new point groups result. Similarly, by adding a reflection perpendicular to the cyclic axis of symmetry, five other point groups are formed. In general, we can say that by adding reflections perpendicular to, or containing the cyclic axis, by adding 2-fold rotations perpendicular to the cyclic axis, by substituting improper for proper rotations, or by any combinations of these three, we obtain new point groups. This is the usual procedure that is used in many solid state or group theory courses. It has the advantage that a good deal of insight is developed into the symmetry operations of the different point groups. One disadvantage of determining the 32 point groups using either of these two general procedures is that after they are obtained we must then determine to which of the seven crystal systems each point group belongs. This is not difficult in practice, but conceptually the result is that the 32 crystallographic groups seem to be quite independent of the seven crystal systems.

Actually, the 32 point groups and the seven crystal systems can both be developed from the same simple symmetry operations discussed in Chapter 1, and this is the approach that we shall take initially. Thus, the crystallographic point groups for each of the crystal systems are derived one at a time. For example, we shall take the tetragonal crystal system and determine which reflections and rotations can be added to the symmetry operations that define the system with the proviso that we remain within the tetragonal crystal system. This will be done for each crystal system and in this way we shall obtain all of the 32 crystallographic point groups. The important advantage of this approach is that we can use a similar procedure to develop the space groups.

Note that the 32 point groups are illustrated in Appendix 5 in two ways: first with diagrams similar to the ones we have used for the symmetry operations, except that for convenience we draw them in the more usual stereograms, and second as scroll-like shapes for which you should be able to determine the symmetry operations. As we discuss each point group, you should consult this appendix.

4.2. THE POINT GROUPS FOR EACH CRYSTAL SYSTEM

4.2.1. Triclinic

In Chapter 2, we discussed the fact that the symmetry operations $1(E)$ and $\bar{1}(i)$ place no restrictions on the unit-cell axes and angles, leading to the triclinic crystal system. Any object, whether it be a crystal, or a molecule or even a set of 'hands' as in Chapter 1, that only possesses the identity element, is said to belong to point group 1 in the International notation or to point group C_1 in the Schoenflies notation. For convenience, we shall continue to write the Schoenflies notation in parentheses after the International notation. Thus, the symbol for the point group will be written $1(C_1)$.

If an object, crystal, set of 'hands', molecule, or whatever, is described by the two symmetry operations $1(E)$ and $\bar{1}(i)$, then the point symmetry is called $\bar{1}(S_2)$. Thus, if we say an object has the point symmetry $\bar{1}(S_2)$, we mean that there are two symmetry operations that take the object into itself, namely 1 and $\bar{1}$ (or E and i). We may also say that the point group of the object is $\bar{1}(S_2)$. Although we shall not dwell on the group theory aspect, this means that the point group $\bar{1}$ has the set of symmetry operations $\{1, \bar{1}\}$, or using the Schoenflies notation we say that the point group S_2 consists of $\{E, i\}$, and these two operations form a group of order 2 ($h = 2$).

Now, within a crystal system, the point group which has the symmetry of the lattice is called the **holohedral** point group, and as such it possesses the largest number of symmetry operations (we might loosely say it has the highest symmetry). As there are only two point groups belonging to the triclinic system, the holohedral point group here is $\bar{1}(S_2)$.

You can see that the notation for some of these point groups is very similar or even the same as for the symmetry operations. This can be troublesome, but bear in mind that a point group is a set, or collection, of symmetry operations obeying certain mathematical laws and that a symmetry operation is a single operation that takes the object into itself. As we encounter point groups with more symmetry operations, this will become less of a problem.

We cannot add any more symmetry operations and remain within the triclinic crystal system. The addition of a 2-fold rotation or a reflection, for instance, results in the monoclinic crystal system. Any other rotations result in other crystal systems. Therefore, we conclude that there are only two point groups compatible with triclinic symmetry. These are the point groups 1 and $\bar{1}$ in the International notation (or C_1 and S_2 in the Schoenflies notation) and we write these as $1(C_1)$ and $\bar{1}(S_2)$ to bring out both notations. The stereograms for these two point groups are shown in Appendix 5. For the point group $1(C_1)$, only a single point is shown; for point group $\bar{1}(S_2)$, there are two points corresponding to the two symmetry operations in the point group. These points are obtained in the following way. We start with a general point above the plane of the paper, indicated by a dot at the bottom-right of the stereogram. Application of the identity operator naturally leaves it unchanged, but the inversion operator, on the other hand, moves it through the center of the stereogram. This means that the general point is taken to the top-left of the diagram but below the plane of projection as represented by the open circle. Note that since we took a general point, that is a point *not* positioned directly on a symmetry element, there are as many points on the stereogram as there are symmetry operations in the point group.

4.2.2. Monoclinic

In Chapter 2, we found that the symmetry operations $2(C_2)$ or $\overline{2}(\sigma_h)$ give rise to the monoclinic set of axes and angles (recall from Chapter 1 that $\overline{2}$ is the same as a reflection and so $\overline{2}$ can be written equally well as m). Therefore, any object described by a single 2-fold rotation is compatible with monoclinic symmetry. It is then said to belong to point group 2 (or C_2), which is a group of order $h = 2$; the h symmetry operations are $\{1, 2\}$ or $\{E, C_2\}$. In the same way, objects with just $\overline{2}$ or m symmetry belong to the point group $m(C_{1h})$ of order $h = 2$, whose symmetry operations are $\{1, m\}$ or $\{E, \sigma_h\}$. The stereograms for these two point groups are shown in Appendix 5. In the monoclinic crystal system we normally take the unique axis to be the **c**-axis (the 1st setting) or the **b**-axis (the 2nd setting). In the appendix both the settings are shown; they are merely different views of the same situation. In each we see that the stereograms show two points corresponding to the two symmetry operations. It is also instructive to examine carefully the scroll-like shapes with point symmetry $m(C_{1h})$ and $2(C_2)$. The presence and positions of the planes and axes of symmetry should be evident and one can see how these shapes and the stereograms display the same symmetry.

It is pertinent to ask whether there are other point groups that put no new restrictions on the monoclinic axes and angles. Clearly, if we add another 2-fold rotation we shall have the necessary conditions for the orthorhombic system. However, if we add a reflection operating perpendicular to the 2-fold axis, no new conditions on the crystal axes or angles will be imposed. This can be checked by the methods used in Chapter 2, but it is also intuitively clear: if the **c**-axis is a 2-fold axis perpendicular to the **ab**-plane, and if the **ab**-plane is a plane of mirror symmetry, then no new conditions can be put on the lengths of the **a**, **b** or **c**-axes and no new condition can be put on the angle between the **a** and **b**-axes. What is new is that the addition of the mirror plane generates the symmetry operation of inversion $\overline{1}(i)$. This can readily be seen by first applying the symmetry operation $2(C_2)$ followed by the reflection operation $m(\sigma_h)$ perpendicular to the 2-fold axis. To show this explicitly, we can write, using the matrices in Appendix 1.

$$\{m[001]\}\{2[001]\} = \begin{bmatrix} 1 & 0 & 0 \\ 0 & 1 & 0 \\ 0 & 0 & -1 \end{bmatrix} \begin{bmatrix} -1 & 0 & 0 \\ 0 & -1 & 0 \\ 0 & 0 & 1 \end{bmatrix}$$

$$= \begin{bmatrix} -1 & 0 & 0 \\ 0 & -1 & 0 \\ 0 & 0 & -1 \end{bmatrix} = \{\bar{1}\} \qquad [4.1]$$

and so we find that $m[001]\ 2[001] = \bar{1}$ (or $\sigma_h\ C_2 = i$). In the International notation the point group is written as $2/m$ and has symmetry operations $\{1, 2, \bar{1}, m\}$ (or in the Schoenflies notation the point group is C_{2h} and has symmetry operations $\{E,\ C_2,\ i,\ \sigma_h\}$). The stereogram for $2/m(C_{2h})$ shows these symmetry operations. Start with the dot at the bottom-right, which represents the effect of the identity operation $1(E)$. Operate on this point with $2(C_2)$ and the dot at the top-left is obtained. Then, operate on the starting point with $\bar{1}(i)$ and the circle at the top-left is obtained (remember, the circle indicates that the general point now lies below the plane of projection). Finally, operate on the starting point with $m(\sigma_h)$ and the circle at the bottom-right is obtained. Thus, we have four points on the stereogram corresponding to the four symmetry operations of the point group. If we operate on any point of the stereogram with any of these four symmetry operations or products of these operations, then one of the other four points on the stereogram will be obtained. This is a consequence of the fact that these four operations form a group. Similarly, if we operate on all four points with any of these operations, the diagram appears exactly the same as before the operation, and hence the operation is a *symmetry* operation.

The **notation** is now becoming clear for the monoclinic as well as for the other point groups. In the International notation n/m means a reflection perpendicular to an n-fold rotation axis, while nm means a reflection in a plane (or really n such reflections) containing an n-fold axis. In the Schoenflies notation C_n is a point group which contains the symmetry operations C_n, C_n^2, and so on. The symbol C_{nh} means that the point group also contains a reflection perpendicular to the C_n-axis. Because of operations such as in Eqn [4.1], the point group C_{nh} will have twice as many symmetry operations as the point group C_n. The symbol C_{nv} means that the mirror plane contains the C_n-axis (as opposed to being perpendicular to it, as in C_{nh}) and, again, there will be twice as many symmetry operations in the point group C_{nv} as in C_n. Recall, from Chapter 1, that in the Schoenflies notation the reflection σ_h acts perpendicularly to the rotation axis while σ_v contains it (*h*, horizontal and *v*, vertical.)

Appendix 4 has all the point groups listed in the International and Schoenflies notations together with the symmetry operations for each point group. Note that in the International notation we define both a **full symbol** and a **short symbol**. For the three point groups the symbols are as follows:

Full symbol (1st setting)	Short symbol	Schoenflies
1 1 2	2	C_2
1 1 *m*	*m*	C_{1h}(or C_s)
1 1 2/*m*	2/*m*	C_{2h}

Here, $2/m(C_{2h})$ is centrosymmetric and holohedral; the other point groups in the crystal system are subgroups of the holohedral group. Within a crystal system this must always be true. In the full symbols, the symmetry operations are indicated with respect to the crystallographic axes, **a**, **b** and **c**, in turn. Thus, we see that there is no symmetry element along **a** and **b** (except for 1), while the important symmetry elements 2 and *m*, respectively, lie along or are perpendicular to **c**. For the second setting, where the **b**-axis is the 2-fold axis, the full symbol of the holohedral group is 1 2/*m* 1.

We shall see that in other crystal systems, particularly the orthorhombic system, the full symbol is more important. Normally, however, the full symbol contains redundant information. The short symbol has sufficient information to enable us to construct all of the symmetry operations of the point group. The same general comments about full and short symbols apply to the space group symbols. Finally, examine the scroll-like shape with point symmetry $2/m(C_{2h})$ in Appendix 5. It is apparent that the addition of a mirror plane to the $2(C_2)$ shape creates the $2/m(C_{2h})$ shape. Moreover, the addition of a 2-fold axis to the $m(C_{1h})$ shape again creates the $2/m(C_{2h})$ shape.

4.2.3. Orthorhombic

For this crystal system two 2-fold axes, or two mirror planes, perpendicular to each other determine the characteristic restrictions on the unit-cell axes and angles (Chapter 2). The two 2-fold symmetry operations imply a third 2-fold axis, perpendicular to both, as can be shown by multiplying the appropriate matrices, as in Eqn [4.1] or by using stereograms. The point group is called 222 with symmetry operations {1, 2[100], 2[010], 2[001]} or in the Schoenflies notation it is D_2 and has the following symmetry operations {*E*, C_2[100], C_2[010], C_2[001]}, and is of order $h = 4$.

Similarly, two perpendicular mirror planes imply the presence of a 2-fold axis along the intersection of the two planes; if we call the axes perpendicular to the mirror planes **a** and **b**, the 2-fold axis is along **c** and the point group is *mm*2 with operations {1, 2[001], *m*[010], *m*[100]}, or C_{2v} with operations {*E*, C_2[001], σ_v[010], σ_v[100]}. The symbol *mm*2 is the **standard form** by which this point group is denoted in the *ITA*. This, however, is only a matter of convenience, since the choice of axes in a crystal may be dictated by other considerations apart from the types of symmetry operations and may necessitate referring to the crystal by a non-standard symbol. If the **a**-axis is the 2-fold axis, the symbol is 2*mm* and if the **b**-axis is the 2-fold axis, the symbol is *m*2*m*.

Again, we ask if there are any other combinations of symmetry operations that will not impose new restrictions on the unit-cell axes and angles of the orthorhombic system. It is clear that a mirror plane can be placed normal to any of the three mutually perpendicular 2-fold axes without affecting the unit-cell geometry. On trying this, it is found that there must be mirror planes perpendicular to all of the 2-fold axes and not to just one or two. The full symbol for the point group thus generated is $2/m\ 2/m\ 2/m$, for which the short symbol *mmm* can be used, or D_{2h} in the Schoenflies notation. In the Schoenflies notation, the point group C_2 means the principal axis has 2-fold symmetry and if there are 2-fold axes perpendicular to the principal axes, then the symbol becomes D_2 as above. The *h* in D_{2h} refers to σ_h, a reflection through a mirror plane perpendicular to the *n*-fold axis. The point group *mmm*(D_{2h}) has eight symmetry operations, the six operations in the full symbol plus the identity operation and a generated center of inversion. That the center of inversion is generated can be seen by multiplying the matrices for the three mutually perpendicular reflection operations:

$$\{m[100]\}\{m[010]\}\{m[001]\} = \begin{bmatrix} -1 & 0 & 0 \\ 0 & 1 & 0 \\ 0 & 0 & 1 \end{bmatrix} \begin{bmatrix} 1 & 0 & 0 \\ 0 & -1 & 0 \\ 0 & 0 & 1 \end{bmatrix} \times \begin{bmatrix} 1 & 0 & 0 \\ 0 & 1 & 0 \\ 0 & 0 & -1 \end{bmatrix} = \begin{bmatrix} -1 & 0 & 0 \\ 0 & -1 & 0 \\ 0 & 0 & -1 \end{bmatrix} = \{\bar{1}\} \qquad [4.2]$$

Another way of seeing this is by means of the stereogram shown in Appendix 5. The general point is taken by the three mirror reflections to form seven other points, making four enantiomorphic pairs. Inspection of the stereogram reveals the presence of a center of inversion.

The orthorhombic point groups are summarized in the following table. Of these, $mmm(D_{2h})$ is centrosymmetric and holohedral. Appendix 5 shows stereograms for these three point groups as well as three shapes having these symmetry operations.

Full symbol	Short symbol	Schoenflies
222	222	D_2
$mm2$	$mm2$	C_{2v}
$\frac{2}{m}\frac{2}{m}\frac{2}{m}$	mmm	D_{2h}

4.2.4. Tetragonal

This crystal system is defined by the presence of a single 4 or $\overline{4}$ axis in the crystal. This results in point groups $4(C_4)$ and $\overline{4}(S_4)$ both of which are of order $h = 4$ (see Appendix 4).

Which symmetry operations can we add and still remain within the crystal system? The obvious choices are mirror reflections and 2-fold rotations placed in strategic orientations. For example, if a 2-fold rotation is added about a direction perpendicular to the 4-fold axis, let us say along **a**, consider what other symmetry operations, if any, will be generated. We can find this out by multiplying the appropriate matrices:

$$\{4[001]\}\{2[100]\} = \begin{bmatrix} 0 & -1 & 0 \\ 1 & 0 & 0 \\ 0 & 0 & 1 \end{bmatrix} \begin{bmatrix} 1 & 0 & 0 \\ 0 & -1 & 0 \\ 0 & 0 & -1 \end{bmatrix} = \begin{bmatrix} 0 & 1 & 0 \\ 1 & 0 & 0 \\ 0 & 0 & -1 \end{bmatrix}$$
$$= \{2[110]\} \qquad [4.3]$$

This shows us that another 2-fold rotation is produced, this time about the face-diagonal of the **ab**-plane. For this reason, the resulting point group is called $422(D_4)$. Notice that because the 4-fold direction is such an important direction (it is the principal axis or unique direction in the tetragonal crystal), it is conventional to place the symbol '4' first, unlike the procedure used in the orthorhombic system where there is no unique direction. In the International notation (Appendix 6), the first '2' after the '4' refers to the 2-fold rotation about **a** (and to an equivalent 2-fold rotation about **b** as a consequence of the 4-fold rotation) while the second '2' refers to the diagonal 2-fold rotation about [110] (and another equivalent 2-fold rotation about $[1\bar{1}0]$). As before, in the Schoenflies notation the '*D*' implies that perpendicular to the principal axis there are 2-fold axes. The occurrence of these symmetry operations can be seen by reference to the stereogram for point group $422(D_4)$ given in Appendix 5.

There are other ways of adding symmetry operations to point groups 4 and $\bar{4}$. For example, a reflection through a mirror plane placed normal to the 4-fold axis results in point group $4/m(C_{4h})$, which is centrosymmetric, as given in Appendix 5. You should now be able to show this by multiplying the relevant matrices. Another thing that can be done is to make a reflection through a plane containing the 4-fold axis and lying in the **ac**-plane. By similar arguments as those used above, we find another mirror plane diagonally between the **a** and **b**-axes (see Appendix 5) giving the point group $4mm(C_{4v})$. Also, we can now add a mirror plane and 2-fold axes perpendicular to **c**. The result is another centrosymmetric point group $4/m\ 2/m\ 2/m$, or $4/mmm$ for short (D_{4h}). Finally, to point group $\bar{4}(S_4)$ a 2-fold axis can be added perpendicular to **c**, say along **a**. If we multiply the appropriate matrices

$$\{\bar{4}[001]\}\{2[100]\} = \begin{bmatrix} 0 & 1 & 0 \\ -1 & 0 & 0 \\ 0 & 0 & -1 \end{bmatrix}\begin{bmatrix} 1 & 0 & 0 \\ 0 & -1 & 0 \\ 0 & 0 & -1 \end{bmatrix}$$

$$= \begin{bmatrix} 0 & -1 & 0 \\ -1 & 0 & 0 \\ 0 & 0 & 1 \end{bmatrix} = \{m[110]\} \qquad [4.4]$$

we obtain a mirror plane along the diagonal of the **ab**-plane. The point group is, therefore, $\bar{4}2m$ (D_{2d}). An equivalent point group is $\bar{4}m2$ (D_{2d}) in which the mirror planes are now perpendicular to **a** and to **b** (Fig. 4.1). Notice how the International symbol shows this change in orientation whereas the Schoenflies symbol fails to do so.

For convenience we summarize the tetragonal point groups below. Note that the point group $4/m(C_{4h})$ is centrosymmetric and $4/mmm(D_{4h})$ is both centrosymmetric and holohedral: the point group $\bar{4}(S_4)$ is a subgroup

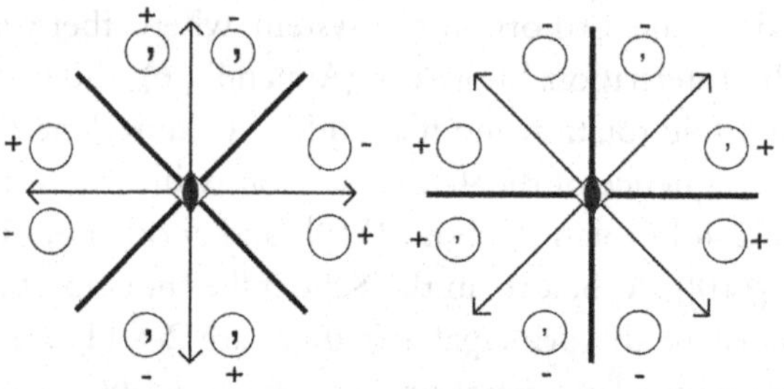

Figure 4.1 Difference between $\bar{4}2m$ and $\bar{4}m2$ point groups.

of $\bar{4}2m(D_{2d})$ and naturally all these point groups are subgroups of $4/mmm(D_{4h})$.

Full symbol	Short symbol	Schoenflies
4	4	C_4
422	422	D_4
$4/m$	$4/m$	C_{4h}
$4mm$	$4mm$	C_{4v}
$\frac{4}{m}\frac{2}{m}\frac{2}{m}$	$4/mmm$	D_{4h}
$\bar{4}$	$\bar{4}$	S_4
$\bar{4}2m$	$\bar{4}2m$	D_{2d}

4.2.5. Trigonal

For this crystal system, we require a 3 or $\bar{3}$ operation. Again, we can successively add other symmetry operations as was done for the tetragonal system and arrive at the five possibilities listed below. $\bar{3}(S_6)$ is centrosymmetric and $\bar{3}m$ (D_{3d}) is both centrosymmetric and holohedral. Notice that by placing a plane of symmetry perpendicular to the 3-fold axis we do not remain within this crystal system, since $3/m = \bar{6}$, which belongs to the hexagonal system.

Full symbol	Short symbol	Schoenflies
3	3	C_3
32	32	D_3
$3m$	$3m$	C_{3v}
$\bar{3}$	$\bar{3}$	S_6 or C_{3i}
$\bar{3}\frac{2}{m}$	$\bar{3}m$	D_{3d}

4.2.6. Hexagonal

This crystal system requires a 6 or $\overline{6}$-operation as discussed in Chapter 2. The resulting seven point groups are quite analogous to those in the tetragonal system, and we list them in the following table.

Full symbol	Short symbol	Schoenflies
6	6	C_6
622	622	D_6
$6/m$	$6/m$	C_{6h}
$6mm$	$6mm$	C_{6v}
$\frac{6}{m}\frac{2}{m}\frac{2}{m}$	$6/mmm$	D_{6h}
$\overline{6}$	$\overline{6}$	C_{3h}
$\overline{6}m2$	$\overline{6}m2$	D_{3h}

$6/m(C_{6h})$ is centrosymmetric and $6/mmm(D_{6h})$ is both centrosymmetric and holohedral. Note how the number 3 appears in the Schoenflies symbols for the $\overline{6}$ and $\overline{6}m2$ point groups (also recall that S_6 is the Schoenflies symbol for the $\overline{3}$ point group in the trigonal crystal system). Unfortunately, these symbols can confuse the unwary into placing these point groups in the incorrect crystal system. Appendix 5 shows scroll shapes and stereograms that display these point symmetries.

4.2.7. Cubic

The cubic crystal system is a little more difficult to treat than the above systems because there is no single principal axis. The crystal system is defined by four 3-fold axes all making equal angles with one another (109° 28′). We want to see which symmetry operations may be added and still remain within the cubic crystal system. Remember that the 3-fold operation implies the symmetry operation $3(C_3)$ as well as $3^2(C_3^2)$.

The first question is do we have a point group using just $1(E)$ and the eight $3(C_3)$ and $3^2(C_3^2)$ operations about the four axes? We realize quickly that the answer is no. This is because for a point group we want to be able to repeat the operations of the set, in any order, and have the object in an equivalent position. However,

$$\{3[\bar{1}11]\}\{3[111]\} = \begin{bmatrix} 0 & -1 & 0 \\ 0 & 0 & 1 \\ -1 & 0 & 0 \end{bmatrix} \begin{bmatrix} 0 & 0 & 1 \\ 1 & 0 & 0 \\ 0 & 1 & 0 \end{bmatrix} = \begin{bmatrix} -1 & 0 & 0 \\ 0 & 1 & 0 \\ 0 & 0 & -1 \end{bmatrix} = \{2[010]\} \qquad [4.5]$$

and so, at the very least, we must include in the set of operations comprising the point group, 2-fold rotations about each of the **a**, **b** and **c**-axes. By repeated multiplications it is found that there are no new operations that are required and that the 2-fold axes keep the symmetry within the cubic crystal system. This leads to the cubic point group 23(T) which consists of 3(C_3) and $3^2(C_3^2)$ symmetry operations about the four <111> directions and 2(C_2) symmetry operations about the three <100> directions. Adding the identity operation, there is a total of 12 symmetry operations in this cubic point group (and note that none of them is a 4-fold operation). This and the other cubic point groups to follow can best be visualized by reference to the diagrams in Appendix 5 (do not confuse this point group with the trigonal point group 32 in the International notation).

To the point group 23(T) we may add reflections through planes perpendicular to the 2-fold axes. From these mirror planes and by taking various products, as in Eqn [4.5], we obtain an inversion operation $\bar{1}(i)$ as well as $\bar{3}$ (S_6^5) and $\bar{3}^5(S_6)$ about the four <111> directions. This cubic point group is denoted $2/m\,\bar{3}$ or m$\bar{3}$ (T_h) and has 24 symmetry operations. Again notice that there are no 4-fold axes in this cubic point group (prior to *ITA* this point group was called simply *m*3 and one had to be clear not to confuse it with the trigonal point group 3*m*, in the International notation).

Now add mirror planes containing both the 3-fold and the 2-fold axes. For example, $m[1\bar{1}0]$ contains the 2[001] axis and the 3[111] and $3[\bar{1}\bar{1}1]$ axes, while *m*[110] also contains the 2[001] axis and the other two 3<111> axes. There are three distinct pairs of such mirror planes, the line of intersection of each pair being coincident with the **a**, **b** and **c**-axes. In the Schoenflies notation, these reflection operations are denoted σ_d to indicate that their mirror planes lie diagonally between the unit-cell axes. The inclusion of such planes gives rise to three axes of $\bar{4}(S_4^3)$ and $\bar{4}^3(S_4)$ symmetry about the **a**, **b** and **c**-axes. The resulting cubic point group, called $\bar{4}3m(T_d)$, has 24 symmetry operations. Again, there are no proper 4-fold operations in this cubic point group.

If we take the cubic point group $23(T)$ and allow 4-fold and $4^3(C_4^3)$ operations about the three 2-fold axes then, using techniques similar to those used in Eqn [4.5], six new 2-fold axes are obtained, that is $2[110]$, $2[1\bar{1}0]$, and so on. This is the cubic point group $432(O)$ which again has 24 symmetry operations. Note that, as in the point group $23(T)$, in this point group there are only proper rotation operations and no inversions, mirrors, or improper rotations.

The last cubic point group is obtained by allowing mirror planes perpendicular to the 4-fold axes. The point group is $4/m\,\bar{3}\,2/m$ or $m\bar{3}m$ for short ($m3m$ prior to the *ITA*) (or O_h) and has 48 symmetry operations. It can be seen from Appendix 5 that all the operations of the other cubic point groups are subgroups of this point group.

Note that only two of the five cubic point groups have proper 4-fold symmetry operations, and also that the cubic point group $23(T)$ has only 12 symmetry operations, not a very large number, and that there are other point groups in the hexagonal, trigonal and tetragonal crystal systems having the same or more symmetry operations. Therefore, cubic point groups should not be associated necessarily with a large number of symmetry operations. The cubic point groups can be summarized as shown in the following table.

Full symbol	Short symbol	Schoenflies
23	23	T
$\frac{2}{m}\bar{3}$	$m\bar{3}$	T_h
432	432	O
$\bar{4}3m$	$\bar{4}3m$	T_d
$\frac{4}{m}\bar{3}\frac{2}{m}$	$m\bar{3}m$	O_h

where $m\bar{3}$ (T_h) is centrosymmetric and $m\bar{3}m$ (O_h) is both centrosymmetric and holohedral. Appendix 4 lists all the symmetry operations of the cubic point groups and Appendix 5 has the point group diagrams. It is worthwhile spending some time testing the various symmetry operations on these diagrams.

4.3. THE 32 POINT GROUPS FROM HOLOHEDRIES

Probably the most elegant way of deriving the 32 point groups is to start from the symmetry of the seven holohedries. These are simply the point symmetries of the lattices for the seven crystal systems, and such a derivation highlights the central importance of the lattice itself. To do this, we just write down the holohedral point groups and list the subgroups that remain in the crystal system.

Thus, starting in the triclinic system, the holohedral group is $\bar{1}$ and its subgroups are 1 and $\bar{1}$ itself. These are the two point groups of the triclinic system. In the monoclinic system, the holohedral group is $2/m$ and its subgroups are 2, *m* and $2/m$ (we do not add 1 and $\bar{1}$, since they are in the triclinic system). In the orthorhombic system, the subgroups of *mmm* are 222, *mm*2, *m*2*m*, 2*mm* and *mmm* itself. In this case, *mm*2, *m*2*m* and 2*mm* are equivalent since they only arise from an interchange of axes, and so in the orthorhombic system there are just three point groups. In the tetragonal system, we start from $4/mmm$ and obtain the subgroups 4, $\bar{4}$, 422, $\bar{4}2m$, $\bar{4}m2$, 4*mm*, $4/m$ and $4/mmm$. Here again, $\bar{4}2m$ and $\bar{4}m2$ are equivalent and so there are seven unique tetragonal point groups. Table 4.1 summarizes this derivation.

Table 4.1 Derivation of 32 point groups from holohedries. The groups bracketed together are equivalent and only differ with respect to the choice of axes

Crystal system	Holohedral point group	Subgroups
Triclinic	$\bar{1}$	1
		$\bar{1}$
Monoclinic	$2/m$	2
		m
		$2/m$
Orthorhombic	*mmm*	222
		[*mm*2
		*m*2*m*
		2*mm*]
		mmm
Tetragonal	$4/mmm$	4
		$\bar{4}$
		$4/m$
		422
		4*mm*
		[$\bar{4}2m$
		$\bar{4}m2$]
		$4/mmm$

Table 4.1 Derivation of 32 point groups from holohedries. The groups bracketed together are equivalent and only differ with respect to the choice of axes—cont'd

Crystal system	**Holohedral point group**	**Subgroups**
Trigonal	$\bar{3}m$	3
		$\bar{3}$
		32
		$3m$
		$\bar{3}m$
Hexagonal	$6/mmm$	6
		$\bar{6}$
		$6/m$
		622
		$6mm$
		$\bar{6}m2$
		$\bar{6}2m$
		$6/mmm$
Cubic	$m\bar{3}m$	23
		$m\bar{3}$
		432
		$\bar{4}3m$
		$m\bar{3}m$

4.4. LAUE CLASSES AND GROUPS

It is a feature of diffraction from a single crystal that the diffraction patterns appear to a good approximation to be centrosymmetric (**Friedel's Law**), even when the crystal is non-centrosymmetric. Thus, if a series of diffraction images is taken in order to determine the symmetry of the crystal, it is not always easy to determine whether a crystal is centrosymmetric or not, although *X*-ray anomalous absorption and other considerations can be used to distinguish between the two cases. Diffraction, therefore, has the effect of adding a center of inversion to the point group of the crystal. This means that one can only distinguish directly between the 11 centrosymmetric groups which, for historical reasons, are also known as the **11 Laue groups**. We may, therefore, classify each of the 32 point groups as members of **Laue classes**. For instance, point groups $4(C_4)$ and $\bar{4}(S_4)$ belong in the same Laue class denoted by the Laue group $4/m(C_{4h})$, that is the diffraction pattern from a crystal having $4(C_4)$ or $\bar{4}(S_4)$ symmetry will look as if it had $4/m(C_{4h})$ symmetry.

Table 4.2 The 11 Laue classes (the Laue groups are marked in bold)

Triclinic	Monoclinic	Orthorhombic	Tetragonal	Trigonal	Hexagonal	Cubic
			International notation			
1	2	222	4	3	6	23
	m	$mm2$	$\bar{4}$		$\bar{6}$	
$\mathbf{\bar{1}}$	$\mathbf{2/m}$	$\mathbf{mmm}$	$\mathbf{4/m}$	$\mathbf{\bar{3}}$	$\mathbf{6/m}$	$\mathbf{m\bar{3}}$
			422	32	622	432
			$4mm$	$3m$	$6mm$	
			$\bar{4}2m$		$\bar{6}m2$	$\bar{4}3m$
			$\mathbf{4/mmm}$	$\mathbf{\bar{3}m}$	$\mathbf{6/mmm}$	$\mathbf{m\bar{3}m}$
			Schoenflies notation			
C_1	C_2	D_2	C_4	C_3	C_6	T
	C_{1h}	C_{2v}	S_4		C_{3h}	
$\mathbf{C_i}$	$\mathbf{C_{2h}}$	$\mathbf{D_{2h}}$	$\mathbf{C_{4h}}$	$\mathbf{S_6}$	$\mathbf{C_{6h}}$	$\mathbf{T_h}$
			D_4	D_3	D_6	O
			C_{4v}	C_{3v}	C_{6v}	
			D_{2d}		D_{3h}	T_d
			$\mathbf{D_{4h}}$	$\mathbf{D_{3d}}$	$\mathbf{D_{6h}}$	$\mathbf{O_h}$

In Table 4.2 are listed the 11 Laue classes with all the point groups associated with them; in each case the Laue group is the last point group in each box. Note that in the 1965 edition of the International Tables the term 'Laue groups' was defined as 'groups of point groups that become identical when a center of symmetry is added to those that lack it.' However, the collections of point groups associated in each Laue class taken together do *not* strictly constitute groups in the mathematical sense. The more recent editions of the International Tables do not, however, make this mistake.

In two dimensions the six Laue classes are associated with (1, 2), (m, $2mm$), (4), ($4mm$), (3, 6), ($3m$, $6mm$), the last group in parentheses being the Laue group symbol for the Laue class.

4.5. POINT GROUP NOTATION

Figure 4.2a is a flow chart for determining the point group of any object in the Schoenflies notation. The cubic point groups should be determined separately by noting the existence of four 3-fold rotations and then checking carefully all of the other symmetry operations.

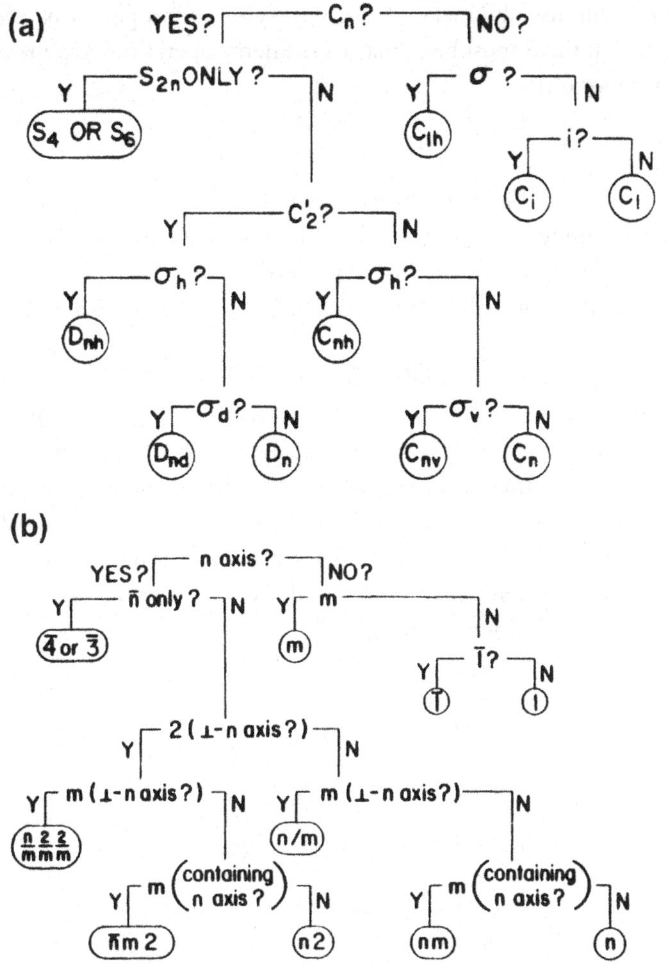

Figure 4.2 A systematic way to determine point groups. Cubic point groups are not included. Note that in (b) the $\bar{6}$ point group would be found as $3/m$, which is equivalent.

Figure 4.2b shows how to obtain the point group of an object using the International notation. This is similar to Fig. 4.2a, although sometimes the International symbol may look slightly different for a particular point group from that given on the chart, because of contraction to the short forms and non-standard directions for the symmetry operations.

Given the point group symbol, one knows the symmetry operations, that is the flow charts in Fig. 4.2 go either way. This is a nice aspect of the point group symbol, namely that the symmetry operations involved can be derived in an instant (drawing a stereogram of several of the symmetry operations will usually make the other symmetry operations clear. Do remember that there must be as many symmetry operations as general points on the stereogram).

Problems

1. $2\bar{2}2$, $\bar{3}/m2$ and $43\bar{2}$ are unusual descriptions of the symmetry of three crystallographic point groups. What are the conventional symbols? What do these point groups have in common?
2. For the point groups $\bar{4}2m$ (D_{2d}) and $622(D_6)$ write out the coordinates of the general points.
3. To which point groups do the following belong? (a) a molecule of H_2O, (b) a brick, (c) an octahedron, (d) a tetrahedron, (e) a propeller, (f) this book (ignoring lettering), (g) each of the letters of the alphabet.
4. An object of point symmetry $m\bar{3}m$ (O_h) distorts under a uniaxial stress along [001]. What is the point group of the distorted object? What is it when the stress is along [111]? An electric field applied to these two distorted objects removes the center of inversion. To which point groups do the resulting objects belong?

CHAPTER 5

Development of Space Groups

Contents

The final frontier
These are the voyages of the Starship, Enterprise
Its 5 year mission to explore strange new worlds
To seek out new life and new civilizations
To boldly go where no man has gone before.

Star Treck (Title sequence)

Space Groups for Solid State Scientists
http://dx.doi.org/10.1016/B978-0-12-394400-9.00005-8

SPACE GROUP OPERATORS

A crystal may be described in differing degrees of detail according to one's requirement. We might assign it to one of the seven crystal systems or else to one of the 14 Bravais lattices, in which case we automatically know the crystal system as well as the type of lattice. If we wish to go deeper, we might then ask to which of the 32 point groups the crystal belongs; this tells us the crystal system and the relationship between the tensor components that describe the macroscopic behaviour and properties of the crystal. We can go further than this by taking into account how the electron density (atoms) is distributed spatially. We have already dealt with the symmetry with respect to a fixed point in the crystal and with the symmetry of space or Bravais lattices. In order to describe the entire crystal (assuming it to be of infinite extent) we must combine the two and also allow for two new types of symmetry operations (which will be discussed below). A space group determined in this way describes the spatial symmetry of the crystal. Thus, a crystallographic **space group** is the set of geometrical symmetry operations that take a three-dimensional periodic object (a crystal) into itself. The set of operations that make up the space group must form a group in the mathematical sense, and certainly must include the primitive lattice translations as well as other symmetry operations.

In order to discuss the set of symmetry operations more simply, consider the lattice to be finite with **periodic boundary conditions**. This is easily understood with reference to a finite one-dimensional lattice of points labelled 1, 2, ..., n, $n + 1$, ..., N, in which case the translational symmetry operation takes each point n into point $n + 1$. We can write the periodic boundary condition as

$$\mathbf{t}_1 = \mathbf{t}_{N+1} \qquad [5.1]$$

where **t** is a unit of translation. This is a useful trick for simulating an infinite lattice and can be applied to two-dimensional or three-dimensional lattices equally well.

Let N be the number of symmetry operations in a finite lattice. Apart from the N operations that translate one primitive unit cell into every other one, there are h symmetry operations that transform the contents of the primitive unit cell into themselves and that are symmetry operations of the entire crystal. Thus, the space group has hN symmetry operations in the set obtained by taking products of the h and N operations. It is convenient to distinguish one product in particular, that is the product of the h symmetry operations with the identity of the translation group, namely, zero translation. These h symmetry operations we shall call **essential space group operations**. The other products of the h operations with primitive lattice translations we shall call **non-essential space group operations**. Operations obtained by the product of the identity of the h operations with the primitive lattice translations are simply the translation symmetry operations. These non-essential operations are perfectly acceptable symmetry operations and are included in the set of hN operations; it is just that they can always be obtained from products of the essential ones with the translations. These ideas are discussed in detail in the following, where we shall see that the operations that we take as the essential ones can often vary.

All space group operations can be described conveniently by means of the **Seitz operator** $\{R|\mathbf{t}\}$ defined by a point operation R, followed by a translation $\mathbf{t}$. R operates on a general position vector $\mathbf{r}$ and can be written as

$$\{R|\mathbf{t}\}\,\mathbf{r} = R\mathbf{r} + \mathbf{t} \qquad [5.2]$$

As discussed in previous chapters, R can refer to either proper or improper rotations or reflections. When $\mathbf{t}$ is any translation, not necessarily one associated with a periodic crystal, the set of operations defined by Eqn [5.2] forms a group known as the **real affine group**, of which space groups are subgroups. An affine group is defined by transformations in space that leave straight lines as straight lines. We can easily check, in the following way, that the set of operations defined by Eqn [5.2] forms a group (see the introduction to Chapter 4 for the definition of a group).

(i) Consider multiplication of any two operations $\{R|\mathbf{t}\}$ and $\{S|\mathbf{u}\}$ where both operations are in the set.

$$\{R|\mathbf{t}\}\,\{S|\mathbf{u}\}\mathbf{r} = \{R|\mathbf{t}\}\,(S\mathbf{r} + \mathbf{u}) = RS\mathbf{r} + R\mathbf{u} + \mathbf{t} = \{RS|R\mathbf{u} + \mathbf{t}\}\mathbf{r} \qquad [5.3]$$

We see then that, since $R\mathbf{u}$ is still a translation, the product of any two operators is another operator that is also a member of the set.

(ii) There is obviously an identity element, $\{1|\mathbf{0}\}$ or $\{E|\mathbf{0}\}$, in the International and Schoenflies notation, respectively.

(iii) The inverse operator $\{R|\mathbf{t}\}^{-1}$ is $\{R^{-1}|-R^{-1}\mathbf{t}\}$ since

$$\{R|\mathbf{t}\}\{R^{-1}| - R^{-1}\mathbf{t}\} = \{1|\mathbf{0}\} \qquad [5.4]$$

as can be checked by using Eqn [5.3].

(iv) From the law of combinations in Eqn [5.3], associativity is obeyed.

These results show that Eqn [5.2] defines a group. This equation is useful for studying the different space groups. Also remember that for crystals the lattice can be described by the Seitz operator $\{1|\mathbf{t}_n\}$, or $\{E|\mathbf{t}_n\}$ in the Schoenflies notation, where $\mathbf{t}_n = n_1\mathbf{a} + n_2\mathbf{b} + n_3\mathbf{c}$.

Space-group symmetry operations may involve a translation, $\boldsymbol{\tau}$, smaller than a primitive lattice translation, coupled with a rotation or reflection. These symmetry operations are known as 'screw rotations' or 'glide reflections'. This leads to an important classification of space groups into two types: symmorphic and non-symmorphic space groups. **A symmorphic space group** is one which may be entirely specified by symmetry operations all acting at a common point and which do not involve one of these translations $\boldsymbol{\tau}$. When it is necessary to specify a space group, with respect to any choice of origin, by at least one operation involving a translation $\boldsymbol{\tau}$, the space group is said to be **non-symmorphic**. Screw or glide operations can still be symmetry operations in symmorphic space groups; it is just that they can also be generated by appropriate combinations of symmorphic operations with translation operations.

5.1. THE SYMMORPHIC SPACE GROUPS

You may think that in order to obtain the space groups one simply combines the 32 point groups with the 14 Bravais lattices. Let us examine this idea more closely and see how a large number of space groups can be generated. This approach does generate the symmorphic space groups, which account for 73 out of a total of 230. These symmorphic space groups also serve as a useful starting point while generating the non-symmorphic space groups.

Consider an orthorhombic primitive, *oP*, lattice (Appendix 3). Now, take any object whose symmetry belongs to one of the orthorhombic point groups and, with the appropriate orientation, relate it to a lattice point (actually to every lattice point by translational symmetry). For example, suppose we start with an object having point symmetry $mm2(C_{2v})$. It might, for example, be one of the special shapes drawn in Appendix 5 or a collection of atoms arranged as in the $mm2(C_{2v})$ stereogram. If the object consists of an arrangement of atoms or molecules, then the result of repeating it in sympathy with the lattice is to produce the **crystal structure**. We have to be careful on how we relate our $mm2(C_{2v})$ object to the lattice. Obviously, to preserve the orthorhombic symmetry at each lattice point the object must be placed so that its mirror planes and 2-fold axes are lined up with the unit-cell axes. The resulting crystal structure has translational symmetry and, in addition, about *any* lattice point it has point symmetry $mm2(C_{2v})$. These two types of symmetry operations are all that are necessary to describe the symmetry of the entire structure.

Figure 5.1a shows an example of how the space groups can be described diagrammatically. The unit cell is outlined with the origin taken, arbitrarily but by convention, at the upper left-hand corner. As usual, a **right-hand convention** is used for the axes with **a** drawn down the page, **b** to the right and **c** out of the page (the same convention as that used throughout the *ITA*). The effect of the symmetry at each lattice point is shown in the left-hand side of the diagram by the arrangement of circles drawn in the same way as discussed in Chapter 1; and it can be seen in our example that this arrangement has symmetry $mm2(C_{2v})$, produced as a result of four operations 1, 2[001], m[100], m[010] (E, C_2, σ_v, σ'_v). The circles on this diagram represent the effect of the symmetry operations on any general point at (x, y, z), the point specified by vector **r** of Eqn [5.2]. The symmetry-equivalent positions thus obtained are known as the **general equivalent positions**. Recall that, in a crystal, each circle may represent a single atom *or* a collection of atoms. The diagram on the right-hand side shows the symbols for the relevant symmetry elements in the locations about which the operations act (Appendix 6). The 2-fold axes are shown to be directed along **c** at the origin together with the two mirror planes (marked by the heavy lines). By translational symmetry {1|**a**}, {1|**b**} and {1|**a**+**b**}, the arrangement of circles and symmetry operations about positions other than

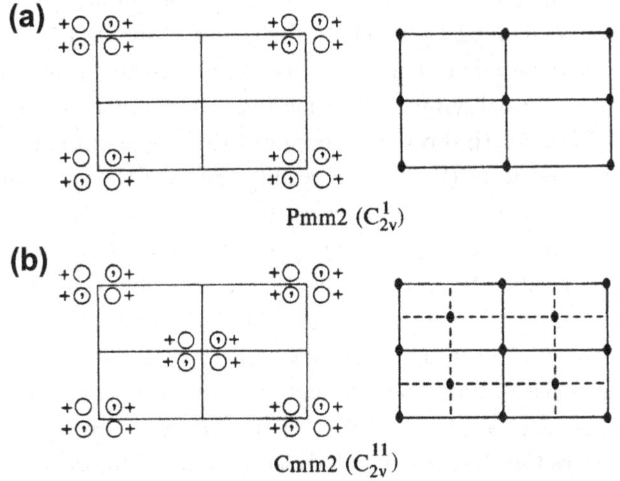

Figure 5.1 Diagrams for two symmorphic orthorhombic space groups.

the origin are shown in the figure. One of the interesting aspects of combining the point symmetry operations with translational symmetry is that many other symmetry operations are generated. Here, we find that halfway between the 2-fold axes, other 2-fold axes occur; and likewise in the case of mirror planes (see Problem 1 in Chapter 3). The symmetry of this space group is given in International notation by the symbol *Pmm*2. *P* stands for the primitive lattice and the *mm*2 describes the other symmetry operations that are **essential** to the space group. By this we mean that the point group symbol *mm*2 allows us to work out the four (in this case) essential symmetry operations mentioned above. The non-essential operations, extra 2-fold rotations and mirror reflections that are generated, are not described in the symbol. The Schoenflies symbol for this space group is C_{2v}^1; it clearly describes the point symmetry used in obtaining the space group but it does not explicitly tell us anything else, since the particular space group is denoted only by the superscript. Here we see the main disadvantage of using Schoenflies notation for space groups.

Returning to the *oP* lattice, appropriately orientated objects could be added having point symmetry $222(D_2)$ or point symmetry $mmm(D_{2h})$. Then two different space groups result whose symmetry operations can be completely described by the space group symbols *P*222 or *Pmmm*, respectively. Diagrams for these two space groups can be drawn similar to those in Fig. 5.1a. Evidently, some space groups can be generated without difficulty.

In the same way, we may consider an orthorhombic base-centered *oC* lattice and add an appropriately orientated object with point symmetry $mm2(C_{2v})$. Again, about *any* lattice point, the point symmetry of the entire crystal structure is $mm2(C_{2v})$. Thus, the space group is labelled *Cmm*2, where the *C* serves to denote the lattice type and the other symbols describe the remaining essential symmetry operations, which are, naturally, the same as in *Pmm*2. Figure 5.1b shows this space group in diagrammatic form. The *C*-centering at position (½, ½, 0) is readily apparent. In fact, without this centering the space group *Cmm*2 reverts to *Pmm*2. Again notice that other non-essential symmetry operations are generated, including special types of reflections, marked by the dashed lines. These will be discussed later in this chapter but note that they have been generated by the combination of point symmetry $mm2(C_{2v})$ with translational symmetry. They are not essential to describe the space group. The Schoenflies symbol has the superscript 11 making it necessary to consult Tables of space groups in order to determine what this means. On the other hand, the beauty of the International symbol is that it contains all the information needed to generate the space group.

We now see how to generate space groups $C222$ and $Cmmm$, and in the same way, other orthorhombic space groups can be formed with oI and oF lattices taken together with the three orthorhombic point groups. The result is that, for the orthorhombic system, when we combine $mm2$, 222 and mmm with the four Bravais arrays, oP, oC, oI and oF, $3 \times 4 = 12$ distinct space groups are obtained.

The same process can be generalized for many other space groups in other crystal systems. Thus, in the triclinic system, there is one Bravais lattice aP and two point groups $1(C_1)$ and $\bar{1}(C_i)$ giving two space groups. These are $P1(C_1^1)$ and $P\bar{1}(C_i^1)$. Similarly, in the monoclinic system there are two Bravais lattices mP and mB and three point groups $2(C_2)$, $m(C_{1h})$ and $2/m(C_{2h})$, giving rise to six space groups labelled $P2(C_2^1)$, $Pm(C_s^1)$, $P2/m(C_{2h}^1)$, $B2(C_2^3)$, $Bm(C_s^3)$ and $B2/m(C_{2h}^3)$. Proceeding in this way for all seven crystal systems, a total of 66 space groups can be found. However, other space groups are obtained when account is taken of the orientation of the point group operations with respect to the Bravais lattice. For example, in the orthorhombic crystal system, oA, oB and oC centered Bravais lattices are equivalent. However, when these lattices are combined with point groups $mm2(C_{2v})$, we can have the 2-fold axis, which we take along **c**, either perpendicular to the centered face or parallel to it. In the former case, we have C-centering, and in the latter A or B-centering. Where the 2-fold axis is either perpendicular or parallel to the centered face, the two cases are physically distinct from each other and lead to space groups $Cmm2$ and $Amm2$. It turns out that by considering the orientation of the point symmetry operations with respect to the Bravais array, a further 7 space groups are obtained giving the **73 symmorphic space groups**. These 73 space groups are listed in Table 5.1 using the International notation for each space group. The exact meaning of this notation will be made clearer at a later stage; however, from this discussion, the notation is obvious. The asterisks in the table mark the seven extra space groups that are generated when the orientation of the point group operations is taken into account with respect to the Bravais unit cell.

Before we go on, a subtle but important point should be brought up. In the development at the beginning of this section, we were careful to place objects belonging to one of the orthorhombic point groups at the lattice points of an orthorhombic lattice. Both the object and the lattice had the symmetry of the same crystal system. This procedure is correct but we should like to question it and see why it is justified. For example, can we put an object with just $1(C_1)$ point symmetry on a cubic Bravais lattice? Conversely, can we put an object with $m\bar{3}m(O_h)$ point symmetry on a triclinic Bravais lattice?

Table 5.1 The 73 symmorphic space groups

Crystal system	Bravais lattice	Space group
Triclinic	*P*	$P1$, $P\bar{1}$
Monoclinic	*P*	$P2$, Pm, $P2/m$
	B or *A*	$B2$, Bm, $B2/m$ (First setting)
Orthorhombic	*P*	$P222$, $Pmm2$, $Pmmm$
	C, *A* or *B*	$C222$, $Cmm2$, $Amm2^*$, $Cmmm$
	I	$I222$, $Imm2$, $Immm$
	F	$F222$, $Fmm2$, $Fmmm$
Tetragonal	*P*	$P4$, $P\bar{4}$, $P4/m$, $P422$, $P4mm$, $P42m$, $P\bar{4}m2^*$, $P4/mmm$
	I	$I4$, $I\bar{4}$, $I4/m$, 1422, $I4mm$, $I\bar{4}2m$, $I\bar{4}m2^*$, $I4/mmm$
Cubic	*P*	$P23$, $Pm\bar{3}$, $P432$, $P\bar{4}3m$, $Pm\bar{3}m$
	I	$I23$, $Im\bar{3}$, $I432$, $I\bar{4}3m$, $Im\bar{3}m$
	F	$F23$, $Fm\bar{3}$, $F432$, $F\bar{4}3m$, $Fm\bar{3}m$
Trigonal	*P*	$P3$, $P\bar{3}$, $P312$, $P321^*$, $P3m1$, $P31m^*$, $P\bar{3}1m$, $P\bar{3}m1^*$
(Rhombohedral)	*R*	$R3$, $R\bar{3}$, $R32$, $R3m$, $R\bar{3}m$
Hexagonal	*P*	$P6$, $P\bar{6}$, $P6/m$, $P622$, $P6mm$, $P\bar{6}m2$, $P\bar{6}2m^*$, $P6/mmm$

The answer to both these questions is no. The reason is that, physically, the potential energy of the crystal compatible with a cubic Bravais lattice can be expanded about a lattice point in combinations of spherical harmonics that have cubic symmetry. There are no combinations that have lower symmetry. Or putting it in another way, since the symmetry operations must act through each lattice point, addition of a non-cubic object automatically forbids overall cubic symmetry. Thus, in a cubic Bravais lattice, the object must have one of the cubic point symmetries. Similarly, for a triclinic Bravais lattice, the potentials expanded about a lattice point have triclinic symmetry and so these potentials would distort the object of higher symmetry and make it adopt triclinic point symmetry. But, remember that the distortion may be small and difficult to observe experimentally.

5.2. NON-SYMMORPHIC OPERATIONS

Space groups that *must* be specified by at least one operation involving a non-primitive translation, τ, are called **non-symmorphic space groups**. Now, if we take such operations into account, it is found that there are a total of 157 non-symmorphic space groups. These taken together with the 73 symmorphic space groups produce 230 in all. We now consider in some detail what happens when rotations and reflections are combined with non-primitive translations to produce the so-called screw and glide operations.

The non-symmorphic operations can conveniently be represented about some origin by the Seitz operator $\{R|\boldsymbol{\tau}\}$ as

$$\{R|\boldsymbol{\tau}\}\mathbf{r} = R\mathbf{r} + \boldsymbol{\tau} \qquad [5.5]$$

R operates first on the position vector $\mathbf{r}$ and is then followed by the non-symmorphic translation $\boldsymbol{\tau}$, where $\boldsymbol{\tau}$ is a translation by a *fraction* of a lattice translation. $\boldsymbol{\tau} = 0$ for the symmorphic operations. Let us now deal with these two non-symmorphic symmetry operations in turn.

5.2.1. Screw Operation

The **screw rotation** is a symmetry operation derived from the coupling of a proper rotation with a non-primitive translation parallel to the axis of rotation, called the **screw axis**. The order in which the two operations are performed is unimportant, as $R\mathbf{r}$ and $\boldsymbol{\tau}$ are commutative under addition.

Consider the rotational part of the screw operation. The rotation R operates on the general vector $\mathbf{r}$ separately from the translation. Since the vectors have magnitude and direction but do not have different origins, for crystals the only allowed proper rotations associated with non-symmorphic operations are the same as in the symmorphic operations, namely, 1, 2, 3, 4 and 6. This is an important point and so we shall explain it more fully.

(i) With reference to Fig. 5.2, consider the point A. At this point there is a screw rotation about an axis perpendicular to the plane of the page that takes the circle marked with a + into the one marked τ +. The symmetry operation is $\{n|\boldsymbol{\tau}\}$ and so the projected angle between the circles is $360/n$, as shown in the diagram.

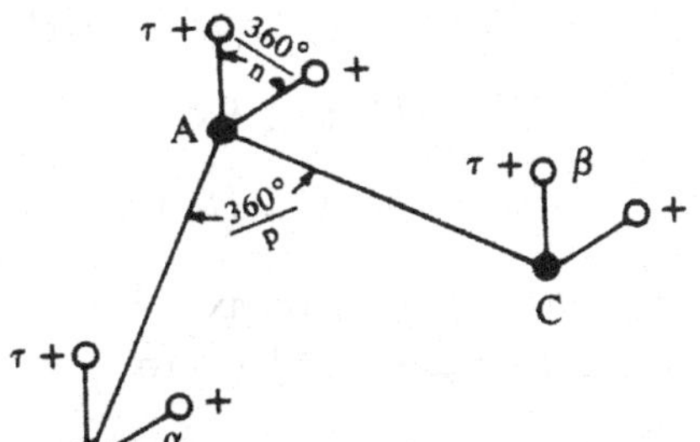

Figure 5.2 The effect of a general screw operation.

(ii) Now consider the effect of the lattice. By translational symmetry, the same arrangement is obtained about the lattice points, B and C, as shown in the figure. These points are chosen so that they are equidistant from the lattice point A, something that can always be done in every crystal system, provided that we choose the angle BAC appropriately. For a particular crystal class, there are restrictions placed on this angle, as discussed in Section 2.4. These conditions can be expressed in terms of the symmetry operations for the lattice $\{p|\mathbf{0}\}$, with $p = 1, 2, 3, 4$ or 6, which implies that angle $BAC = 360°/p$.

(iii) Now reconsider the operation $\{n|\boldsymbol{\tau}\}$ at the point A. If we want this operation to be a symmetry operation of the crystal it must take the entire crystal into itself. This operation takes the circle labelled α with height denoted by the $+$ sign into the one labelled β at height $\tau +$. This operation makes the angle projected on the plane of the page $\alpha A\beta = 360°/n$. However, the symmetry operation of the lattice makes the projected angle $\alpha A\beta = 360°/p$. Thus, if we demand that the screw operation $\{n|\boldsymbol{\tau}\}$ be a symmetry operation of the crystal, $n = p$. Thus, the conditions placed on the unit-cell axes and angles for proper rotations must also be true for screw rotations.

Another point to realize is that the choice of rotation affects the value of $\boldsymbol{\tau}$. This can be demonstrated in the following way. Consider an n-fold screw operation $\{R|\boldsymbol{\tau}\}$ performed n times on a position vector $\mathbf{r}$. Just as with the proper rotations, this must rotate the object that is being operated on through a complete revolution of 360°. However, in contrast to proper rotations, the object will also be moved *along* the rotation axis, in general, through an integral number N of unit lattice translations. Thus we may write

$$\{R|\boldsymbol{\tau}\}^n = \{1|\mathbf{t}_N\} = \mathbf{t}_N \quad [5.6]$$

Now consider what $\{R|\boldsymbol{\tau}\}^n$ is, by writing it explicitly:

$$\begin{aligned}\{R|\boldsymbol{\tau}\}^n\,\mathbf{r} &= \{R|\boldsymbol{\tau}\}\{R|\boldsymbol{\tau}\}\ldots\{R|\boldsymbol{\tau}\}\{R|\boldsymbol{\tau}\}\mathbf{r} \\ &= \{R|\boldsymbol{\tau}\}\{R|\boldsymbol{\tau}\}\ldots\{R|\boldsymbol{\tau}\}(R\mathbf{r}+\boldsymbol{\tau}) \\ &= \{R|\boldsymbol{\tau}\}\{R|\boldsymbol{\tau}\}\ldots(R^2\mathbf{r}+R\boldsymbol{\tau}+\boldsymbol{\tau})\end{aligned} \quad [5.7]$$

Since the translation $\boldsymbol{\tau}$ is *along* the rotation axis, $R\boldsymbol{\tau}$ is equal to $\boldsymbol{\tau}$. Equation [5.7] then, on continuing the process, reduces to

$$\{R|\boldsymbol{\tau}\}^n\mathbf{r} = R^n\mathbf{r} + n\boldsymbol{\tau} = \{R^n|n\boldsymbol{\tau}\}\mathbf{r} \quad [5.8]$$

Thus, from Eqns [5.6] and [5.8]

$$\{1|\mathbf{t}_N\} = \{R^n|n\boldsymbol{\tau}\} \qquad [5.9]$$

In other words, performing the rotation operation n times results in one or more unit cell translations given by $\mathbf{t}_N$. Therefore, the amount of translation is $(1/n)$th of one or more unit repeat distances.

In the example shown in Fig. 5.3, a 4-fold screw axis is shown, with each rotation of **r** through 90° and translations $\boldsymbol{\tau}$ of magnitude **c**/4. The diagram marked 4_1 in Fig. 5.4 shows the same screw rotation drawn in the conventional manner looking down the **c**-axis. The general point at (x, y, z) is taken by the first screw operation, which is denoted by 4_1, to the point $(\bar{y}, x, ¼+z)$. The next screw operation 4_1^2 takes this point to $(\bar{x}, \bar{y}, ½+z)$. The third screw operation 4_1^3 takes it to $(y, \bar{x}, ¾+z)$. Finally, the fourth operation 4^4 returns the general point to its initial position, but translated along **c** through a unit repeat distance.

Note that in designating the screw axes a subscript is used. In the International notation, the screw rotation is denoted R_q, where it is to be understood that the fractional unit of translation is q divided by the order n of the rotation R.

Consider the 4_2 operation shown in Fig. 5.4. For rotation through $2\pi/4 = 90°$ the general point is displaced along **c** through a distance of (2/4)**c**.

Thus, the initial point at (x, y, z) is taken by 4_2 to $(\bar{y}, x, ½+z)$. The operation 4_2^2 then takes it to $(\bar{x}, \bar{y}, 1+z)$ which by translational invariance is equivalent to the point $(\bar{x}, \bar{y}, z)$. The third operation 4_2^3 results in $(y, \bar{x}, 1½+z)$, which is equivalent, by the same token, to $(y, \bar{x}, ½+z)$.

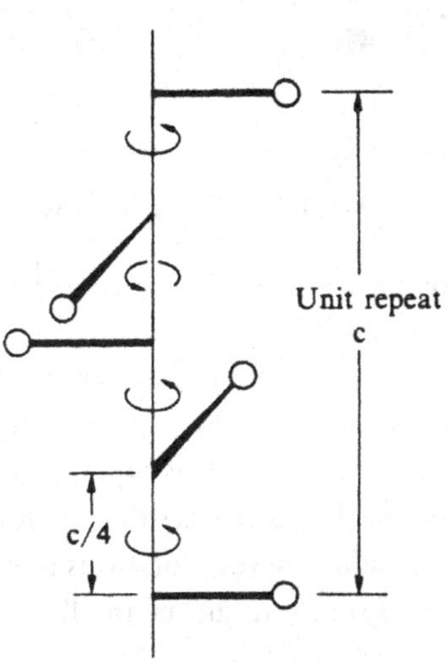

Figure 5.3 A 4_1 screw axis.

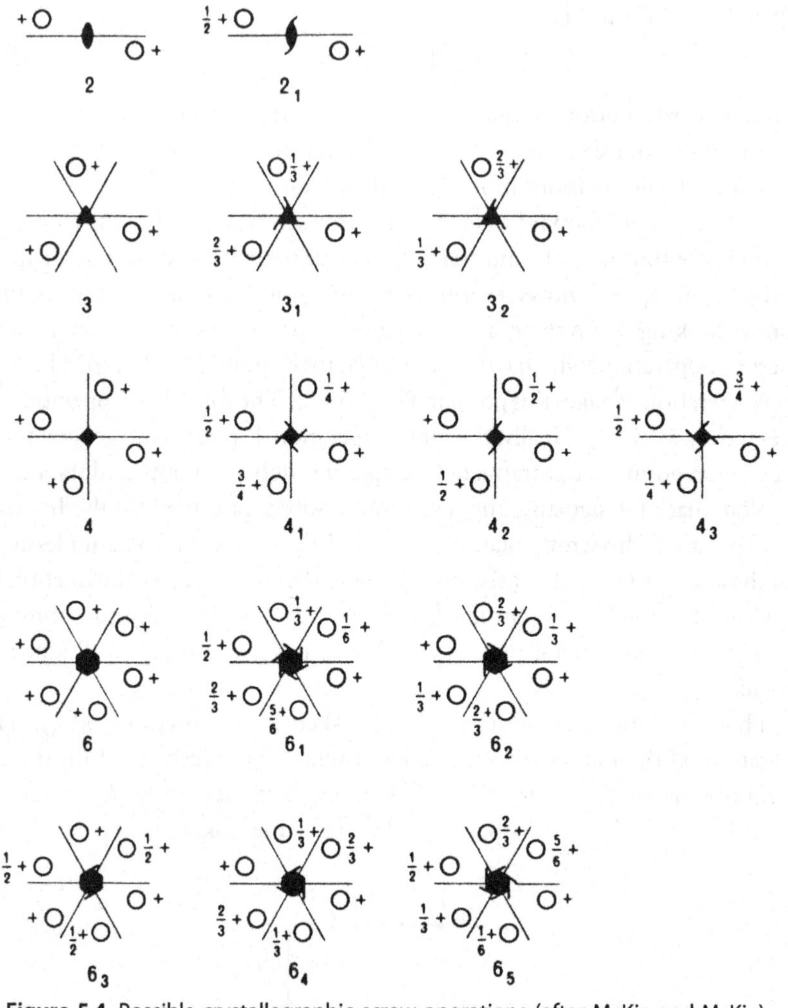

Figure 5.4 Possible crystallographic screw operations (after McKie and McKie).

Finally, carrying out the operation four times, 4_2^4, results in the point at $(x, y, 2 + z)$, which is equivalent, by translational symmetry, to the starting position. The important thing to note is that performing this operation four times corresponds to a total translation of two unit repeat distances. The basic translation, as given by Eqn [5.9], is equal to one-fourth of the total translation of 2 unit repeats, that is it is $2/4 = ½$ along **c**.

It should now be clear which screw rotations are possible. There are 11 crystallographically allowed screw rotations in all

$$2_1,\ 3_1,\ 3_2,\ 4_1,\ 4_2,\ 4_3,\ 6_1,\ 6_2,\ 6_3,\ 6_4,\ 6_5$$

and all these are shown in Fig. 5.4.

It is easy to work out the effect of applying a screw rotation to the general point at (x, y, z) by applying the rotation matrix R from Appendix 1 and then adding the appropriate translation, $\boldsymbol{\tau}$, as expressed by Eqn [5.6]. Take for example, the operation 3_2 about [001]:

$$3_2[001] = \{3[001]|0,\ 0,\ \tfrac{2}{3}\} \qquad [5.10]$$

Here, the vector $\boldsymbol{\tau} = \left(0,\ 0,\ \tfrac{2}{3}\right)$ and means zero translations along **a** and **b** with a translation along [001] of $\tfrac{2}{3}\mathbf{c}$. Now operate on a position vector **r**

$$\{3\,[001]|0,\ 0,\ \tfrac{2}{3}\}\mathbf{r} = 3[001]\mathbf{r} + \left(0,\ 0,\ \tfrac{2}{3}\right) \qquad [5.11]$$

which in matrix form is equal to

$$\begin{bmatrix} 0 & -1 & 0 \\ 1 & -1 & 0 \\ 0 & 0 & 1 \end{bmatrix} \begin{bmatrix} x \\ y \\ z \end{bmatrix} + \begin{bmatrix} 0 \\ 0 \\ \tfrac{2}{3} \end{bmatrix} = \begin{bmatrix} -y \\ x-y \\ \tfrac{2}{3}+z \end{bmatrix} \qquad [5.12]$$

Thus, the general point (x, y, z) is taken by the $3_2[001]$ operation to $(\bar{y}, x-y, \tfrac{2}{3}+z)$.

These new operations are represented by conventional symbols, as shown in Appendix 6 as well as in Fig. 5.4.

5.2.2. Glide Operations

Just as we considered the combination of proper rotations with translation to obtain the screw rotations, it is natural to try the same approach with improper rotations instead. It turns out that only one improper rotation can be combined with translation to result in a symmetry operation, namely, $\bar{2}$ or m (see Problem 3). Thus, the only appropriate combination is reflection plus translation to give what is known as a **glide reflection**. There are, however, different types of glide reflections in crystals: prior to the *ITA* there were three, namely the **axial glide**, the **diagonal glide** and the **diamond glide**. However, more recently a fourth type of glide, the **double glide** has been introduced. In all of these glides one reflects across a plane, the so-called **glide plane** and translates by some distance that is a fraction of the unit cell. The order in which these two steps are carried out is unimportant.

In the **axial glide**, the magnitude of the translation vector **τ** is one half of a unit-cell translation *parallel* to the reflection plane. We refer to the axial glide as an *a*, *b* or *c*-glide according to the axis along which the translation is carried out. The reflection that accompanies this translation can be across any of the planes **ab, bc** or **ca**, that is (001), (100) or (010). Precisely, which plane is involved in a particular axial glide depends on the space group being considered; it is usually clear from the space group symbol. This point will be discussed more fully in the next section.

As an example of an axial glide, Fig. 5.5 shows a *b*-glide with a reflection across a plane perpendicular to the **a**-axis. As in reflection operations, the glide reflection produces the enantiomorph of the original object; thus the comma in the figure. As with the screw rotation, Eqn [5.5] can be used to work out the fractional coordinates of a general point after applying the axial glide operation. For example, this *b*-glide, assuming that the glide plane is at $x = 0$, gives

$$\{m[100]|0,½,0\}\mathbf{r} = m[100]\mathbf{r} + (0,\ ½,\ 0) \qquad [5.13]$$

$$= \begin{bmatrix} -1 & 0 & 0 \\ 0 & 1 & 0 \\ 0 & 0 & 1 \end{bmatrix} \begin{bmatrix} x \\ y \\ z \end{bmatrix} + \begin{bmatrix} 0 \\ ½ \\ 0 \end{bmatrix} = \begin{bmatrix} -x \\ ½+y \\ z \end{bmatrix} \qquad [5.14]$$

so that the point at (x, y, z) is taken to $(\bar{x}, ½+y, z)$. As can be seen by this result, or by Fig. 5.5, applying this operation twice is equivalent to a unit cell translation. As already noted, the glide plane cannot be perpendicular to the glide direction. For example, if you try to draw a diagram of an *a*-glide with the reflection plane perpendicular to **a**, the result will be an ordinary symmorphic mirror operation.

The **diagonal glide**, usually called the ***n*-glide**, involves translations along two or three directions. In general, the translations are $(\mathbf{a}+\mathbf{b})/2$,

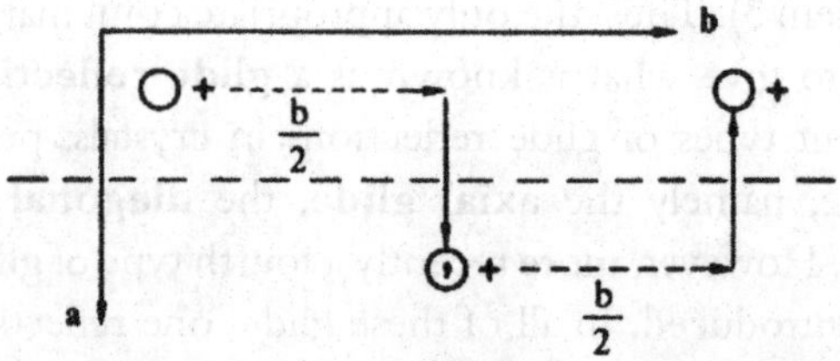

Figure 5.5 The effect of two successive *b*-glide operations. The dashed line represents the glide plane.

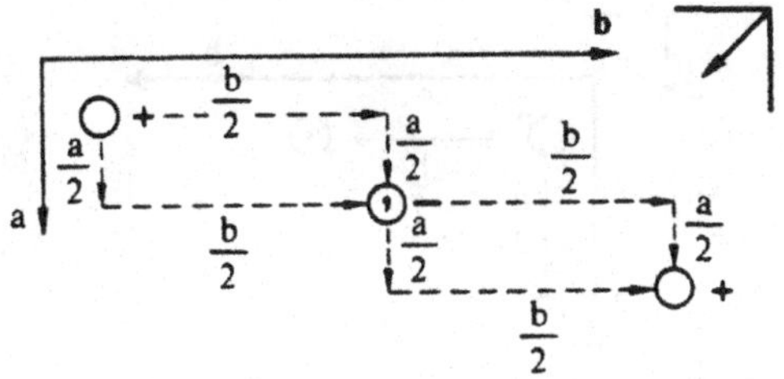

Figure 5.6 The effect of two *n*-glide operations. The symbol at top right indicates the *n*-glide with reflection in the plane of projection.

(**b**+**c**)/2 or (**c**+**a**)/2. However, in tetragonal and cubic crystal systems the *n*-glide can also include a translation of (**a**+**b**+**c**)/2. This occurs only in tetragonal and cubic crystals.

Figure 5.6 illustrates an *n*-glide with a reflection across the plane perpendicular to **c**. The point at (x, y, z) is translated through $(\mathbf{a}/2 + \mathbf{b}/2)$ followed by reflection across the (001) glide plane. Using Eqn [5.5], we can write this as (assuming that the glide plane is at $z = 0$)

$$\{m[001]|½, ½, 0\}\mathbf{r} = m[001]\mathbf{r} + (½, ½, 0) \qquad [5.15]$$

$$= \begin{bmatrix} 1 & 0 & 0 \\ 0 & 1 & 0 \\ 0 & 0 & -1 \end{bmatrix} \begin{bmatrix} x \\ y \\ z \end{bmatrix} + \begin{bmatrix} ½ \\ ½ \\ 0 \end{bmatrix} = \begin{bmatrix} ½+x \\ ½+y \\ -z \end{bmatrix} \qquad [5.16]$$

The point at (x, y, z) is taken by this operation to $(½+x, ½+y, \bar{z})$. As before, the column vector representing $\boldsymbol{\tau}$ is obvious once one knows which n-glide is involved. The position of the symbol '*n*' in the space group symbol (see later) determines the reflection plane and translation directions, as with the axial glide.

The next type of glide is the **diamond** or ***d*-glide**. Here the translations are (**a**±**b**)/4 or (**b**±**c**)/4 or (**a**±**c**)/4 and (**a**±**b**±**c**)/4 in tetragonal and cubic crystals. The same general comments apply to the *d*-glide as to the others, and they will be discussed when encountered in one of the space group examples that follow.

Finally, the recently introduced **double glide** or ***e*-glide** (Fig. 5.7) is defined by two perpendicular glide vectors related by a centering translation and so only occurs in certain centered unit cells.

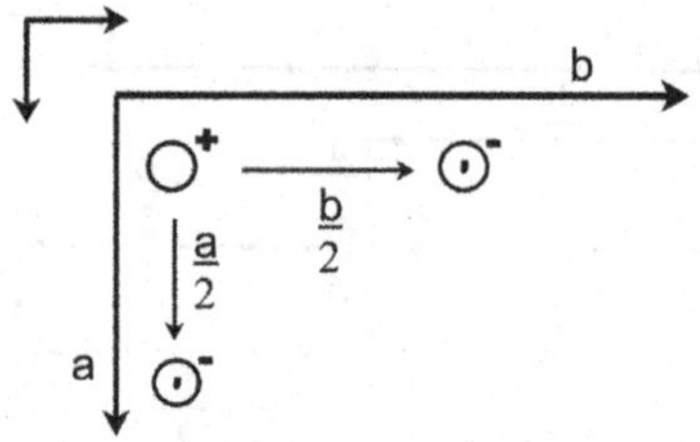

Figure 5.7 The effect of a double glide operation. The symbol at top left indicates the *e*-glide with reflection in the plane of the paper.

The glide operations and their symbols are conveniently summarized in Appendix 6.

5.3. POINT GROUP OF A SPACE GROUP

We can define the **point group of a space group** in terms of the set of symmetry operations obtained when all the translations in the space group symmetry operations are taken to be zero. As usual, we take $\mathbf{t}_n$ as the primitive lattice translation vectors and $\boldsymbol{\tau}$ as a fraction of a primitive lattice vector. Thus, $\boldsymbol{\tau}$ with a point operation R allows one to describe the glide, screw or any general non-symmorphic symmetry operation. We may symbolically describe the method of obtaining the operations of the point group of a space group as

$$\{R_i|\boldsymbol{\tau}_i + \mathbf{t}_n\} \rightarrow \{R_i|\mathbf{0}\} \qquad [5.17]$$

where we have added the subscript i (= 1 to h) to make it clear that a certain value of $\boldsymbol{\tau}$ goes with a certain R and that the identical R_i appears on both sides of this expression.

For a symmorphic space group, these h symmetry operations $\{R|\mathbf{0}\}$ are always symmetry operations of the space group. However, for a non-symmorphic space group, at least one of the h operations cannot be a symmetry operation of the space group. Thus, you might wonder why we have made this definition and of what use it is.

As always, when talking about a symmorphic crystal, the h essential space group symmetry operations $\{R_i|\mathbf{0}\}$ can be considered about one origin. We know that these operations can, and usually do, imply other symmetry operations, such as glides or screws, but these are not essential in describing the space group symmetry. These other symmetry operations are obtained by the product $\{R|\mathbf{0}\}\{1|\boldsymbol{\tau}_n\} = \{R|R\boldsymbol{\tau}_n\}$. A non-symmorphic space group is a space group in which some operations of the form $\{R|\boldsymbol{\tau}\}$ are *required* no matter which origin one uses.

There are several important reasons for defining the point group of a space group. In particular, it is intimately bound up with the relationship between macroscopic physical properties and symmetry (**Neumann's 'principle'**). Possibly, a more important consequence of defining the point group of a space group is that it leads to a fundamental theorem in solid-state science that allows one to handle the symmetry operations of a space group in terms of a point group. First, we consider the connection between macroscopic physical properties and symmetry and then go on to discuss this fundamental theorem at some length.

Neumann's 'principle' states that the macroscopic (tensor) properties of a crystal have *at least* the symmetry of the point group of the space group. The physical reason for this is that in measuring a macroscopic property one would not expect to be able to detect the effect of a translation that is only a fraction of a primitive unit cell. In other words, no distinction could be made between $\{R|\boldsymbol{\tau}\}$ and $\{R|\mathbf{0}\}$ using macroscopic physical properties alone. On the other hand, the rotational part of the symmetry operation will relate points within the crystal that are separated by macroscopic distances and hence, will have an effect on the macroscopic properties. Actually, we are not strictly correct here, as certain physical properties, such as optical activity, can distinguish between enantiomorphic groups. This is because such a chiral property adds in phase from one unit cell to the next to result in an overall macroscopic effect.

Note the use of the words *at least* in the statement of Neumann's principle. It is important to understand that the point group of the space group is generally of lower symmetry than the symmetry of the physical properties. For example, if we consider properties specified by second-rank tensors, we find that cubic crystals (those having cubic point groups) are isotropic, that is the diagonal elements of the tensor are equal and non-zero, while the non-diagonal elements are zero. This means that such properties are the same in all directions and therefore have spherical symmetry. Clearly, all of the symmetry operations of the cubic point groups are contained by such spherical symmetry, so that we can say that the cubic point groups are subgroups of a spherically symmetric group. With properties that are described by tensors of higher rank, the situation is more involved. Several books that discuss the use of Neumann's principle in considerable detail are listed in the Bibliography.

Note that there is an **isomorphism** (a one-to-one correspondence) between the operations of the point group of a space group and the operations of the factor group (see below) of a space group, where the factor group is taken with respect to the primitive lattice translation group.

In order to understand and prove this fundamental theorem, we review briefly some concepts that are normally taught at the beginning of a group theory course. For a more detailed discussion of these concepts, refer to any of the various group theory books in the Bibliography, in particular, the book by Burns, Chapter 2.

Subgroup. The group S is called a subgroup of the group G if all the elements of S are also contained in G. If there are g elements in G and s in S, then one can prove that g/s equals an integer known as the **index** of S in G.

Complex. A complex is a set of elements of a group and this set need not be a subgroup. If α and β are two complexes, then the product $\alpha\beta$ is the product of each element in α with every element in β, where the resulting product of two elements, if it occurs more than once, is only counted once.

Coset. If p is an element of the group G and is not contained in the subgroup S, then the complexes pS and Sp are the left and right cosets of S. Cosets must be formed with respect to a subgroup. Thus, cosets can never be subgroups themselves, since the identity $1(E)$ is contained in S, and therefore, pS cannot contain an identity. If p and q are different elements of G and are not in S, then the cosets pS and qS are distinct (have no overlapping elements). This can be proved by the so-called rearrangement theorem.

Factored set. A finite group can be written in terms of a finite number of distinct cosets, $G = S + pS + qS + \ldots$, where p and q are in G but not in S and q is not in pS. In fact, the number of distinct cosets is g/s.

Conjugate elements. Let a, b, c, …, x, … be the elements of a group. Elements a and b are conjugate if, for some element x in the group, $a = x^{-1}bx$. Every element is conjugate with itself because we may take $x = 1(E)$. It is very easy to prove that: if a is conjugate to b, then b is conjugate to a; if a is conjugate to b and to c, then b is conjugate to c.

Class. The set of elements of a group that are conjugate are said to form a class. The importance of the class concept is that the character (trace) is identical for the matrix representation of all the members of the same class. Thus, in character tables of group representations, the group elements are gathered into the different classes.

Invariant subgroup. A subgroup that consists of complete classes is called an invariant subgroup (or sometimes a **normal divisor** or **normal subgroup**). By this, we mean that T is an invariant subgroup of G if it is a subgroup of G *and* if, for any g in G and t in T, $g^{-1}tg$ is an element of T. From this we see that $g^{-1}Tg = T$ or $Tg = gT$. T consists of complete classes belonging to G. Thus, for an invariant subgroup, the left and right cosets are the same. We may also say that an invariant subgroup is self-conjugate. For example, consider the relationship between point groups $4mm(C_{4v})$ and $mm2(C_{2v})$ (Fig. 5.8).

The operations in point group $4mm(C_{4v})$ are:

$$\{1,\ 4,\ 4^3,\ 2,\ m[100],\ m[010],\ m[110],\ m[\bar{1}10]\}$$

or

$$\{E,\ C_4,\ C_4^3,\ C_2,\ \sigma_v,\ \sigma_v',\ \sigma_d,\ \sigma_d'\}$$

These are arranged in five classes:

1	2[001]	4[001] 4^3[001]	m[100] m[010]	m[110] $m[\bar{1}10]$
E	C_2[001]	C_4[001] C_4^3[001]	σ_v[100] σ_v[010]	σ_d[110] $\sigma_d[\bar{1}10]$

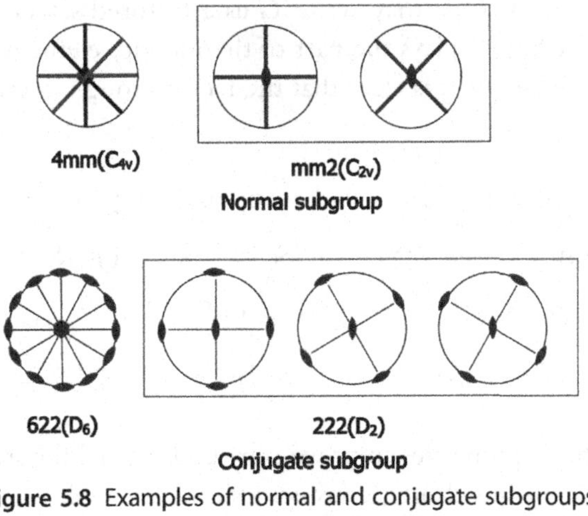

Figure 5.8 Examples of normal and conjugate subgroups.

Point group $mm2(C_{2v})$ is a subgroup that contains four classes that are derived from $4mm$ in two ways, as follows:
Subgroup 1:

1	2[001]	$m[100]$	$m[010]$
E	$C_2[001]$	$\sigma_v[100]$	$\sigma_v[010]$

Subgroups 2:

1	2[001]	$m[110]$	$m[110]$
E	$C_2[001]$	$\sigma_v[110]$	$\sigma_d[\bar{1}10]$

Note that both these subgroups consist of complete classes of $4mm$ and no operation in $4mm$ relates these two. Therefore, $mm2$ is a normal subgroup of $4mm$.

On the other hand, $222(D_2)$ is a subgroup of $622(D_6)$ that can be formed in three ways (Fig. 5.8), depending on the set of 2-fold rotations that one chooses. Note that each of these choices of subgroup is related by the 6-fold symmetry of the supergroup 622. Subgroups like this are called **conjugate subgroups**: point group 222 is said to be conjugate in 622.

Factor group. A factor group of G is a set of elements obtained by considering an invariant subgroup S as one element and all the cosets of this invariant subgroup as the remaining elements. In order to form the factor group, take p, g, t, ... which are distinct elements of G not contained in S. Then, the factor group of G with respect to S has elements $\{S, pS, gS, tS, \ldots\}$. One may write G as a factored set $G = S + pS + qS + tS + \ldots$, where S plays the part of the identity element of the factor group, as we see below. It is clear that the factor group is a group from the following:

Associativity – same as the original group

Closure – $(pS)(qS) = pSSq = pSq = (pq)S = tS$

Identity – $S(pS) = pSS = (pS)$

Inverse – $(pS)^{-1}(pS) = S^{-1}p^{-1}pS = S^{-1}S = S$

We shall use these properties to prove our fundamental theorem. To make some of these concepts clearer, consider a simple example of a factor group of the point group $4mm(C_{4v})$. Thus, we may take $S = \{1, 2[001]\}$ or $\{E, C_2[001]\}$, and since S is a subgroup and is made up of complete classes it is an invariant subgroup. Then

$4S = (4[001], 4^3[001])$	$C_4S = (C_4[001], C_4^3[001])$
$m[100]S = (m[100], m[010])$	$\sigma_vS = (\sigma_v[100], \sigma_v[010])$
$m[110]S = (m[110], m[\bar{1}10])$	$\sigma_dS = (\sigma_d[110], \sigma_d[\bar{1}10])$

as can be verified by the use of a group multiplication table, stereograms or the matrix multiplication method. Thus, the factor group with respect to the invariant subgroup $S = \{1, 2\}$ or $\{E, C_2\}$ can be written as $\{S, 4S, m[100]S, m[110]S\}$ or $\{S, C_4S, \sigma_vS, \sigma_dS\}$. We see the **homomorphism** (many-to-one correspondence) between the point group and the factor group. There is a two-to-one correspondence, where the elements $1(E)$ and $2(C_2)$ of the point group correspond to the element S of the factor group, and so on. Thus, the multiplication table of the point group G_p and the factor group G_p/S are identical so long as this correspondence is recognized. The invariant subgroup S always plays the role of the identity, as shown above. Also, we see that this factor group is smaller (has fewer elements) than the original, in fact one half the size in our example. Thus, with the factor group concept, we may deal with larger groups in terms of smaller groups.

5.3.1. The Factor Group of a Space Group

We now prove the fundamental theorem. Let G be the space group, and let T (**the translation group**) be the subgroup of G that consists of all the pure primitive lattice translations $\{1|\mathbf{t}_n\}$. It is clear that T is a subgroup of G; in fact, in the space group $P1(C_1^1)$ it is the same as G. It is easy to show that T is an invariant subgroup of G. In order to do this, take a typical element of T, $\{1|\mathbf{t}_n\}$, and a typical element of G, $\{R|\mathbf{t}_n + \boldsymbol{\tau}\}$. We find the elements conjugate to $\{1|\mathbf{t}_n\}$ by

$$\begin{aligned}\{R|\mathbf{t}_n + \boldsymbol{\tau}\}^{-1}\ \{1|\mathbf{t}_n\}\ \{R|\mathbf{t}_n + \boldsymbol{\tau}\} &= \{R^{-1}| - R^{-1}\mathbf{t}_n - R^{-1}\boldsymbol{\tau}\}\ \{R|2\mathbf{t}_n + \boldsymbol{\tau}\} \\ &= \{1|R^{-1}\mathbf{t}_n\}\end{aligned} \quad [5.18]$$

The final result is just a pure translation since $R^{-1}\mathbf{t}_n$ is still a primitive lattice vector. Thus, the subgroup T is conjugate to itself and it is an invariant subgroup.

Now, form the factor group of G with respect to the invariant subgroup T, that is G/T. The space group is written in a factored set

$$G = \{1|\mathbf{0}\}T + \{R_2|\boldsymbol{\tau}_2\}T + \cdots + \{R_g|\boldsymbol{\tau}_g\}T \qquad [5.19]$$

where we have taken the $i = 1$ term to be the identity. The corresponding factor group has the elements T, $\{R_2|\boldsymbol{\tau}_2\}T$, ..., $\{R_g|\boldsymbol{\tau}_g\}T$.

Let us examine this equation more closely. So far in this book, we have always taken the lattice to be infinite (an infinite number of symmetry operations in T). We could just as well have taken the lattice to be finite with periodic boundary conditions. For the sake of a better understanding of this equation, let us do just that and assume that the group T has some huge number of symmetry operations equal to m. This equation, then, is a short-hand way of writing the gm symmetry operations of the space group G. Clearly, each one of the elements of this factor group is 'hiding' a huge number of operations. As we shall see below, Eqn [5.19] contains, at most, 48 (= g) entries on the right-hand side.

In order to prove that G/T and the point group G_p are isomorphic, we must show two things.

(i) First, all elements of one coset have R_i as the rotational part of the element. This is fairly easy to show since a typical element of one coset is $\{R_i|\boldsymbol{\tau}\}$ $\{1|\mathbf{t}_n\} = \{R_i|R_i\mathbf{t}_n + \boldsymbol{\tau}\}$. Clearly, the rotational part of the result is R_i.

(ii) Second, which is the converse, all the elements of G that have R_i as the rotational part must be contained in the particular coset $\{R_i|\boldsymbol{\tau}\}T$. Consider an element of G, $\{R_i|\mathbf{t}'\}$, and assume for the moment that it is not contained in the coset $\{R_i|\boldsymbol{\tau}\}T$. Then

$$\{R_i|\boldsymbol{\tau}_i\}\{R_i|\mathbf{t}'\}^{-1} = \{R_i|\boldsymbol{\tau}_i\}\ \{R_i^{-1}| - R_i^{-1}\mathbf{t}'\} = \{1| - \mathbf{t}' + \boldsymbol{\tau}_i\} \qquad [5.20]$$

Because the result must be a member of G, it must also be a pure primitive lattice translation (the rotational part is the identity) and so $\mathbf{t}'$ may differ from $\boldsymbol{\tau}$ only by a pure primitive lattice vector. If we take $\mathbf{t}' = \mathbf{t}_n + \boldsymbol{\tau}_i$, then the term $\{R_i|\mathbf{t}'\}$ is in the coset $\{R_i|\boldsymbol{\tau}_i\}T$ because the general term of this coset is $\{R_i|\boldsymbol{\tau}_i\}$ $\{1|\mathbf{t}_n\} = \{R_i|\mathbf{t}_n + \boldsymbol{\tau}_i\}$. Thus, our assumption is contradicted since $\{R_i|\mathbf{t}'\}$ must be contained in the product of $\{R|\boldsymbol{\tau}\}$ and a primitive translation of T.

5.4. SPACE GROUPS

We are now in a position to describe the space groups using the symmetry operations already discussed. This can be done either by drawing diagrams or by using mathematical techniques. The method that we shall adopt involves both techniques. We shall see throughout that it is possible to explain the space group diagrams taken from the *ITA* by using the matrix operators discussed in the preceding chapters. There are two main features to explain: the coordinates of the general equivalent positions and the positions and occurrence of the symmetry operations, essential and non-essential, with respect to any particular choice of origin, both of which can be obtained from the International space group symbol alone.

The procedure, then, will be to consider the point group of any space group, as discussed in the last section, and to ascertain the number of essential symmetry operations. This automatically tells us the number of general equivalent positions in the primitive unit cell (we must be careful, however, since in the *ITA* the conventional Bravais cells are used and this means that the number of general equivalent positions will be a multiple of that for the primitive cell). Having done this, we shall then pick out certain features of the space group examples given and explain them by matrix operator methods. It will be seen that, simply by using the Seitz operator together with the matrices in Appendix 1, the coordinates of all the general equivalent positions can be determined. In so doing, the presence and positions of other symmetry operations, including those that are not given in the space group symbol, are obtained. While all this can be achieved purely by means of diagrams, and this is the way that most crystallographers construct a particular space group, it is useful to do this with some kind of mathematical formalism as well. This does have the benefit of making the process of constructing any space group more automatic and precise and also often gives more insight into the relationships between the various symmetry operations. No attempt will be made in this section to actually derive the space groups; we shall indicate how this was done by Fedorov and Schoenflies in Section 5.5.

We have chosen a few examples of space groups each of which displays certain features that we consider worthwhile discussing in detail. After reading this section, the reader should be able to derive all of the general equivalent positions and symmetry operations of any space group, given only the International space group symbol. It is a remarkable fact that a single symbol can succinctly describe *all* of the features of a space group. The Schoenflies symbol for the space group, on the other hand, is made up from the Schoenflies point group symbol and a serial number as a superscript. The superscript serves only to differentiate between one space group and another; however, it is arbitrary and does not allow one to generate all the symmetry operations or general equivalent positions. Nevertheless it is useful because it shows the point group of the space group at a glance and is independent of the choice of the axes of reference. In order to determine the point group from the International space group symbol, it is necessary to replace all non-symmorphic operations, when they occur, by their symmorphic equivalents, that is replace all screw rotations by proper rotations and all glides by mirrors.

Before going on to the examples, it is worth saying a little more about the International notation (Section 5.1). The space group symbol always takes the form of a letter, *P*, *A*, *B*, *C*, *I*, *F* or *R* for the Bravais lattice followed by one or more symmetry symbols. For example, $Pnma(D_{24}^{16})$ and $P2_1/c(C_{2h}^5)$ are two space groups belonging to different crystal systems (their point groups are $mmm(D_{2h})$ and $2/m(C_{2h})$, respectively) but both have primitive lattices. In the *ITA*, the *standard* space group symbols are listed. However, in the International notation, just as the point group symbols can be different according to our choice of axes of reference, so too can the space group symbols. The standard form adopted by the *ITA* is for convenience only, and it is stated explicitly that 'no official importance is attached to the particular setting of the space group adopted as standard.' There are many reasons why a particular crystal structure may be given a non-standard space group symbol. For example, it may prove easier to visualize a particular structure in relation to some other form, such as a high-temperature phase, by taking a non-standard orientation for the axes of reference.

One final point of warning: by definition, a space group always includes the infinite number of translation operations $\{1|\mathbf{t}_n\}$. This is so fundamental that it is sometimes ignored while attention is focused on the other essential symmetry operations. Please remember that the translational symmetry is always understood to exist.

5.4.1. General Equivalent Positions and Coset Representatives

The space group can be defined in terms of the **general equivalent positions**. These positions are obtained by starting with a general point at (x, y, z) and applying the h essential symmetry operations $\{R|\boldsymbol{\tau}\}$: we now drop the subscript in order to conform to normal practice. Thus, the set of general equivalent positions consists of h positions and any essential or non-essential symmetry operation of the space group operating on this set of positions will not produce any more. Naturally, we may always apply the translational symmetry operations to gather these h positions about a convenient origin.

Note that in the *ITA*, h, $2h$, $3h$ or $4h$ general equivalent positions are given when the Bravais lattice is primitive, base or body-centered rhombohedral referred to hexagonal axes or face-centered respectively. This is because the conventional centered unit cells are used in these tables (Appendix 3 shows diagrams of these unit cells). However, it is more fundamental to think in terms of the primitive cell of any lattice, which has only h general equivalent positions corresponding to the h symmetry operations $\{R|\boldsymbol{\tau}\}$.

Note that the space group may be completely described by giving the lattice type (P, C, F, I) and a set of h general equivalent positions. The latter tells us all the symmetry operations of $G/T$and hence the crystal system, and so on, while the lattice type tells us all the symmetry operations of T. Consider the example of space group $Aba2(C_{2v}^{17})$. The translation subgroup is given by

$$\{1|1,0,0\},\ \{1|0,1,0\},\ \{1|0,0,1\},\ \{1|0,½,½\} \qquad [5.21]$$

and the general coordinates from the *ITA* are

$$x,y,z \quad \bar{x},\bar{y},z \quad x+½,\bar{y}+½,z \quad \bar{x}+½,y+½,z \qquad [5.22]$$

We say that the **coset representatives** of the translation subgroup $T=A$ in the representative space group $G = Aba2$ are:

$$\{1|0,0,0\}\ \{2[001]|0,0,0\}\ \{m[010]|½,½,0\}\ \{m[100]|½,½,0\} \quad [5.23]$$

Thus, the translation group $T = A$ and this set of coset representatives uniquely define the space group $Aba2$.

We now consider examples of the space groups in different crystal systems.

5.4.2. Triclinic System

We have already seen that in the triclinic crystal system there is only one unique space lattice, the primitive lattice aP. Furthermore, as discussed in Section 5.2, because there are only two triclinic point groups, $1(C_1)$ and $\bar{1}$ (C_i), there can only be two symmorphic triclinic space groups, $P1(C_1^1)$ and $P\bar{1}$ (C_i^1). Since reflection operations and rotations of order 2 or more are not compatible with triclinic symmetry, there cannot be any screw or glide operations. This implies that there are no non-symmorphic triclinic space groups.

How do we represent these space groups diagrammatically? Fig. 5.9 shows the way this is done in the *ITA* for space group $P1(C_1^1)$. Two separate unit cells are shown side by side in precisely the same way as in the examples discussed briefly in Section 5.2. The cell on the right illustrates the effect of the symmetry operations on a general point at (x, y, z), represented as usual by a circle, and the cell on the left shows the locations of the symmetry operations. Although this is a trivial example, it does show certain features. Consider the left-hand side cell. Since the point group of $P1$ is $1(C_1)$, the space group is completely specified by the identity operation $1(E)$ and the primitive lattice translations $\{1|\mathbf{t}_n\}$. There is no special diagrammatic symbol for $1(E)$ and so only the basic outline of the unit cell is drawn. Notice that the unit cell has been deliberately drawn with non-orthogonal axes to emphasize that the space group is triclinic. In the right-hand side cell, the general point at the top-left of the cell, which is deliberately shown to be above the plane of projection, as indicated by the + sign, is operated on by the identity operation leaving it unchanged. This is further operated on by the primitive translations to produce other points related to it by unit translations along **a**, **b** and **a** + **b**. Clearly, these belong in neighbouring cells so that there is only one general equivalent position per unit cell, in agreement with the order of the point group. Using Eqn [5.5], we write

$$\{1|\mathbf{0}\}\mathbf{r} = \begin{bmatrix} 1 & 0 & 0 \\ 0 & 1 & 0 \\ 0 & 0 & 1 \end{bmatrix} \begin{bmatrix} x \\ y \\ z \end{bmatrix} = \begin{bmatrix} x \\ y \\ z \end{bmatrix} \qquad [5.24]$$

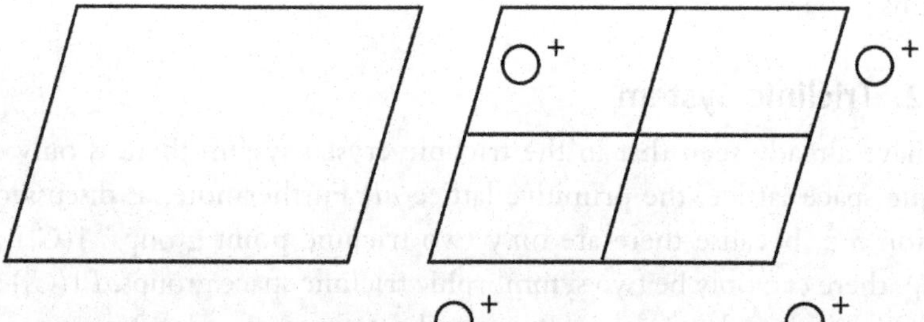

Figure 5.9 $P1(C_1^1)$.

Thus, operating on the general point at (x, y, z) by the identity operator gives (x, y, z) again. This may seem a complicated way of looking at what is really a trivial problem but the approach will be appreciated when we come on to more complex space groups.

The second triclinic space group $P\bar{1}(C_i^1)$ is shown in Fig. 5.10. The point group of this space group is $\bar{1}(C_i)$, and has two symmetry operations $1(E)$ and $\bar{1}(i)$, so that we need two symmetry operations to completely specify the space group. The left-hand cell shows the location of the inversion centers represented by small circles. It is conventional to take an inversion operation to act at the origin of the unit cell since this halves the information needed to specify the electron density in the unit cell, each half being related by the $\bar{1}(i)$ operation. This is particularly important in diffraction studies because it simplifies the phase relationships between scattered waves (for an elementary discussion of this, see any of the standard books on X-ray crystallography).

Notice that having placed a center of inversion at the origin of the unit cell, and hence, by translational invariance, at the origins of all unit cells, new centers of inversion appear at positions halfway along the axes, at the centers of the faces and in the center of the unit cell. This can be understood best by consulting the right-hand side diagram. The point at (x, y, z) is operated on by $\{\bar{1}|\mathbf{0}\}$ to produce a new equivalent point thus:

$$\{\bar{1}|\mathbf{0}\}\mathbf{r} = \begin{bmatrix} -1 & 0 & 0 \\ 0 & -1 & 0 \\ 0 & 0 & -1 \end{bmatrix} \begin{bmatrix} x \\ y \\ z \end{bmatrix} = \begin{bmatrix} -x \\ -y \\ -z \end{bmatrix} \quad [5.25]$$

We see from this result that the general point at (x, y, z) has been taken to ($\bar{x}$, $\bar{y}$, $\bar{z}$) by the $\bar{1}$ operation at (0, 0, 0), forming an enantiomorphically

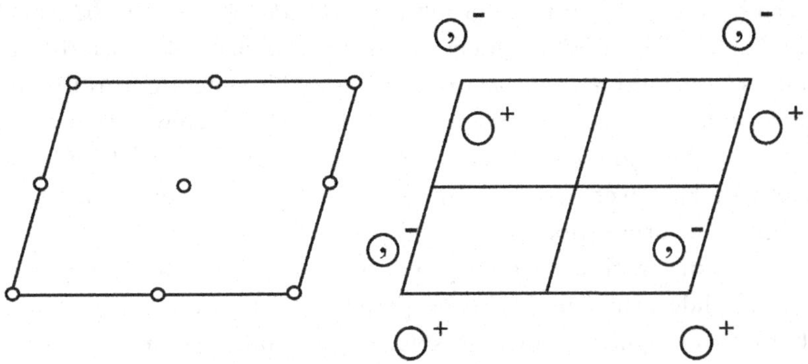

Figure 5.10 $P\bar{1}(C_i^1)$.

related point in the process. The same occurs at other places in the diagram separated by the unit translation distances. The result of this is that there are two general equivalent positions per unit cell (equal to the number of operations in the point group of the space group). The point at (x, y, z) is also related to that at $(\bar{x}, 1-y, \bar{z})$ by a center of inversion at (0, ½, 0). These and other centers of inversion result from the combination of the point symmetry at the origin with the translation symmetry of the lattice. The point at $(\bar{x}, 1-y, \bar{z})$ is, of course, equivalent to the point at $(\bar{x}, \bar{y}, \bar{z})$ by the translational symmetry along **b**. We can write a similar expression to that used in Eqn [5.25] in order to relate the two general positions by the center of inversion at (0, ½, 0). This is done by including a translation vector that takes care of the shift of origin of this symmetry operation. The expression is

$$\{\bar{1}|2(0,½,0)\}\mathbf{r} = \begin{bmatrix} -1 & 0 & 0 \\ 0 & -1 & 0 \\ 0 & 0 & -1 \end{bmatrix} \begin{bmatrix} x \\ y \\ z \end{bmatrix} + \begin{bmatrix} 0 \\ 1 \\ 0 \end{bmatrix} = \begin{bmatrix} -x \\ 1-y \\ -z \end{bmatrix} \quad [5.26]$$

It is evident from this that in order to relate the two positions, we have to add a vector translation equal to twice the shift of origin of the symmetry operation. This is simply a consequence of the fact that the inversion center must lie midway between the points related by it. The same general rule applies to 2-fold rotations and reflections but more complicated rules are needed for other symmetry operations.

5.4.3. Monoclinic System

In the monoclinic crystal system, there are two unique space lattices that are conventionally used, *mP* and *mB* (first setting) or *m*C (2nd setting), and three point groups, 2, *m* and 2/*m* (C_2, C_s and C_{2h}). This means that we can have 2 × 3 = 6 space groups, *P*2, *B*2, *Pm*, *Bm*, *P*2/*m*, *B*2/*m* (C_2^1, C_2^3, C_s^1, C_s^3, C_{2h}^1, C_{2h}^3) as discussed in Section 5.2. The fact that the symmetry operations 2 and *m* are allowed in this system means that it is now possible to generate space groups containing 2_1 screw axes and glide planes. In the monoclinic system, both axial and diagonal glides are permitted. However, it turns out that they become equivalent if one redefines the orientation of the unit cell. We take for this discussion the '1st setting' with the mirror planes perpendicular to the **c**-axis. Thus the axial glide planes must also be perpendicular to the **c**-axis, leading only to *a* and *b*-glides. However, since the assignment of **a** and **b**-axes is arbitrary, there is only one unique glide direction. The *a*-glide can be called a b-glide simply by interchanging the labels of the **a** and **b**-axes. In the *ITA*, the convention is to call it a b-glide (c-glide in the second setting). Thus, the possible space groups involving screw axes and glide planes are $P2_1$, $B2_1$, Pb, Bb, $P2_1/m$, $B2_1/m$, $P2/b$, $B2/b$, $P2_1/b$, $B2_1/b$. However, as we shall see, some of these are equivalent.

Before considering some examples in detail, it is worth saying something about the space group notation. The symbols that we have used are known as the **standard short symbols**; this is because they contain sufficient information to determine the space group. There is a **full symbol** which is quoted sometimes. This takes the form Γ …, where Γ represents the symbol for the Bravais lattice, P, A, B, C, I, F or R, and the dots denote the positions of the symbols for the symmetry operators, which in the monoclinic and orthorhombic systems are taken to be relative to the **a**, **b** and **c**-axes, respectively. Thus, a 2-fold axis parallel to **a** would be indicated by the figure '2' in the first place. An a-glide with the plane of reflection perpendicular to **c** would be denoted by 'a' in the third place, and so on. In the monoclinic system, the '1st setting' convention is to place the 2 or 2_1 axis parallel to **c** and the m or b-glide plane perpendicular to **c**. Thus the full symbol for $P2$, say, would be $P112$; this means that there is '1-fold' (trivial) symmetry along **a** and **b** and a 2-fold axis parallel to **c**. Similarly, the space group $P2_1/b$ would be written as $P112_1/b$. It can be seen why the full symbol is not normally used in the monoclinic system; it gives no information that we do not know already. In the orthorhombic system, however, the full symbols are often necessary. In Appendix 7, all the short and full symbols are listed.

We shall consider three examples, one corresponding to each point group of the monoclinic system. Each example illustrates certain features that can be applied to the other monoclinic space groups not covered here. The space groups chosen are $B2$, Bb and $P2_1/b$.

(1) Space group $B2(C_2^3)$, number 5 in the *ITA*. Consider Fig. 5.11a, where the unit cells are drawn in the usual orientation. The **c**-axis is out of the paper and perpendicular to **a** (down the page) and **b**. The point group for $B2$ is obtained by removing all translations and so we have point group $2(C_2)$, which has two symmetry operations, 1 and 2 (E and C_2). Thus, these two operations together with the primitive lattice translations completely specify the space group. However, since in the *ITA* the conventional Bravais cell is used, we take this into account in determining the general equivalent positions. Thus, in the B-centered cell, every point specified with respect to

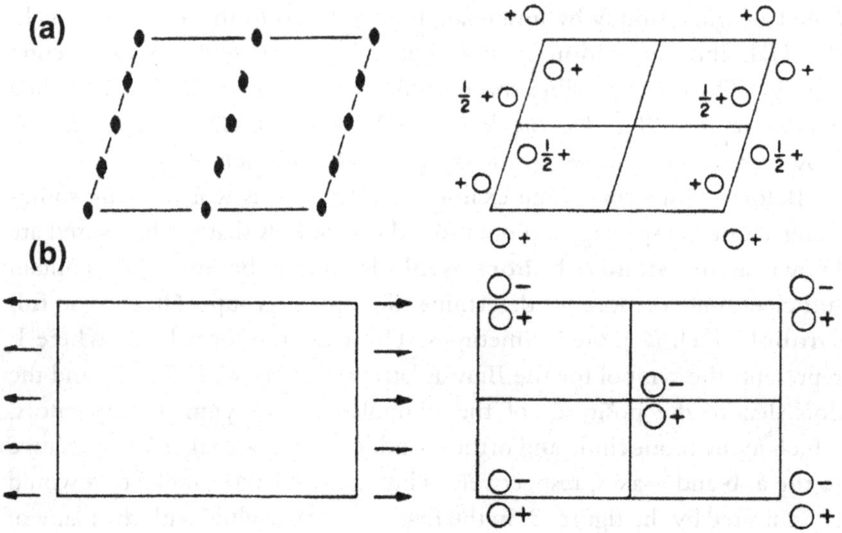

Figure 5.11 (a) $B2(C_2^3)$ (b) $C2(C_2^3)$.

an origin at (0, 0, 0) is repeated with respect to the lattice point occurring at (½, 0, ½).

This means that there will be twice as many general equivalent positions listed in these tables as symmetry operations in the point group, that is there are four in all. Returning to Eqn [5.5], set the vector $\boldsymbol{\tau} = 0$ since we are considering only the 2-fold operation. The effect of this 2-fold rotation on the general point is then given by

$$\{2[001]|\mathbf{0}\}\mathbf{r} = \begin{bmatrix} -1 & 0 & 0 \\ 0 & -1 & 0 \\ 0 & 0 & 1 \end{bmatrix} \begin{bmatrix} x \\ y \\ z \end{bmatrix} = \begin{bmatrix} -x \\ -y \\ z \end{bmatrix} \qquad [5.27]$$

Thus, the point at (x, y, z) is taken to $(\bar{x}, \bar{y}, z)$, as shown in the right-hand diagram in Fig. 5.11a. We can now add the centering translation (½, 0, ½) to each of these equivalent points thus generating two new points situated at $(½+x, y, ½+z)$ and at $(½-x, \bar{y}, ½+z)$. Again, these are marked on the right-hand diagram of Fig. 5.11a.

There are other symmetry operations that occur in this space group. For example, consider the two points at (x, y, z) and $(½-x, \bar{y}, ½+z)$. There must be some symmetry operation given by $\{R|\tau\}$ linking them. Therefore we can write

$$R\begin{bmatrix} x \\ y \\ z \end{bmatrix} + \tau = \begin{bmatrix} \tfrac{1}{2} - x \\ -y \\ \tfrac{1}{2} + z \end{bmatrix} = \begin{bmatrix} -x \\ -y \\ z \end{bmatrix} + \begin{bmatrix} 0 \\ 0 \\ \tfrac{1}{2} \end{bmatrix} + \begin{bmatrix} \tfrac{1}{2} \\ 0 \\ 0 \end{bmatrix} \qquad [5.28]$$

The last two vectors are reminiscent of non-symmorphic translations. The first vector on the right side suggests a 2-fold rotation through an axis at the origin of the unit cell operating on the point at (x, y, z), so that we must consider the possibility of having a 2_1 screw axis in this space group. In order to be consistent with monoclinic symmetry (and point group 2), this 2_1 axis must lie along the [001] direction, and hence its translation vector would be consistent with the second vector on the right-hand side in Eqn [5.28]. The third vector on the right-hand side of this equation is then seen to arise from a shift of origin for this screw axis. Since this must lie halfway between the points situated at (x, y, z) and ($\tfrac{1}{2}-x$, $\bar{y}$, $\tfrac{1}{2}+z$), this additional translation vector (½, 0, 0) arises from the 2_1 axis passing through ½(½, 0, 0) = (¼, 0, 0). This and other screw axes that are present are marked on the left-hand diagram of Fig. 5.11a. It can be seen that the space group $B2$ contains non-symmorphic operations. Nevertheless, it is still a symmorphic space group, since the space group can be specified completely by symmorphic operations alone. The screw rotations are non-essential symmetry operations and arise from combining the $2(C_2)$ point operation with translational symmetry. Of course, we are quite at liberty to call this space group $B2_1$ should we wish to do so, rather than $B2$. Either would do; it is just conventional to use the latter symbol, which has the merit of revealing that the space group is symmorphic. Therefore, crystals of point group $2(C_2)$ belong to one of the three possible space groups: $P2(C_2^1)$, $P2_1(C_2^2)$ and $B2(C_2^3)$ or $C2(C_2^3)$. Fig. 5.11b shows the space group diagrams for $C2$ (2nd setting).

(2) Space group $Bb(C_s^4)$, number 9 in the *ITA*. Crystals having this space group belong to the point group $m(C_s$ or $C_{1h})$, because when the translational part of the b-glide is removed a mirror reflection is obtained. Point group $m(C_{1h})$ has two symmetry operations, 1 and m (E and σ_h). When these operations are combined with the B-centering, we expect to obtain four general equivalent positions. Figure 5.12a shows the space group diagrams. Proceeding as before, apply the b-glide to the general point at (x, y, z) to obtain a new point at (x, $\tfrac{1}{2}+y$, $\bar{z}$) and then apply the B-centering to get the points ($\tfrac{1}{2}+x$, y, $\tfrac{1}{2}+z$) and ($\tfrac{1}{2}+x$, $\tfrac{1}{2}+y$, $\tfrac{1}{2}-z$). These are the four general

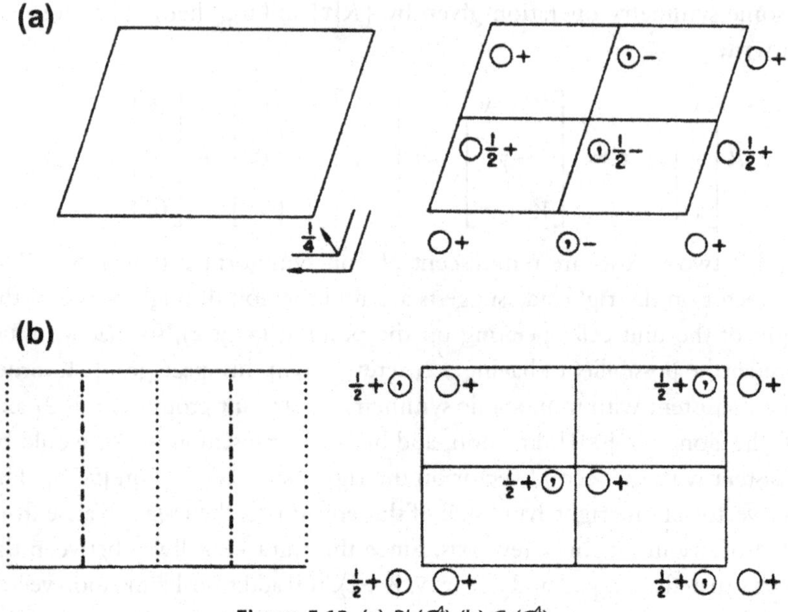

Figure 5.12 (a) $Bb(C_s^4)$ (b) $Cc(C_s^4)$.

equivalent positions in this space group. However, we can find other symmetry operations. Consider the relationship between the two points at (*x*, *y*, *z*) and (½+*x*, ½+*y*, ½−*z*). As before, we can subtract some translations from the latter:

$$R\begin{bmatrix} x \\ y \\ z \end{bmatrix} + \tau = \begin{bmatrix} ½ + x \\ ½ + y \\ ½ - z \end{bmatrix} = \begin{bmatrix} x \\ y \\ -z \end{bmatrix} + \begin{bmatrix} ½ \\ 0 \\ 0 \end{bmatrix} + \begin{bmatrix} 0 \\ ½ \\ 0 \end{bmatrix} + \begin{bmatrix} 0 \\ 0 \\ ½ \end{bmatrix} \quad [5.29]$$

The first vector suggests that *R* is a reflection acting on (*x*, *y*, *z*), and so we should suspect the presence of a glide. The second and third vectors correspond to glides along both **a** and **b** simultaneously; this is equivalent to an *n*-glide perpendicular to **c**. The plane can be located precisely by considering the fourth vector. Its magnitude represents the fractional distance along **c** that any general point is taken through on reflection across the *n*-glide plane. Now this plane must lie halfway between the two mirror-related points, so that the position is given by one-half of the fourth vector given by (0, 0, ½). The *n*-glide plane, therefore passes through (0, 0, ¼), as shown on the left-hand diagram of Fig. 5.12a. Space group Bb can, of course, be labelled Bn. We could equally well have started with the n-glide plane passing through the origin and lying perpendicular to **c** and then have generated the b-glide, this time through (0, 0, ¼). It is simply the normal custom to use the former symbol and origin. Figure 5.12b shows the same space group but in the 2nd setting, that is Cc. Proceeding in this way for all the possible space groups with point group $m(C_{1h}$ or $C_s)$, we find that the only uniquely different space groups are $Pm(C_s^1)$, $Pb(C_s^2)$, $Bm(C_s^3)$ and $Bb(C_s^4)$.

(3) Space group $P2_1/b(C_{2h}^5)$, number 14 in the *ITA*. This is an important space group, particularly for organic crystal structures, as it is the most common space group encountered in organic crystals.[1] The point group of this space group is $2/m(C_{2h})$ which has four operations, 1, 2[001], m[001] and $\overline{1}$ (E, C_2, σ_h and i). We expect therefore to generate four general equivalent positions.

Figure 5.13a shows the space group diagrams. Start with the 2_1 axis and b-glide plane through the origin. The 2_1 operation generates from the point (x, y, z) a new point at $(\overline{x}, \overline{y}, ½+z)$ and the b-glide takes these two points to the positions $(x, ½+y, \overline{z})$ and $(\overline{x}, ½-y, ½-z)$, respectively. This can be verified by consulting the diagram or by applying the matrices in the way done previously. Now, consider the two points located at (x, y, z) and at $(\overline{x}, ½-y, ½-z)$ which are enantiomorphs of one another. The symmetry operation relating these points is given by

$$R\begin{bmatrix} x \\ y \\ z \end{bmatrix} + \boldsymbol{\tau} = \begin{bmatrix} -x \\ ½ - y \\ ½ - z \end{bmatrix} = \begin{bmatrix} -x \\ -y \\ -z \end{bmatrix} + \begin{bmatrix} 0 \\ ½ \\ ½ \end{bmatrix} \qquad [5.30]$$

Thus, we recognize that R is the inversion operator acting midway between the two points related by it; in this case, the inversion center is positioned at ½(0, ½, ½) = (0, ¼, ¼). Its existence is a direct consequence of having the 2_1 axis perpendicular to the b-glide plane. This is hardly surprising when we recall that the point group of this space group is $2/m(C_{2h})$, with the 2_1 axis becoming $2(C_2)$ and b becoming $m(\sigma)$ after the non-symmorphic translations are removed. Since $2/m$ is a centrosymmetric point group, any space group with this point group will also be centrosymmetric.

It is convenient to place the center of inversion at the origin of the unit cell (as explained in Section 5.4.2). Figure 5.13b shows the conventional

[1] It is not clear if this is a real effect or is simply due to a reluctance of many crystallographers to work with crystals with any different space group!

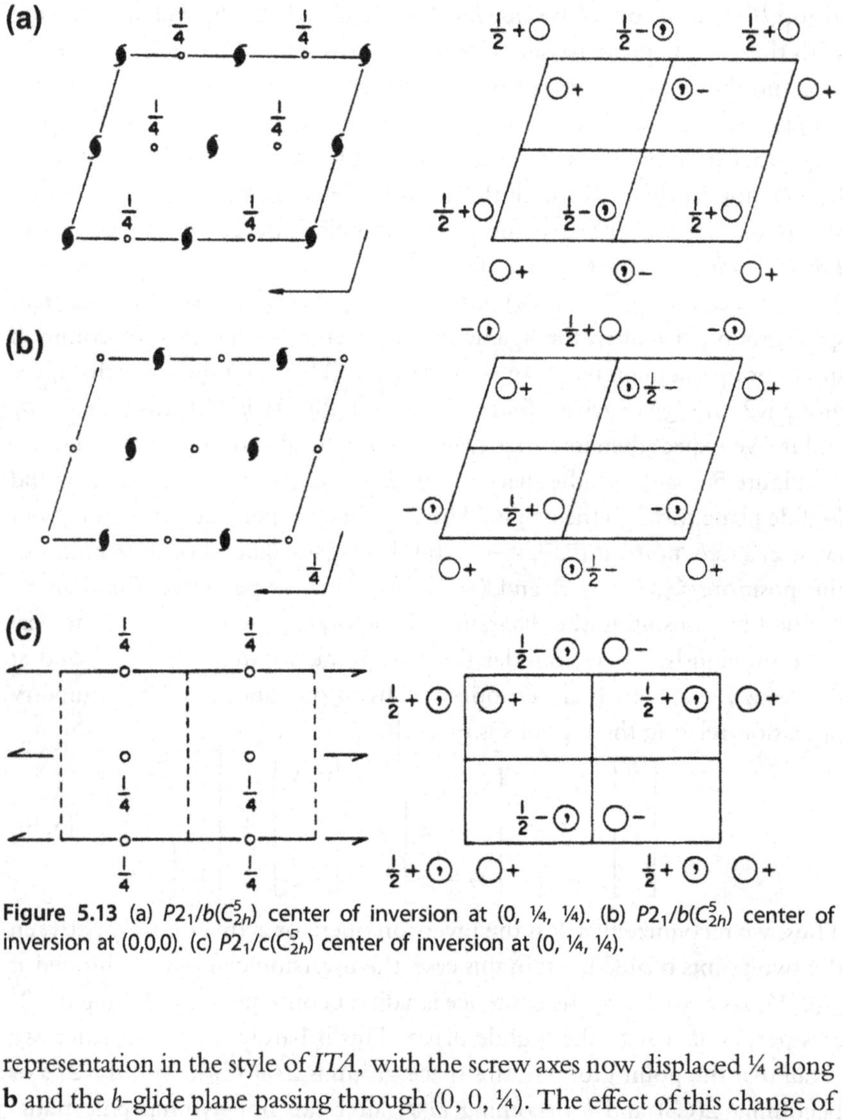

Figure 5.13 (a) $P2_1/b(C_{2h}^5)$ center of inversion at (0, ¼, ¼). (b) $P2_1/b(C_{2h}^5)$ center of inversion at (0,0,0). (c) $P2_1/c(C_{2h}^5)$ center of inversion at (0, ¼, ¼).

representation in the style of *ITA*, with the screw axes now displaced ¼ along **b** and the *b*-glide plane passing through (0, 0, ¼). The effect of this change of origin is to give four general equivalent positions with coordinates:

$$(x, y, z) \quad (\bar{x}, \tfrac{1}{2}-y, \tfrac{1}{2}+z)$$
$$(x, \tfrac{1}{2}+y, \tfrac{1}{2}-z) \quad (\bar{x}, \bar{y}, \bar{z})$$

Figure 5.13c shows the space group diagrams for the second setting, $P2_1/c(C_{2h}^5)$. Note that while the International symbol is different, the Schoenflies symbol is the same.

5.4.4. Orthorhombic System

In this system, there are three point groups and four different Bravais lattices. These are as follows:

$222(D_2)$

$mm2(C_{2v})$ with oP, oC(or oA or oB), oF and oI

$mmm(D_{2h})$

In addition, 2_1 rotations and axial, diagonal and diamond glides are possible. When these are taken into account, making allowance for some of the possible combinations being equivalent, it is found that there are 9 space groups with point group $222(D_2)$, 21 with point group $mm2(C_{2v})$ and 28 with point group $mmm(D_{2h})$. These are all listed in Appendix 7. We shall discuss four examples from the orthorhombic system.

(1) Space group $P2_12_12_1(D_2^4)$, number 19 in the *ITA*. The space group symbol $P2_12_12_1$ means that the lattice is primitive, as indicated by the letter *P*; one 2_1 axis is parallel to **a** (indicated by the 2_1 in the first place after *P*), one 2_1 axis is parallel to **b** (indicated by the 2_1 in the second place) and one 2_1 axis is parallel to **c** (indicated by the 2_1 in the third place). Figure 5.14 shows the space group diagrams.

The point group of this space group is $222(D_2)$, which has four operations, 1, 2[001], 2[010] and 2[100] (E, $C_2[001]$, $C_2[010]$ and $C_2[100]$), and so there are four general equivalent positions per unit cell. Notice also that the three screw axes have been deliberately drawn so that they do not intersect one another. Consider the effect of these operations on the general point (x, y, z). First, consider the 2_1 rotation about [001]. If it is taken, for the moment, to pass through the unit cell origin, we get

$$\{2[001]|(0,0,½\}\mathbf{r} = \begin{bmatrix} -1 & 0 & 0 \\ 0 & -1 & 0 \\ 0 & 0 & 1 \end{bmatrix} \begin{bmatrix} x \\ y \\ z \end{bmatrix} + \begin{bmatrix} 0 \\ 0 \\ ½ \end{bmatrix} = \begin{bmatrix} -x \\ -y \\ ½+z \end{bmatrix} \quad [5.31]$$

Now, if we place the $2_1[001]$ operation through the point (¼, 0, 0) as shown in Fig. 5.14, we must add to the coordinates a translation of 2(¼, 0, 0) = (½, 0, 0), since the 2_1 axis must lie midway between the points that it relates. The general equivalent position then becomes (½$-x$, $\bar{y}$, ½$+z$) as shown on

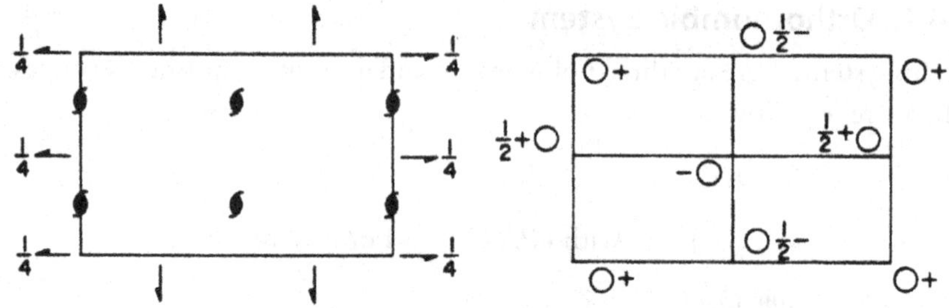

Figure 5.14 $P2_12_12_1(D_2^4)$.

the left-hand side of the figure. We can apply the same procedure with the other screw rotations to give the four general positions

$$(x, y, z) \quad (½-x, \bar{y}, ½+z) \quad (½+x, ½-y, \bar{z}) \quad (\bar{x}, ½+y, ½-z)$$

You may wonder why the three screw axes were made non-intersecting in the first place. In fact, why were they displaced from one another by a factor of ¼ of a unit cell repeat? Why not ⅛ or ⅓ or any other value? Let us consider what happens if any two mutually perpendicular 2_1 screw axes are displaced from each other through a coordinate distance $n/2$, where n has to be determined. For example, suppose we have a $2_1[100]$ axis passing through the origin of the unit cell and a $2_1[010]$ axis passing through the point (0, 0, ½n). The second operator is displaced through a distance $n/2$ along **c**, and will contain a translation vector 2(0, 0, ½n) = (0, 0, n) in addition to its normal non-symmorphic translation of (0, ½, 0), since the $2_1[010]$ element lies midway between the points that it relates. Consider now the product of these two screw operations acting on a position vector **r**. We can express this as follows:

$$\begin{aligned}&\{2_1[100]\}\{2_1[010]\}\mathbf{r}\\&=\{2[100]|½,0,0\}\{2[010]|(0,½,0)+(0,0,n)\}\mathbf{r}\\&=\{2[100]|½,0,0\}(2[010]\mathbf{r}+(0,½,n))\\&=\{2[100]\}\{2[010]\}\mathbf{r}+\{2[100]\}(0,½,n)+(½,0,0)\end{aligned} \qquad [5.32]$$

Writing this out in matrix form, we get

$$\begin{bmatrix}1&0&0\\0&-1&0\\0&0&-1\end{bmatrix}\begin{bmatrix}-1&0&0\\0&1&0\\0&0&-1\end{bmatrix}\begin{bmatrix}x\\y\\z\end{bmatrix}+\begin{bmatrix}1&0&0\\0&-1&0\\0&0&-1\end{bmatrix}\begin{bmatrix}0\\½\\n\end{bmatrix}+\begin{bmatrix}½\\0\\0\end{bmatrix} \qquad [5.33]$$

which reduces to

$$\begin{bmatrix} -x \\ -y \\ z \end{bmatrix} + \begin{bmatrix} 0 \\ 0 \\ -n \end{bmatrix} + \begin{bmatrix} \frac{1}{2} \\ -\frac{1}{2} \\ 0 \end{bmatrix} \qquad [5.34]$$

We have deliberately written the result in three parts. The first term clearly corresponds to a 2-fold rotation about the **c**-axis. The second term determines whether this operation is symmorphic or non-symmorphic. Obviously, there are only two possible values for n if we wish to have a symmetry operation at all. If $n = \frac{1}{2}$, the generated operation is $2_1[001]$, consistent with space group $P2_12_12_1$. We see therefore that, for this space group, the first two screw axes have to be non-intersecting with a displacement between them of $(\frac{1}{2})(0, 0, \frac{1}{2}) = (0, 0, \frac{1}{4})$. The other case, when $n = 0$, has two intersecting screw axes and gives space group $P2_12_12$. Note that in both space groups, the third symmetry operation does not intersect the first two. This is shown by the third column matrix in Eqn [5.34], which represents the point through which the third operation acts, that is $(\frac{1}{2})(\frac{1}{2}, -\frac{1}{2}, 0) = (\frac{1}{4}, -\frac{1}{4}, 0)$. If in the case of $P2_12_12$, we do not take the point of intersection of the two screw axes to be at the origin but, instead, at $(-\frac{1}{4}, \frac{1}{4}, 0)$; then the 2-fold axis directed along **c** passes through the point $(\frac{1}{4}, -\frac{1}{4}, 0) + (-\frac{1}{4}, \frac{1}{4}, 0) = (0, 0, 0)$. This is the choice of origin used in the *ITA* (Fig. 5.15).

Similar considerations apply when body-centering is added to $P2_12_12_1$ and $P2_12_12$. In the former case, additional 2-fold axes along **a**, **b** and **c** are generated (see Fig. 5.16), which again do not intersect with one another. In the latter case, intersecting 2-fold axes are obtained along **a** and **b** together

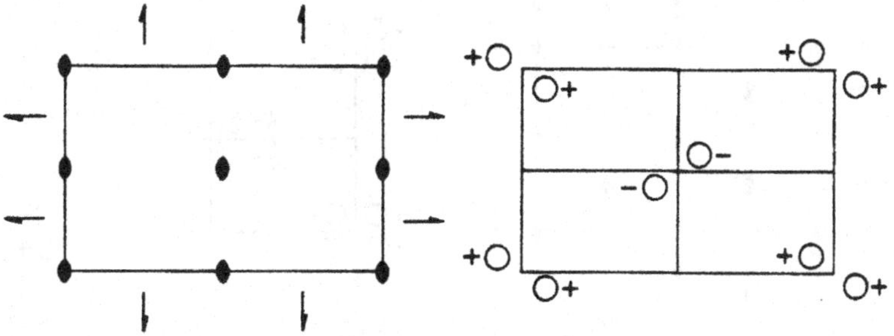

Figure 5.15 $P2_12_12(D_2^3)$.

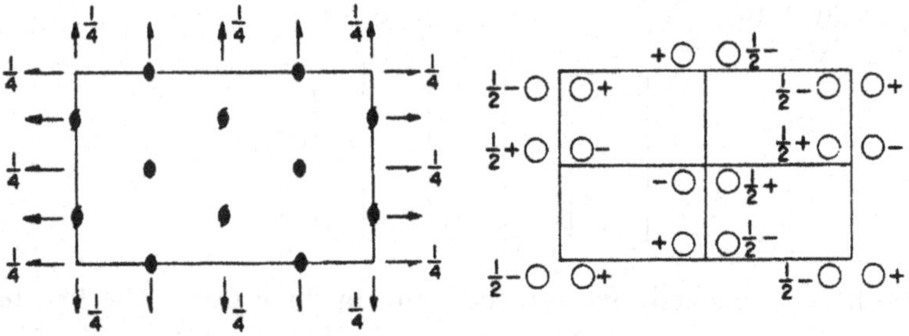

Figure 5.16 $I2_12_12_1(D_2^9)$.

with a screw axis along **c** intersecting those already along **a** and **b** (see Fig. 5.17).

Strangely enough both space groups could equally well be called by the same name, $I222$ or $I2_12_12_1$, since both have three mutually perpendicular 2-fold and 2_1-screw axes. The difference between them lies in whether or not the 2-fold axes intersect (this sort of ambiguity, fortunately, only occurs in two other space groups, namely, the cubic space groups I23 and $I2_13$). When they intersect, the space group is symmorphic and when they do not it is non-symmorphic.

It might be tempting to think that as the space group in Fig. 5.16 could be completely described by 2-fold rotations rather than 2-fold screws it should be considered to be symmorphic. However, it should be realized that these 2-fold rotations are *displaced* from one another and therefore have $\boldsymbol{\tau} \neq 0$ (they are not what are normally called screw rotations since in this case $\boldsymbol{\tau}$ is not a vector along the 2-fold axes).

(2) Space group $Pna2_1(C_{2v}^9)$, number 33 in the *ITA*. Proceeding as before we note that the point group of $Pna2_1$ is $mm2(C_{2v})$, which has

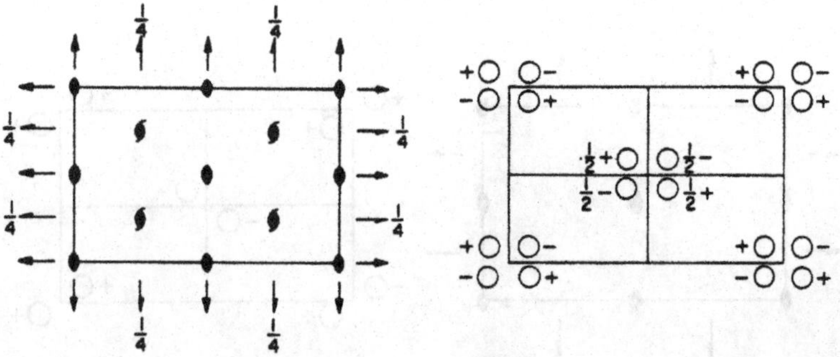

Figure 5.17 $I222(D_2^8)$.

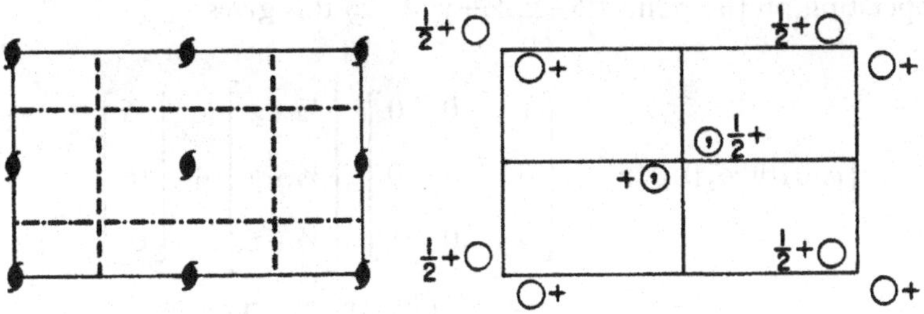

Figure 5.18 $Pna2_1(C_{2v}^9)$.

four operations, namely, 1, 2[001], m[010] and m[100] (E, C_2, σ_v[010] and σ_v[100]). Thus, we expect to have four general equivalent positions. Let us deal with each symmetry operation separately. The n-glide symbol occurs in the first position of the space group symbol indicating that the glide plane is *perpendicular* to **a**. The translations therefore must be along **b** and **c**. Thus, this n-glide operating on the general point (x, y, z) gives

$$\{m[100]|(0,½,½)\}\mathbf{r} = \begin{bmatrix} -1 & 0 & 0 \\ 0 & 1 & 0 \\ 0 & 0 & 1 \end{bmatrix}\begin{bmatrix} x \\ y \\ z \end{bmatrix} + \begin{bmatrix} 0 \\ ½ \\ ½ \end{bmatrix} = \begin{bmatrix} -x \\ ½+y \\ ½+z \end{bmatrix} \quad [5.35]$$

The space group diagrams are shown in Fig. 5.18. The n-glide plane is taken to be at ¼ along [100] in this diagram (denoted by the dot-dash lines), so that in order to be consistent with this choice 2(¼, 0, 0) must be added to the point that has just been generated. The general equivalent position will then be (½−x, ½+y, ½+z). The a-glide symbol is in the second place of the space group symbol; this means that its reflecting plane is *perpendicular* to **b**. Therefore, operating with the a-glide on the general point at (x, y, z), gives

$$\{m[010]|½,0,0\}\mathbf{r} = \begin{bmatrix} 1 & 0 & 0 \\ 0 & -1 & 0 \\ 0 & 0 & 1 \end{bmatrix}\begin{bmatrix} x \\ y \\ z \end{bmatrix} + \begin{bmatrix} ½ \\ 0 \\ 0 \end{bmatrix} = \begin{bmatrix} ½+x \\ -y \\ z \end{bmatrix} \quad [5.36]$$

and operating on the point (½−x, ½+y, ½+z) this gives

$$\{m[010]|½,0,0\}\mathbf{r} = \begin{bmatrix} 1 & 0 & 0 \\ 0 & -1 & 0 \\ 0 & 0 & 1 \end{bmatrix} \begin{bmatrix} ½-x \\ ½+y \\ ½+z \end{bmatrix} + \begin{bmatrix} ½ \\ 0 \\ 0 \end{bmatrix}$$

$$= \begin{bmatrix} 1-x \\ ½-y \\ ½+z \end{bmatrix} = \begin{bmatrix} -x \\ ½-y \\ ½+z \end{bmatrix} \quad [5.37]$$

In Fig. 5.18, the *a*-glide plane is shifted from the origin by ¼ along [010], so that 2(0, ¼, 0) must be added to the coordinates generated. This gives the points (½+x, ½−y, z) and ($\overline{x}$, $\overline{y}$, ½+z). Now consider the relationship between (x, y, z) and ($\overline{x}$, $\overline{y}$, ½+z).

$$R\begin{bmatrix} x \\ y \\ z \end{bmatrix} + \tau = \begin{bmatrix} -x \\ -y \\ z \end{bmatrix} + \begin{bmatrix} 0 \\ 0 \\ ½ \end{bmatrix} \quad [5.38]$$

which is a 2_1 axis along [001] passing through the unit-cell origin. This is the 2_1 axis in the third place of the space group symbol.

(3) Space group *Fddd* (D_{2h}^{24}), number 70 in the *ITA*. The point group of *Fddd* is *mmm*(D_{2h}) which has eight operations, namely 1, 2[001], 2[010], 2[100], $\overline{1}$, *m*[001], *m*[010] and *m*[100] (*E*, C_2[001], C_2[010], C_2[100], *i*, σ[001], σ[010], σ[100]). Therefore, there are eight general equivalent positions. The *ITA* use the conventional centered unit cell for this face-centered lattice. Thus, for every one of the eight general points generated for the primitive unit cell by the symmetry operations of the space group, we must add (0, ½, ½), (½, 0, ½) and (½, ½, 0), giving 4 × 8 = 32 general equivalent positions for the all-face-centered unit cell.

Figure 5.19 shows the space group diagrams. Although they look complicated, they are really simple if you consider each symmetry operation in turn. We shall not deal with all these operations here; rather we concentrate on certain features. First, since the point group is *mmm*(D_{2h}), it is centrosymmetric. We are therefore not surprised to find a center of

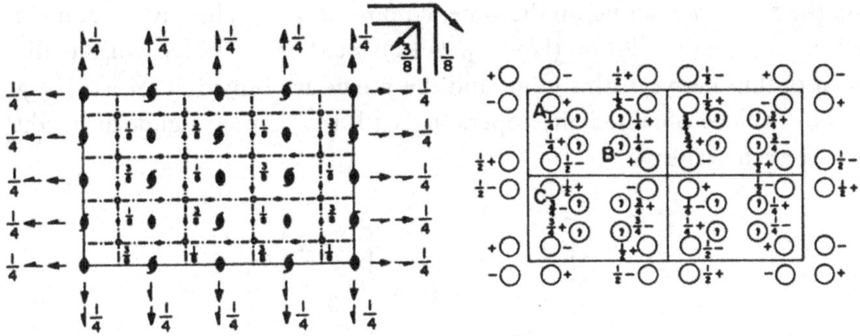

Figure 5.19 *Fddd*(D_{2h}^{24}).

inversion, which in this case has been placed at (⅛, ⅛, ⅛). Second, consider the diamond glides. The space group symbol tells us that there are diamond glide planes perpendicular to the **a**, **b** and **c**-axes, since the symbol *d* appears in all three positions after the Bravais lattice symbol. Consider the diamond glide perpendicular to **c** as an example. The effect of this on a general point at (*x*, *y*, *z*) is

$$\{m[001]|\tfrac{1}{4},\tfrac{1}{4},0\}\mathbf{r} = \begin{bmatrix} 1 & 0 & 0 \\ 0 & 1 & 0 \\ 0 & 0 & -1 \end{bmatrix} \begin{bmatrix} x \\ y \\ z \end{bmatrix} + \begin{bmatrix} \tfrac{1}{4} \\ \tfrac{1}{4} \\ 0 \end{bmatrix} = \begin{bmatrix} \tfrac{1}{4}+x \\ \tfrac{1}{4}+y \\ -z \end{bmatrix} \quad [5.39]$$

This means that the point at (*x*, *y*, *z*), marked A in Fig. 5.19, has been taken to (¼+*x*, ¼+*y*, $\bar{z}$), by this diamond glide at zero height. If we wish to be consistent with Fig. 5.19, however, we must place this glide plane at height 1/8 and therefore, we add the vector 2(0, 0, 1/8) to the coordinate (¼+*x*, ¼+*y*, $\bar{z}$) to obtain this position (¼+*x*, ¼+*y*, ¼−*z*), marked B on the diagram. This glide means a movement of the general point out of the plane of projection and towards the lower right, as indicated by the diagrammatic symbol at the top right, the arrow showing the general direction of movement. The value 1/8 next to it gives the height of the glide plane.

Notice that there is another 'diamond' glide symbol with the arrow pointing to the lower left and with height marked 3/8. This second 'diamond' glide plane may be thought of as a consequence of the face-centering. This can be seen in the following way. Consider the centering on the (010) face acting on the general point (x, y, z). This gives a point at $(x, y, z) + (½, 0, ½) = (½+x, y, ½+z)$, marked C. Now, what is the relationship between this point and the previous point B at $(¼+x, ¼+y, ¼-z)$ that we obtained after operating with the diamond glide at height ⅛? We can write

$$R\begin{bmatrix} ¼+x \\ ¼+y \\ ¼-z \end{bmatrix} + \boldsymbol{\tau} = \begin{bmatrix} ½+x \\ y \\ ½+z \end{bmatrix} \qquad [5.40]$$

This can be rewritten as

$$R\begin{bmatrix} x \\ y \\ -z \end{bmatrix} + R\begin{bmatrix} ¼ \\ ¼ \\ ¼ \end{bmatrix} + \boldsymbol{\tau} = \begin{bmatrix} x \\ y \\ z \end{bmatrix} + \begin{bmatrix} ½ \\ 0 \\ ½ \end{bmatrix} \qquad [5.41]$$

which, on comparing rotations, gives

$$R = \begin{bmatrix} 1 & 0 & 0 \\ 0 & 1 & 0 \\ 0 & 0 & -1 \end{bmatrix}$$

$$\boldsymbol{\tau} = \begin{bmatrix} ½ \\ 0 \\ ½ \end{bmatrix} - \begin{bmatrix} 1 & 0 & 0 \\ 0 & 1 & 0 \\ 0 & 0 & -1 \end{bmatrix}\begin{bmatrix} ¼ \\ ¼ \\ ¼ \end{bmatrix} = \begin{bmatrix} ¼ \\ -¼ \\ ¾ \end{bmatrix} = \begin{bmatrix} ¼ \\ -¼ \\ 0 \end{bmatrix} + \begin{bmatrix} 0 \\ 0 \\ ¾ \end{bmatrix} \qquad [5.42]$$

We see immediately that R is a reflection, $m[001]$. The operation connecting these two points is therefore

$$\{m[001]|¼, -¼, 0\}\mathbf{r} + (0, 0, ¾) \qquad [5.43]$$

The first term is a diamond glide towards the lower left of the space group diagram in Fig. 5.19. The vector (0, 0, ¾) tells us that this diamond glide plane is at height ½(0, 0, ¾), that is at ⅜ along **c**. This is then ¼**c** above the previous diamond glide plane. Such an alternation of glide directions with successive glide planes is characteristic of diamond glides. Although we have shown it to result from the all-face-centering combined with the diamond operation, we could equally well show that it results from having two mutually perpendicular diamond glides. These give the face-centering; either approach is valid. It is difficult to give a ready explanation of where diamond glides are possible, although it is worth noting that they only occur in orthorhombic, tetragonal and cubic space groups, and are always found in conjunction with *F* or *I*-centering.

Finally, notice that the three diamond glides automatically produce three mutually perpendicular 2-fold rotations. Note too that all the operations of the point group are found in the space group. The full International space group symbol for *Fddd* is *F* 2/*d* 2/*d* 2/*d*, signifying that the 2-fold axes are perpendicular to the glide planes. This suggests another way to visualize the *Fddd* space group.

Consider the space group $F222(D_2^7)$ to which a center of inversion has been added. If we place the center at the unit cell origin to be coincident with the three intersecting 2-fold axes, then the space group *F* 2/*m* 2/*m* 2/*m* equivalent to $Fmmm(D_{2h}^{23})$ is generated. If, on the other hand, the inversion center is placed at the point (⅛, ⅛, ⅛) with respect to the intersecting 2-fold axes, then the space group $Fddd(D_{2h}^{24})$ is generated. We suggest that, as an exercise, you should verify this for yourself.

(4) Space group *Aem*2 (C_{2v}^{15}), number 39 in the *ITA*.

This space group illustrates the recently defined **double** or ***e*-glide** operation. This new operation has resulted in a change in the space group symbol for five space groups:

No.	39	41	64	67	68
New	*Aem*2	*Aea*2	*Cmce*	*Cmme*	*Ccce*
Old	*Abm*2	*Aba*2	*Cmca*	*Cmma*	*Ccca*

Currently, the old symbols are still much in use, especially in computer programs that use space groups, but it is likely that the new symbols will become more evident, especially in crystal structure descriptions in the literature. Figure 5.20 shows the diagrams for space group $Aem2(C_{2v}^{15})$.

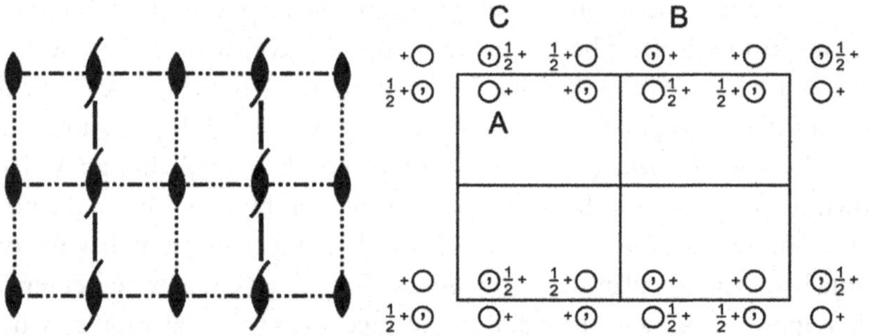

Figure 5.20 *Aem*2(C_{2v}^{15}).

Consider the general position (x, y, z) marked by the letter A. The double glide results in two new general positions B and C via the equations:

$$\{m[100]|0,½,0\}\mathbf{r} = \begin{bmatrix} -1 & 0 & 0 \\ 0 & 1 & 0 \\ 0 & 0 & 1 \end{bmatrix} \begin{bmatrix} x \\ y \\ z \end{bmatrix} + \begin{bmatrix} 0 \\ ½ \\ 0 \end{bmatrix} = \begin{bmatrix} -x \\ ½+y \\ z \end{bmatrix}$$

$$\{m[100]|0,0,½\}\mathbf{r} = \begin{bmatrix} -1 & 0 & 0 \\ 0 & 1 & 0 \\ 0 & 0 & 1 \end{bmatrix} \begin{bmatrix} x \\ y \\ z \end{bmatrix} + \begin{bmatrix} 0 \\ 0 \\ ½ \end{bmatrix} = \begin{bmatrix} -x \\ y \\ ½+z \end{bmatrix}$$

[5.44]

Thus, the point B at ($\bar{x}$, ½+y, z) is obtained by a reflection across the (100) plane, since the letter *e* is in the first place of the space group symbol, plus a half translation along **b**. The point C found at ($\bar{x}$, y, ½+z) is obtained by reflection again across the (100) plane, but this time with a half translation along **c**. Note that in addition to the 2-fold axes along **c** (third place in the space group symbol) passing through the unit cell origin, 2-fold screw axes have now been generated in between. The mirror plane perpendicular to **b** (second place in the space group symbol) passes through the screw axes rather than through the unit cell origin.

5.4.5. Tetragonal System

In this system there are seven point groups and two Bravais lattices.

$4(C_4)$	
$\bar{4}(S_4)$	
$4/m(C_{4h})$	*tP* and *tI* (or *tC* and *tF* with **a** and **b**
$422(D_4)$	axes rotated through 45° about **c**)
$4mm(C_{4v})$	
$\bar{4}2m(D_{2d})$	
$4/mmm(D_{4h})$	

Combining the seven point groups with the two Bravais lattices, and taking into account possible non-symmorphic operations, one finds that there are 68 tetragonal space groups. We shall just take four space group examples, the first three belonging to point group $4(C_4)$.

(1) Space group $P4_1(C_4^2)$, number 76 in the *ITA*. Figure 5.21 shows the space group diagrams. These diagrams are just the 4_1 diagrams of Fig. 5.4 with translational symmetry added. The point group of this space group is $4(C_4)$ which has four operations, 1, 4, 2 ($=4^2$) and 4^3 (E, C_4, C_2 and C_4^3), so that there are four general equivalent positions in the unit cell. You should be able to find for yourself the coordinates of these general equivalent positions by applying successive operations of $\{4[001]|0, 0, ¼\}$ to the point at (x, y, z). Notice that the general points form a helical arrangement about the **c**-axis through the unit cell origin (0, 0, 0) and the face center at (½, ½, 0). This helix is anticlockwise out of the paper along the +**c** direction. A crystal having this space group would have a 'handedness' or **chirality** and could, in principle, show optical rotation of plane-polarized light. This helical arrangement forms a right-handed

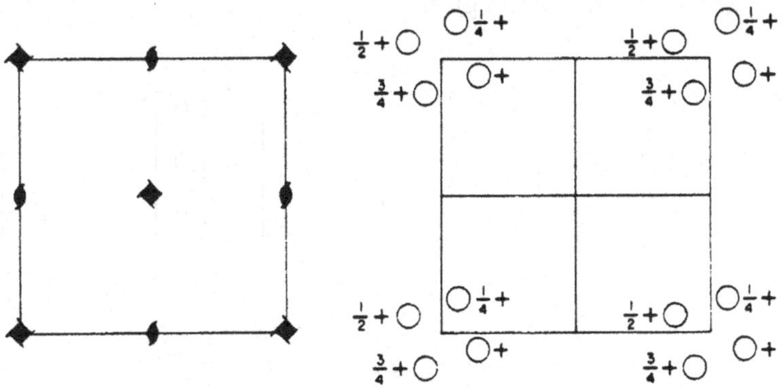

Figure 5.21 $P4_1(C_4^2)$.

screw – we normally define the hand of a helix looking down on it. In this case, the points are rotated to the right away from you. This is a potential source of confusion, since the wrong usage is sometimes seen in the literature. One of us, during a study of the connection between crystal structure and optical activity,[2] found many such errors in the literature dealing with the link between optical rotation of crystals and crystal structure. The space group $P4_1$ is an example of an **enantiomorphic** or **chiral** space group.

Note that we could have defined a different unit cell with new **a** and **b**-axes set at 45° to the original **a** and **b**-axes. These new axes would be longer and the resulting unit cell would then be *C*-centered. The space group symbol would then become $C4_1$. Sometimes this unit cell may be preferred over the former. For example, when tracing the relationship between different phases of a particular material it may be visually more convenient to have the larger unit cell.

(2) Space group $P4_2(C_4^3)$, number 77 in the *ITA*. As with the previous space group, the number of equivalent positions is 4. Figure 5.22 shows the space group diagrams which again are like the 4_2 diagram of Fig. 5.4 with translational symmetry added.

The point (x, y, z) operated on by the 4_2 operation gives the result

$$\{4[001]|0,0,½\}\mathbf{r} = \begin{bmatrix} 0 & -1 & 0 \\ 1 & 0 & 0 \\ 0 & 0 & 1 \end{bmatrix} \begin{bmatrix} x \\ y \\ z \end{bmatrix} + \begin{bmatrix} 0 \\ 0 \\ ½ \end{bmatrix} = \begin{bmatrix} -y \\ x \\ ½+z \end{bmatrix} \qquad [5.45]$$

Applying this operation again gives

$$\{4[001]|0,0,½\}\mathbf{r} = \begin{bmatrix} 0 & -1 & 0 \\ 1 & 0 & 0 \\ 0 & 0 & 1 \end{bmatrix} \begin{bmatrix} -y \\ x \\ ½+z \end{bmatrix} + \begin{bmatrix} 0 \\ 0 \\ ½ \end{bmatrix} = \begin{bmatrix} -x \\ -y \\ 1+z \end{bmatrix} \qquad [5.46]$$

[2] A.M. Glazer & K. Stadnicka, *J. Appl. Cryst.*, **19**, 108 (1986).

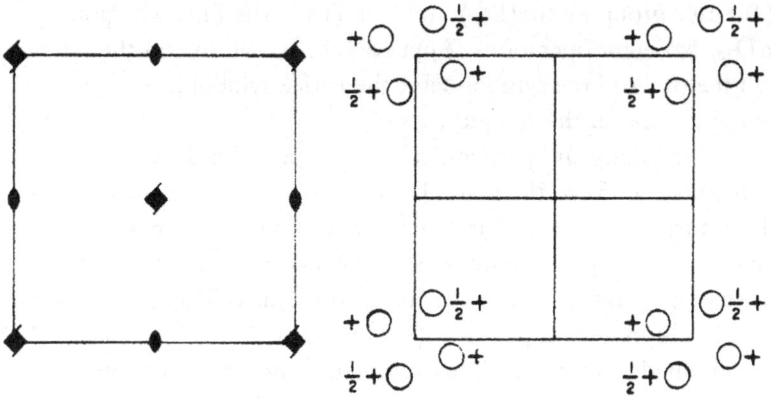

Figure 5.22 $P4_2(C_4^3)$.

The result ($\overline{x}$, $\overline{y}$, 1+z) is, by translational invariance, equivalent to ($\overline{x}$, $\overline{y}$, z). Proceeding in this way, the four general equivalent positions are (x, y, z), ($\overline{y}$, x, ½+z), ($\overline{x}$, $\overline{y}$, z) and (y, $\overline{x}$, ½+z). Note that there is no sense of handedness about this space group.

(3) Space group $P4_3(C_4^4)$, number 78 in the *ITA*. Figure 5.23 shows the space group diagram of $P4_3$. The important point is that it is very similar to $P4_1$, except that now the helix of general positions is clockwise along the +**c** direction. Obviously, the chirality of this space group is enantiomorphically related to $P4_1$, all other features being the same. This makes it possible for a particular substance to crystallize equally well in either space group, the differences between the crystals being revealed by some chiral property such as optical rotation.

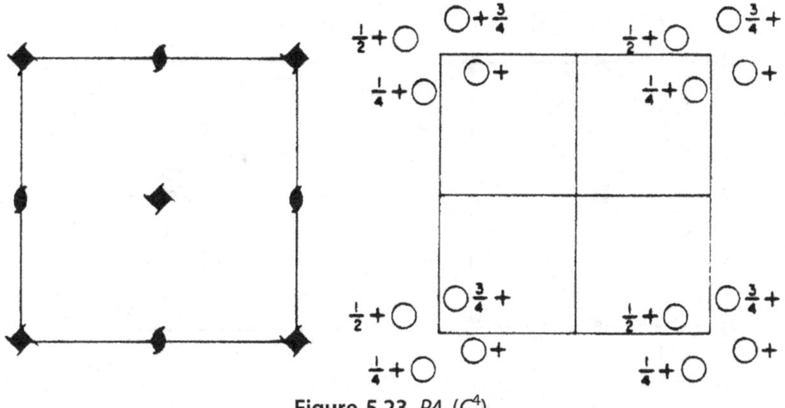

Figure 5.23 $P4_3(C_4^4)$.

(4) Space group $P\bar{4}2_1m(D^3_{2d})$, number 113 in the *ITA*. The point group $\bar{4}2m(D_{2d})$ has eight operations (Appendix 4). Recall that in the tetragonal system the order of the symbols after the lattice symbol is:

Symmetry about the unique **c**-axis

Symmetry along and perpendicular to the **a** and **b**-axes

Symmetry at 45° to the **a** and **b**-axes

Therefore, we expect to find eight general equivalent points generated by the space group operations. Start by placing the axis of the $\bar{4}(S^3_4)$ symmetry operation parallel to **c** and at the unit cell origin as shown in Fig. 5.24. This results in points at (x, y, z), $(\bar{y}, x, \bar{z})$, $(\bar{x}, \bar{y}, z)$ and $(y, \bar{x}, \bar{z})$. This combined with translational symmetry gives rise to another axis of $\bar{4}(S^3_4)$ symmetry through the center of the unit cell and parallel to **c**. Notice that in Fig. 5.24, screw axes have been placed perpendicular to the **c**-axis, just as with the 2-fold axes in point group $\bar{4}2m(D_{2d})$, and that they do not intersect the $\bar{4}$ axes, that is they are along lines such as x, ¼, 0. The result of this is that four other general equivalent points are generated, at (½+x, ½+y, z), (½−x, ½+y, $\bar{z}$), (½−y, ½-x, z) and (½+x, ½−y, $\bar{z}$). The last four points are related to the first four by mirror planes parallel to **c** and 45° to the screw axes (just like the mirror planes at 45° to the 2-fold axes in the point group). Note, from the diagram, that glide planes parallel to the diagonals of the unit cell along <110> directions are also generated. They are not axial glide planes, since the glide direction is not along an axis of the unit cell, nor are they diagonal glides, since there is no component along **c**. There is no standard notation for such glide planes. There was a recommendation in 1992 by the Nomenclature Commission of the International Union of Crystallography to call them *k*-glides, but this has not been

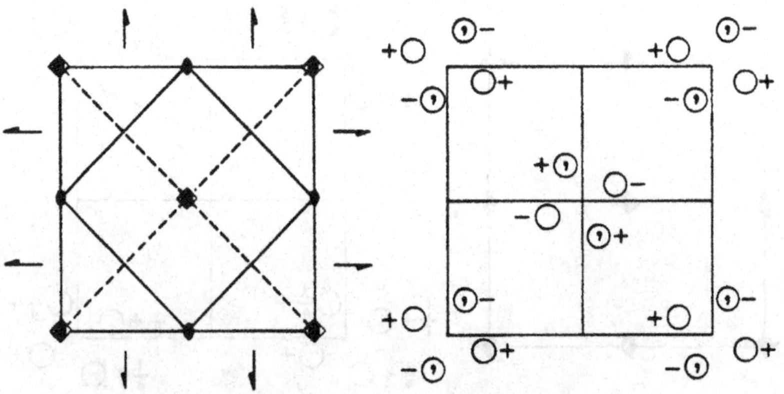

Figure 5.24 $P\bar{4}2_1m(D^3_{2d})$.

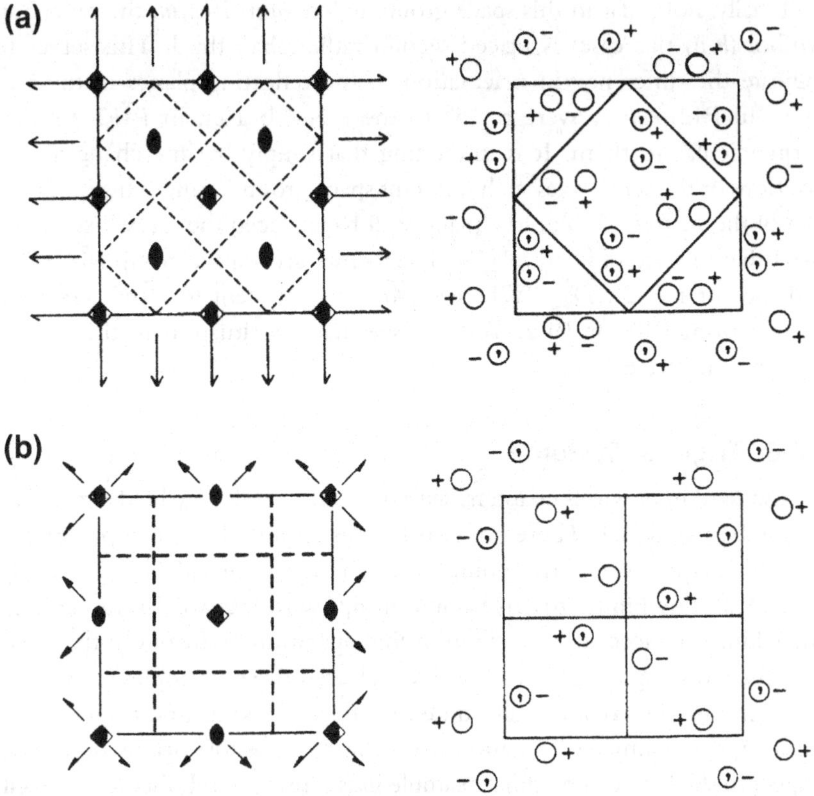

Figure 5.25 $P\bar{4}b2(D_{2d}^{7})$.

adopted in the *ITA*. They are merely symmetry operations that follow as a consequence of the others, and, of course, if the unit cell is redefined by taking new unit cell axes, **a** and **b** at 45° to the old **a** and **b**-axes, a *C*-centered cell is obtained and then these *k*-glides are seen to be axial glides.

Consider what happens if the screw axes are positioned to intersect the $\bar{4}(S_4^3)$ axis. In Fig. 5.25a, we show this with screw axes parallel to **a** and **b** as well, that is along the two lines *x*, 0, 0 and 0, *y*, 0. There are now 16 general equivalent positions. It can be seen that the unit cell is now *C*-centered. However, if the unit cell is redefined 45° about the **c**-axis, a primitive unit cell with eight general equivalent positions results. The conventional space group diagrams for the new cell are shown in Fig. 5.25b, where it can be seen that axial glides are also present. This is space group $P\bar{4}b2_1(D_{2d}^{7})$. Actually, there are also 2-fold axes present parallel to the screw axes but midway between them, so that the space group can also be called $P\bar{4}b2(D_{2d}^{7})$. This is in fact the conventional name for this space group.

Finally, note that in this space group, unlike m in $P\bar{4}2_1m$, the reflection symbol (b in this case) is placed second rather than third. This serves to indicate the difference in orientation of the reflecting planes in the two cases. In $P\bar{4}2_1m$ they were at 45° to the **a** and **b**-axes, in $P\bar{4}b2$ they are perpendicular to them. It is interesting that simply by interchanging the positions of the screw axes such different space groups result. This is not the end of the matter. We could equally well have placed the screw axes along two other lines, x, 0, ¼ or x, ¼, ¼. The former gives space group $P\bar{4}n2(D_{2d}^8)$ and the latter $P\bar{4}2_1c(D_{2d}^4)$. Whereas $P\bar{4}n2$ is equivalent to $P\bar{4}n2_1$, $P\bar{4}2_1c$ is distinct from $P\bar{4}2c$ or $P\bar{4}c2$. You can see that the situation in these space groups is not trivial.

5.4.6. Trigonal System

As discussed in previous chapters, *some* trigonal crystals may be described by a rhombohedral cell. There are two Bravais lattices that are appropriately labelled hP and hR, and five point groups, 3, $\bar{3}$, 32, $3m$ and $\bar{3}m$ (C_3, S_6, D_3, C_{3v}, D_{3d}). Combining the five point groups with the two Bravais lattices, and taking into account non-symmorphic operations, it turns out that there are 25 trigonal space groups. We have selected three examples of trigonal space groups. The first two illustrate the effect of taking a point group and allowing the symmetry operations to act in two possible orientations with respect to the lattice. The third example illustrates a rhombohedral unit cell.

(1) Space group $P312(D_3^1)$, number 149 in the *ITA*. The point group of this symmorphic space group is $32(D_3)$. We start by placing the 3-fold axis parallel to **c** and at the origin of the unit cell, and then find that translational symmetry gives rise to 3-fold axes along ⅔, ⅓, z and ⅓, ⅔, z, as shown in Fig. 5.26. The 2-fold axes can be positioned in two ways. To obtain this space group, the 2-fold axes are placed *perpendicular* to the **a**, **b** and [110] directions. This is why the figure 2 is placed in the third position of the symbol. In order to make it clear that this is the third position, the identity 1 is written in the second position (Appendix 6). The six general equivalent positions are as follows:

$$(x, y, z) \quad (\bar{y}, x-y, z) \quad (y-x, \bar{x}, z) \quad (\bar{y}, \bar{x}, \bar{z}) \quad (x, x-y, \bar{z}) \quad (y-x, y, \bar{z})$$

(2) Space group $P321(D_3^2)$, number 150 in the *ITA*. Notice that in this space group we have placed the figure 1 in the third place. This is simply a convention to distinguish it from the previous one. Here, the 2-fold axes are placed in the second possible orientation, that is they are rotated about **c**

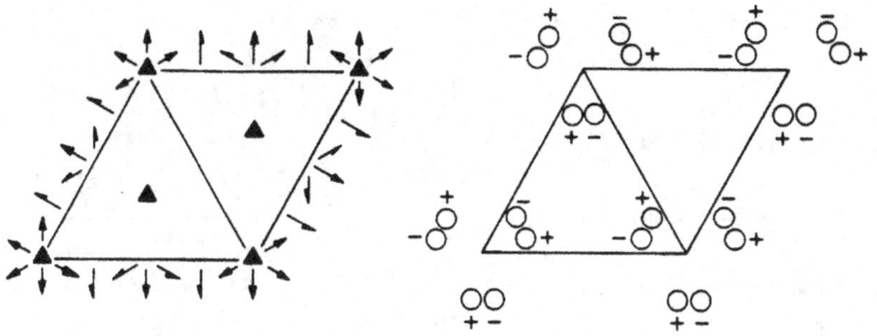

Figure 5.26 $P312(D_3^1)$.

through 30° with respect to the previous case. Now they lie *parallel* to the **a**, **b** and [110] axes (Fig. 5.27). Again, there are six general equivalent positions:

$$(x, y, z) \quad (\bar{y}, x-y, z) \quad (y-x, \bar{x}, z) \quad (y, x, \bar{z}) \quad (\bar{x}, y-x, \bar{z}) \quad (x-y, \bar{y}, \bar{z})$$

(3) Space group $R3c(C_{3v}^6)$, number 161 in the *ITA*. Figure 5.28 shows the space group diagrams for $R3c$. The R indicates a rhombohedral Bravais lattice. Two unit cells are outlined on the right-hand side; one is the trigonal cell and the other is the obverse rhombohedral cell viewed in projection down the [111] direction (see Fig. 2.10). The c-glide planes pass through the threefold axes at the origin and the centered positions. The reason for this goes back to the point discussed in Chapters 2 and 3, namely that the rhombohedral lattice is only obtained from a centered hexagonal lattice. Therefore, as the c-glide plane passes through (0, 0, 0), it must, in order to keep the rhombohedral centering, pass through the centering positions $\pm(⅔, ⅓, ⅓)$. We leave it to the reader to show what

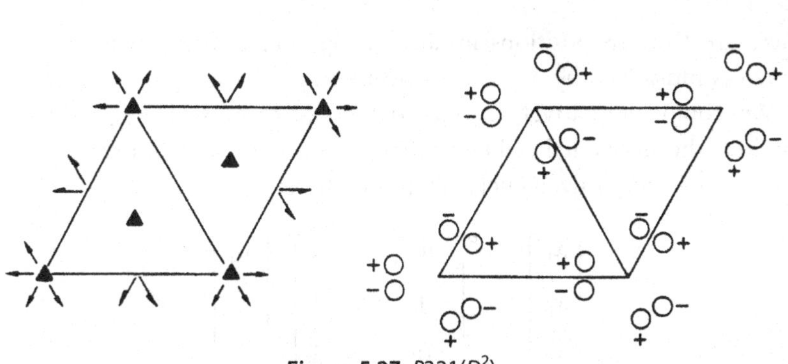

Figure 5.27 $P321(D_3^2)$.

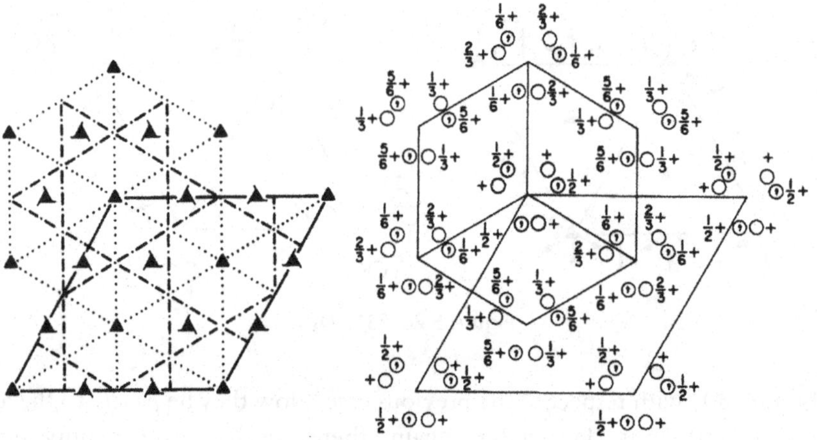

Figure 5.28 $R3c(C_{3v}^{6})$.

the other, redundant, symmetry operations are, and that the general equivalent positions are, with respect to hexagonal axes, given by

$$\left.\begin{array}{c}(0,0,0)\\ (\tfrac{1}{3},\tfrac{2}{3},\tfrac{2}{3})\\ (\tfrac{2}{3},\tfrac{1}{3},\tfrac{1}{3})\end{array}\right\} + \left\{\begin{array}{l}(x,y,z)\\ (\bar{y},x-y,z)\\ (y-x,\bar{x},z)\\ (\bar{y},\bar{x},\tfrac{1}{2}+z)\\ (x,x-y,\tfrac{1}{2}+z)\\ (y-x,y,\tfrac{1}{2}+z)\end{array}\right.$$

There are thus 18 positions in all, corresponding to the 6 point group operators multiplied by 3 because of centering.

We could alternatively specify the coordinates with respect to the primitive rhombohedral cell by applying the transformation matrix of Eqn [3.14] which, for convenience, is repeated here:

$$\begin{bmatrix} x_r \\ y_r \\ z_r \end{bmatrix} = \begin{bmatrix} 1 & 0 & 1 \\ -1 & 1 & 1 \\ 0 & -1 & 1 \end{bmatrix} \begin{bmatrix} x_h \\ y_h \\ z_h \end{bmatrix} \qquad [5.47]$$

where r and h refer to rhombohedral and hexagonal axes of reference, respectively. Equation [5.47] can be written as

$$
\begin{aligned}
x_r &= x_h + z_h \\
y_r &= -x_h + y_h + z_h \\
z_r &= -y_h + z_h
\end{aligned}
\qquad [5.48]
$$

Now apply the transformation matrix to the second general position ($\bar{y}$, $x-y$, z):

$$
\begin{bmatrix} 1 & 0 & 1 \\ -1 & 1 & 1 \\ 0 & -1 & 1 \end{bmatrix}
\begin{bmatrix} -y_h \\ x_h - y_h \\ z_h \end{bmatrix}
=
\begin{bmatrix} -y_h + z_h \\ x_h + z_h \\ -x_h + y_h + z_h \end{bmatrix}
\qquad [5.49]
$$

The result can now be written in terms of the rhombohedral coordinates in Eqn [5.48], and we find that

$$
\begin{aligned}
-y_h + z_h &= z_r \\
x_h + z_h &= x_r \\
-x_h + y_h + z_h &= y_r
\end{aligned}
\qquad [5.50]
$$

Thus, the point ($\bar{y}$, $x-y$, z) is transformed to (z, x, y), with subscripts now dropped for convenience. Proceeding in this way, the 6 general equivalent positions grouped around the origin of the trigonal unit cell are transformed to (x, y, z), (z, x, y), (y, z, x), (½+x, ½+y, ½+z), (½+z, ½+x, ½+y) and (½+x, ½+z, ½+y). The other 12 general positions are related by the rhombohedral centering, but when we apply this transformation we now find that they lie outside the rhombohedral cell, that is in neighbouring unit cells. For example, transform the point (⅓+x_h, ⅔+y_h, ⅔+z_h). The following result is obtained:

$$
\begin{aligned}
1 + x_h + z_h &= 1 + x_r \\
1 - x_h + y_h + z_h &= 1 + y_r \\
-y_h + z_h &= z_r
\end{aligned}
\qquad [5.51]
$$

which, by translational symmetry, is equivalent to (x, y, z). Clearly then, this rhombohedral unit cell has one-third the volume of the trigonal cell. It is a primitive cell, and this is one reason why it is useful. Note that with the rhombohedral cell there are six general equivalent positions in agreement with the number of symmetry operations in the point group $3m(C_{3v})$.

5.4.7. Hexagonal System

The hexagonal space groups are rather like the tetragonal ones, and so we shall not give any examples of them. There are seven point groups and one Bravais lattice (*hP*). The point groups are 6, $\bar{6}$, $6/m$, 622, $6mm$, $\bar{6}m2$ and $6/mmm$ (C_6, C_{3h}, C_{6h}, D_6, C_{6v}, D_{3h} and D_{6h}). From these it is possible to derive 27 hexagonal space groups, which are all listed in Appendix 7.

5.4.8. Cubic System

In the cubic system, there are three Bravais lattices, *cP*, *cF* and *cI*, and five point groups, 23, $m\bar{3}$, 432, $\bar{4}3m$ and $m\bar{3}m$ (T, T_h, O, T_d and O_h). When combined with non-symmorphic elements, 36 cubic space groups result and are listed in Appendix 7.

In the International Tables (up to 1969), diagrams were not given for the cubic space groups. This is because of the complexity introduced by the high symmetry of such space groups. It may seem surprising to the casual reader that the cubic space groups should be described as 'complex'; after all, the symmetry being high should make things simpler. It is true that the symmetry is high, but it is also a fact that the number of general equivalent positions is large. This makes it difficult to keep track of all the symmetry-related points. *ITA* does, however, have space group diagrams for the cubic system. For the cubic point groups, there are 12, 24 or 48 symmetry operations (APPENDIX 4). That is, there are 12, 24 or 48 general equivalent positions in the primitive cell of the cubic space groups. Combine this with lattice centering, such as all-face-centering, and the total number of general positions can be as many as 192. Furthermore, bearing in mind that each general position may describe an assemblage of n atoms, there may be $192n$ atoms in the unit cell. Fortunately, in many of the cubic crystals that are studied the atoms do not usually occupy general positions in the unit cell. Often they lie at **special positions**. For example, certain atoms may lie exactly on mirror planes, and there will be half as many of such atoms in the unit cell as those in general positions. The placing of atoms in special positions means that there will be a reduction in the number of atoms in the unit cell. We shall deal more fully with special positions in the next chapter.

5.5. DERIVATION OF SPACE GROUPS

As mentioned in the Preface, the derivation of the 230 space groups was mainly due to the efforts of three people: a Russian, E. S. Fedorov, a German, A. Schoenflies and an Englishman, W. Barlow, all working roughly at the same time but independently from one another. The derivation of Fedorov was completed first (1890) and followed soon after by that of Schoenflies (1891) and then Barlow (1894). However, because Fedorov's work was published in Russian while that of Schoenflies was in the more (at that time) acceptable scientific language of German, the Schoenflies approach made the most impression on the scientific community. Barlow's approach was not very clearly set out and, in any case, was incomplete.[3]

Fedorov's method, which owed much to the earlier work of Sohnke, Moebius and others, involved the division of space groups into three classes:

(1) Symmorphic class, which Fedorov thought of in terms of symmorphic operations acting at a point.

(2) Hemisymmorphic class, in which only proper symmetry operations act at a point.

(3) Asymmorphic class, in which there are no such points of intersection.

For symmetry operations, Fedorov used rotations, reflections and combined rotation-reflection operations. He disliked the use of inversion operations, preferring to accomplish the same effect by a 2-fold rotation-reflection. Fedorov's greatest accomplishment perhaps was the mathematical technique that he developed for expressing combinations of symmetry operations. With this, every space group was described by a set of three simple equations and it was the generalization of this that led to the discovery that there were 230 space groups in all. Fedorov took matters further and showed that similar results could be obtained by assuming that crystals could be described by convex polyhedra with pairs of equal opposite faces packed together through all space. However, curiously, this approach led Fedorov to reject some space groups as being impossible for real crystal structures. They were $Fdd2(C_{2v}^{19})$, $Fddd(D_{2h}^{24})$, $I\bar{4}3d(T_d^6)$, $P4_332(O^6)$ and $P4_132(O^7)$. In spite of this, we now know that there are several crystal structures with these space groups, and yet the fault in Fedorov's logic is not obvious. The reader is recommended to read the series of his papers translated into English by D. K. Harker.

[3] William Barlow (1845–1934) was an amateur scientist. There is a (apocryphal) story that he used to buy large numbers of gloves from the shops in Cambridge, often confounding the sales people who wanted to know what size he wanted. After his death, apparently, the gloves were found on the walls of his home in the arrangements of the space groups!

Let us turn to the Schoenflies approach, which we should like to discuss in more detail. The basic idea was simple and the systematic approach straightforward; again, it owed much to the pioneering work of Sohnke and others. An appreciation of how this was done leads to a deeper understanding of the space groups themselves.

The Schoenflies approach to the determination of the 230 space groups started from that outlined in Section 5.1 for the symmorphic space groups; that is the easy part. Schoenflies then allowed the possibility of glide planes and screw axes. Finally, he coupled all these possibilities by realizing that to completely describe a space group all that is needed is the lattice (all the translation symmetry operations) and h space group operations, where h is the number of operations in the point group of the space group. These h operations of the space group are isomorphic to the h operations of the point group of the space group, as discussed in Section 5.4. This idea is simple. For example, if for a given lattice and point group the symmorphic space group has a symmetry operation $4(C_4)$, then the corresponding non-symmorphic space groups may have a symmetry operation 4_1 or 4_2 or 4_3 and none other. Similarly, a symmorphic operation $m(\sigma)$ may only be replaced in non-symmorphic space groups by one of the various types of glide reflections. Thus, although the task is very large, it is finite and has well-defined bounds. The examples that we show are meant to give a flavour for the approach as well as make the remarks clearer.

The triclinic space groups are too trivial to clarify the above points and in any case, the techniques of Section 5.1 determine both of them perfectly well. This is because there is only the *aP* lattice to consider; the point group $1(C_1)$ gives the space group that Schoenflies calls C_1^1 and the point group $\bar{1}(C_i)$ gives the space group C_i^1 ($P1$ and $P\bar{1}$, respectively).

For the monoclinic crystal system, it is easier to see how the approach is used. There are only two Bravais space lattices, the *mP* and *mB* lattices, and the point groups that are consistent with monoclinic symmetry are $2(C_2)$, $m(C_s$ or $C_{1h})$, and $2/m(C_{2h})$. As discussed in Section 5.1, there are six symmorphic space groups that one can immediately obtain and they are listed in Table 5.1. Now, starting with the point group $2(C_2)$ we may determine all of the space groups, including the non-symmorphic ones, that have point group $2(C_2)$. We start with the primitive lattice and the point group $2(C_2)$ just as in Section 5.1 and find the first space group C_2^1 ($P2$ in the International notation). Then, if the screw axis 2_1 replaces the 2-fold axis as a symmetry operation, the point group of the resulting space group will still be $2(C_2)$ and so we try it. Doing this, we obtain a new space group which is different from C_2^1, called $C_2^2(P2_1)$. We now have exhausted all the possible non-symmorphic operations that will yield space groups whose point groups are $2(C_2)$. Thus, we proceed to the base-centered lattice and immediately have the symmorphic space group which we call $C_2^3(B2)$. The next possibility is a 2_1 operation with an mB lattice. When we try this we obtain the same space group as C_2^3 but with the origin shifted. Thus, a new space group is not obtained in this case. The three unique space groups we have found in this way are $C_2^1(P2)$, $C_2^2(P2_1)$ and $C_2^3(B2)$.

Now, we proceed to monoclinic space groups whose point groups are $m(C_s)$. Here we know that the only possible non-symmorphic operation is a b-glide. Thus, we start with the mP lattice and the $m(C_s)$ operation and obtain $C_s^1(Pm)$. Furthermore, we find that $C_s^2(Pb)$ is different. For the base-centered lattice, we obtain $C_s^3(Bm)$ and $C_s^4(Bb)$. Hence a fourth unique space group is obtained.

For the monoclinic space groups whose point groups are $2/m(C_{2h})$, there are four possible combinations of symmetry operations: $2/m$, $2_1/m$, $2/b$ and $2_1/b$. We may adjoin these to the two lattices mP and mB; six new space groups are then obtained. These can best be displayed as direct products of two groups; this brings out the simplicity of the approach.

$$C_{2h}^1 = C_2^1 \times \{E, \sigma\} \quad \text{or} \quad P2/m = P2 \times \{1, m\}$$

$$C_{2h}^2 = C_2^2 \times \{E, \sigma\} \quad \text{or} \quad P2_1/m = P2_1 \times \{1, m\}$$

$$C_{2h}^3 = C_2^3 \times \{E, \sigma\} \quad \text{or} \quad B2/m = B2 \times \{1, m\}$$

$$C_{2h}^4 = C_2^1 \times \{E, b\} \quad \text{or} \quad P2/b = P2 \times \{1, b\}$$

$$C_{2h}^5 = C_2^2 \times \{E, b\} \quad \text{or} \quad P2_1/b = P2_1 \times \{1, b\}$$

$$C_{2h}^6 = C_2^3 \times \{E, b\} \quad \text{or} \quad B2/b = B2 \times \{1, b\}$$

This also explains the order of space groups adopted by Schoenflies and by the *ITA*.

It is tempting to stop here because it gets more difficult for the other crystal systems. There is the occasional simple situation, such as in the hexagonal crystal system, where with just a hP lattice, one may deduce the six space groups C_6^1 to C_6^6 ($P6$, $P6_1$, $P6_5$, $P6_2$, $P6_4$, $P6_3$) immediately. However, we discuss one more set of space groups simply to stress that it is nevertheless feasible.

Consider the orthorhombic system and, in particular, let us concentrate on the use of point group $mm2(C_{2v})$ in conjunction with a primitive lattice.

There are 10 space groups that can be generated. This large number is obtained because the various symmorphic or non-symmorphic planes may contain the 2 or 2_1-axis (which we take here as the **c**-axis). Also, the vertical planes may be displaced a distance ¼**a** or ¼**b**. The displacement of ¼ occurs for reasons similar to those found in Section 5.4.4 for the three screw axes in space group $P2_12_12_1$. For conciseness, we can describe these space groups as a direct product of $C_2^1(P2)$ or $C_2^2(P2_1)$ with a group that involves a mirror plane σ perpendicular to the **a**-axis at $x = 0$ or a similar plane σ' displaced to $x = ¼$. The 10 space groups obtained are as follows:

$$Pm-2 = Pmm2 = C_{2v}^1 = C_2^1 \times \{E, \sigma\}$$

$$Pm-2_1 = Pmc2_1 = C_{2v}^2 = C_2^2 \times \{E, \sigma\}$$

$$Pc-2 = Pcc2 = C_{2v}^3 = C_2^1 \times \{E, \{\sigma| c/2\}\}$$

$$Pm'-2 = Pma2 = C_{2v}^4 = C_2^1 \times \{E, \sigma'\}$$

$$Pc'-2_1 = Pca2_1 = C_{2v}^5 = C_2^2 \times \{E, \{\sigma'|c/2\}\}$$

$$Pn-2 = Pnc2 = C_{2v}^6 = C_2^1 \times \{E, \{\sigma'|(b+c)/2\}\}$$

$$Pm'-2_1 = Pmn2_1 = C_{2v}^7 = C_2^2 \times \{E, \sigma'\}$$

$$Pb'-2 = Pba2 = C_{2v}^8 = C_2^1 \times \{E, \{\sigma'|b/2\}\}$$

$$Pn'-2_1 = Pna2_1 = C_{2v}^9 = C_2^2 \times \{E, \{\sigma'|(b+c)/2\}\}$$

$$Pn'-2 = Pnn2 = C_{2v}^{10} = C_2^1 \times \{E, \{\sigma'|(b+c)/2\}\}$$

The standard International and Schoenflies notations are used except for the symbol on the left. This symbol describes the operations that are absolutely necessary to obtain the space group and is just a shorthand notation for the direct product on the right. This symbol describes the oP lattice and 2 or 2_1 along the **c**-axis corresponding the Schoenflies space groups C_2^1 or C_2^2, respectively. The other symbol is a plane perpendicular to the **a**-axis (hence it is put in the first position of the symbol) which is at $x = 0$ or at $x = ¼$; the latter is denoted by a prime. The symbols m, b, c or n describe the direction of the glide in the usual way. The dash in the position of the plane perpendicular to the **b**-axis is a quantity derived from the rest of the information.

The complexity of the problem is apparent, although it is clearly finite. It is instructive to read the original book by Schoenflies, since it gives an appreciation of the ingenuity required to determine the 230 space groups.

5.6. SPACE GROUP CLASSIFICATIONS

The compilers of the *ITA* have established a hierarchy of classifications for the space group symmetries relevant to crystals. We have already met a number of them, and here we briefly describe some of the other classifications that have been adopted. For a more detailed explanation, we refer the reader to the *ITA*. Figure 5.29 shows the classification as a flow diagram.

5.6.1. Enantiomorphic Space Groups and Space Group Types

As we saw, for the space groups $P4_1(C_4^2)$ and $P4_3(C_4^4)$ some space groups are related to each other by inversion or mirror symmetry, thus changing the chirality of the space groups. In total, there are 11 pairs of such space groups (Table 5.2).

Because of this, it has been argued that strictly speaking one should only list one of each of these groups among the total number of space groups, thus making 219 (17 in two dimensions) distinct space group types. The 219 are known as **affine space group types**, while the traditional 230 groups are formally called the **crystallographic space group types**.

5.6.2. Arithmetic Crystal Classes

This refers to those space groups that are described by the Seitz operators $\{R|\mathbf{0}\}$ given a suitable choice of origin. They therefore correspond to the 73 (13 in two dimensions) symmorphic space groups discussed in Section 5.1. These are given special symbols by replacing all screw and glide operations in the symbol by corresponding rotations and reflections and interchanging the lattice letter and the point group part. For instance, the space groups $P2/m$, $P2_1/m$ and $P2_1/c$ all belong to the same arithmetic crystal class called $2/mP$.

5.6.3. Geometric Crystal Classes

We have already met the term ‘crystal class’ when discussing point groups. Geometric crystal classes cover space groups and their point groups together, and correspond to the morphological symmetry of the external shapes of macroscopic crystals. All space groups belong in the same geometric crystal class if the matrix parts of the Seitz operators are identical, provided that one chooses suitable axial bases. Thus, they each are associated with the point groups and so there are 32 (10 in two dimensions) in total.

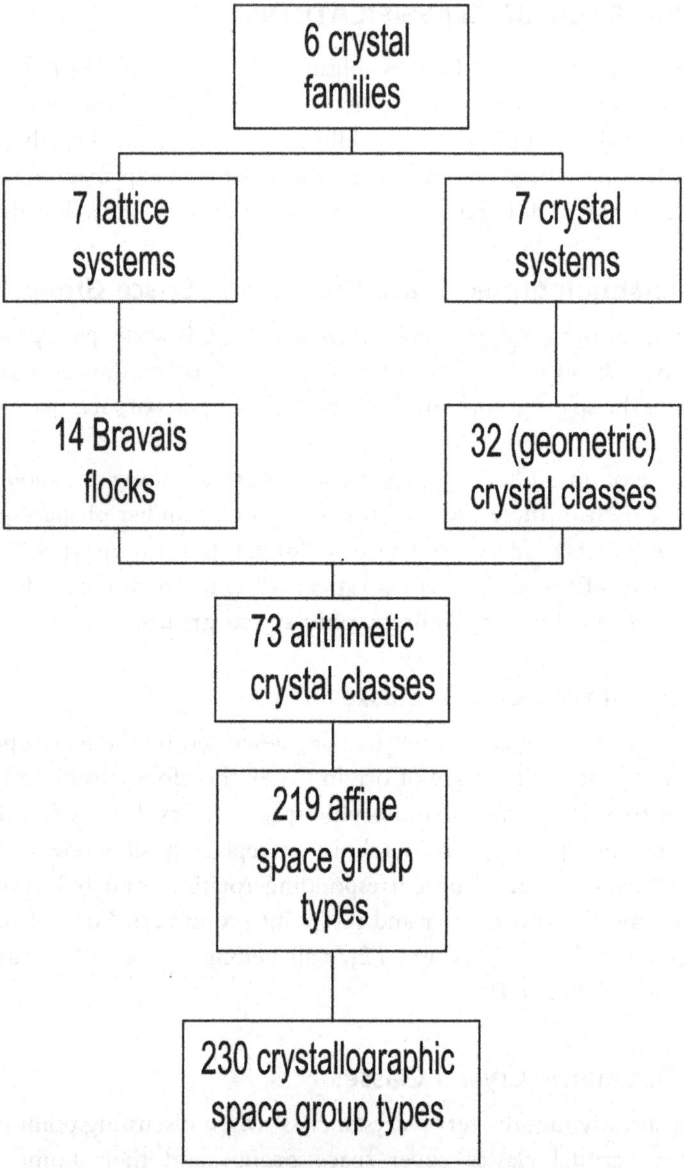

Figure 5.29 Classification of space groups.

5.6.4. Bravais Classes

The formal way to think of the Bravais lattices is in terms of classes defined by matrix operations. Thus, an arithmetic crystal class using matrices defining the lattices forms a Bravais arithmetic class, or simply a Bravais class.

Table 5.2 The 11 Enantiomorphic space group types

$P4_1$	(C_4^2)	$P4_3$	(C_4^4)
$P4_122$	(D_4^3)	$P4_322$	(D_4^7)
$P4_12_12$	(D_4^4)	$P4_32_12$	(D_4^8)
$P3_1$	(C_3^2)	$P3_2$	(C_3^3)
$P3_121$	(D_3^4)	$P3_221$	(D_3^6)
$P3_112$	(D_3^3)	$P3_212$	(D_3^5)
$P6_1$	(C_6^2)	$P6_5$	(C_6^3)
$P6_2$	(C_6^4)	$P6_4$	(C_6^5)
$P6_122$	(D_6^2)	$P6_522$	(D_6^3)
$P6_222$	(D_6^4)	$P6_422$	(D_6^5)
$P4_132$	(O^7)	$P4_332$	(O^6)

For example, the arithmetic crystal class $6/mmmP$ corresponds to the hexagonal lattice and so is one of the Bravais classes. On the other hand, the arithmetic crystal classes, $6/mP$ or $6mmP$ do not correspond to any Bravais lattice. Thus, each lattice is associated with a Bravais class, and so there are 14 (5 in two dimensions) in total.

5.6.5. Bravais Flocks

This classification arises because of the possibility of an accidental metric relationship that is not related to the true symmetry of the underlying crystal structure. We met this idea when discussing crystal systems, only here we apply the concept to space groups. For example, suppose we have a monoclinic crystal described by a particular Bravais lattice and that the angle β is not equal to 90°. Suppose too that at a certain temperature this angle is measured to be 90° to within the precision of the experiment. The lattice then, taken on its own, will appear to have orthogonal axes and might be considered to belong in the orthorhombic system. However, inspection of the crystal structure shows that the arrangement of atoms do not conform to the symmetry restrictions imposed by orthorhombic symmetry. A nice example is provided in the *ITA* and we repeat it here for clarity.

The space group $I4_1$ belongs in the arithmetic crystal class $4I$ (Section 5.6.2). The possible Bravais classes are then (a) $4/mmmI$ or (b) $m\overline{3}mI$, since both contain 4-fold axes and are body-centered. We exclude $m\overline{3}mI$ because $4/mmmI$ is of smaller order. Thus, even if the unit cell is measured to be such that a = b = c as far as we can tell, the space group $I4_1$ is uniquely assigned to the Bravais class $4/mmmI$, despite the fact that the space symmetry of the lattice appears to be $Im\bar{3}m$. Space groups that are assigned to the same Bravais class are said to belong to the same **Bravais flock of space groups**.

5.6.6. Crystal Families

This is the smallest set of space groups containing, for any of its members, all space groups of the Bravais flock and all space groups of the geometrical crystal class to which this member belongs. Thus, $R3$ and $P6_1$ are in the same crystal family because $R3$ and $P3$ belong in the same geometrical crystal class 3, while $P3$ and $P6_1$ are members of the same Bravais flock $6/mmmP$. $P3$ therefore serves as a link between $R3$ and $P6_1$. There are six crystal families (four in two dimensions): see Table 5.3. Notice that there is no trigonal family and so one should not confuse the terms crystal family with crystal system (Appendix 2).

5.6.7. Lattice Systems

These were originally called **Bravais systems** and classify space groups and lattice types, but not crystallographic point groups. A **lattice system** of space groups contains complete Bravais flocks. All Bravais flocks that intersect exactly the same set of geometric crystal classes belong to the same lattice system. There are seven lattice systems (four in two dimensions).

5.7. TWO-DIMENSIONAL SPACE GROUPS

In Chapter 4, we discussed the ten two-dimensional point groups. In order to obtain the **two-dimensional space groups** or **plane groups** (sometimes known as **wallpaper groups**), we must combine these with the five two-dimensional lattices. If we do this to generate the symmorphic plane groups, as we did in Section 5.1, then we obtain a total of 13 plane groups, 2 of which ($p3m1$ and $p31m$) differ only in the

Table 5.3 The six crystal families and their symbols

Triclinic (anorthic)	*a*
Monoclinic	*m*
Orthorhombic	*o*
Tetragonal	*t*
Hexagonal	*h*
Cubic	*c*

Table 5.4 The 17 plane groups (two-dimensional space groups)

Lattice	Point group	Space group full	Symbol short	Number
Oblique	1	*p*1	*p*1	1
	2	*p*211	*p*2	2
Rectangular	*m*	*p*1*m*1	*pm*	3
		*p*1*g*1	*pg*	4
		*c*1*m*1	*cm*	5
	2mm	*p*2*mm*	*pmm*	6
		*p*2*mg*	*pmg*	7
		*p*2*gg*	*pgg*	8
		*c*2*mm*	*cmm*	9
Square	4	*p*4	*p*4	10
	4*mm*	*p*4*mm*	*p*4*m*	11
		*p*4*gm*	*p*4*g*	12
Hexagonal	3	*p*3	*p*3	13
	3*m*	*p*3*m*1	*p*3*m*1	14
		*p*31*m*	*p*31*m*	15
	6	*p*6	*p*6	16
	6*mm*	*p*6*mm*	*p*6*m*	17

orientation of the symmetry elements with respect to the lattice (as in space groups *P*3*m*1 and *P*31*m*). The only non-symmorphic operation permitted in two dimensions is the glide with reflection across a line in the **ab**-plane, given the symbol *g*. All the other non-symmorphic operations we met in three-dimensional space groups involve out-of-plane displacements of the general point. Taking into account the glide operation *g*, a further four non-symmorphic plane groups are generated, making a total of 17 in all. They are listed in Table 5.4.

It should be a simple matter for the reader to draw the space group diagrams for all of the plane groups. The results can be checked by consulting the *ITA*. Note that the order of symbols is not quite the same as in the three-dimensional space groups (see Appendix 6).

5.8. SUBPERIODIC GROUPS

The International Tables Volume E (*ITE*) describes other types of groups under the general heading of **subperiodic groups**. These are 7 **frieze groups**, 75 **rod groups** and 80 **layer groups**. Complete details can be found in the *ITE*. Briefly, these are defined as follows:

Frieze groups – These classify designs on two-dimensional surfaces that repeat in one direction. These are given in Table 5.5.

Table 5.5 The seven frieze groups

ITE	Schoenflies	Common name
$\mathscr{p}1$	C_∞	Hop
$\mathscr{p}211$	D_∞	Spinning hop
$\mathscr{p}1m1$	$C_{\infty v}$	Sidle
$\mathscr{p}11m$	$C_{\infty h}$	Jump
$\mathscr{p}11g$	S_∞	Step
$\mathscr{p}2mm$	$D_{\infty h}$	Spinning jump
$\mathscr{p}2mg$	$D_{\infty d}$	Spinning sidle

In other words, these are the symmetries shown by friezes, for example, on tops of walls or in wallpaper. Figure 5.30 shows some wrought-ironwork with the frieze symmetry $\mathscr{p}2mm$, and Fig. 5.31 shows an example of some brickwork that adopts the pattern of the frieze group $\mathscr{p}_b2gm$ (note the way the lattice symbol is written).

The symmetry operations for the frieze group $\mathscr{p}_b2gm$ are:

(1) 1	(2) 2 0, 0	(3) *m* *x*, ¼	(4) *g* 0, *y*

Most of the bricks are at the 4*c* general positions (1): (x, y), (2): $(\bar{x}, \bar{y})$, (3): $(x, \bar{y}+½)$ and (4): $(\bar{x}, y+½)$, while a few are on the mirror planes at 2*b* sites (site symmetry *.m.*): $(x, ¼)$ and $(\bar{x}, ¾)$ (see Section 6.3.8 for site symmetry and labels for positions).

Rod groups – These are formed by starting with a point group and then adding one translational symmetry axis. This type of symmetry can be useful where linear features in a crystal need to be described. For instance, the crystal structure of α-$CaTeO_3$ (Fig. 5.32) has open channels down $(0, 0, z)$ and $(½, ½, z)$ whose rod group symmetry is symbolized by $\mathscr{p}4_3$.

Figure 5.30 Example of the 'spinning jump' frieze pattern $\mathscr{p}2mm$.

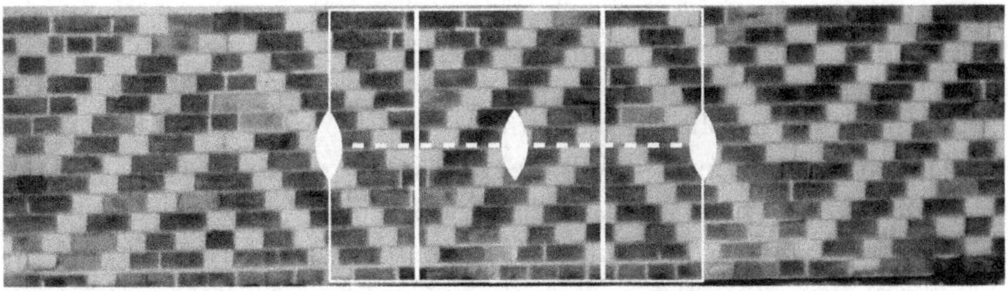

Figure 5.31 'Spinning sidle' frieze pattern on wall of Keble College, Oxford. Frieze group p_b2gm outlined in white.

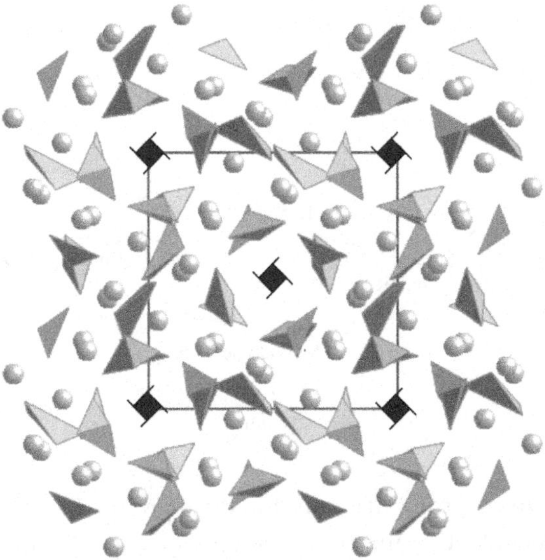

Figure 5.32 Crystal structure of α-$CaTeO_3$ projected on (001).[4] Te-O polyhedra are shown, spheres are Ca atoms.

Layer groups (diperiodic groups) – These are three-dimensional extensions of a two-dimensional plane or wallpaper group. They are formed by adding translational symmetry in two dimensions to a point group. The symmetry group at each lattice point is an axial crystallographic point group. Layer groups can be useful for describing surfaces, polytypes and anywhere that plate-like structural features are important, for example in graphitic compounds.

4 B. Stöger, N. Weil, E. Zobetz & G. Giester. *Acta Cryst.*, **B65**, 167 (2009).

PROBLEMS

1. Within a given crystal system, show that the Bravais lattices always belong to the holohedral point group.
2. The diagram below shows the **hierarchy of the seven crystal systems** in which every crystal system can be produced by an infinitesimal distortion from one above, provided it is connected by a line. Show that this is indeed so. Consider placing an object at a lattice point in any crystal system. Show (a) that all the symmetry operations possessed by this object must be possessed by the lattice (the lattice can, of course, be described by more symmetry operations) and (b) the object must possess *at least* one symmetry operation that is not found in the lattice that is next lower in the hierarchy of the crystal systems.

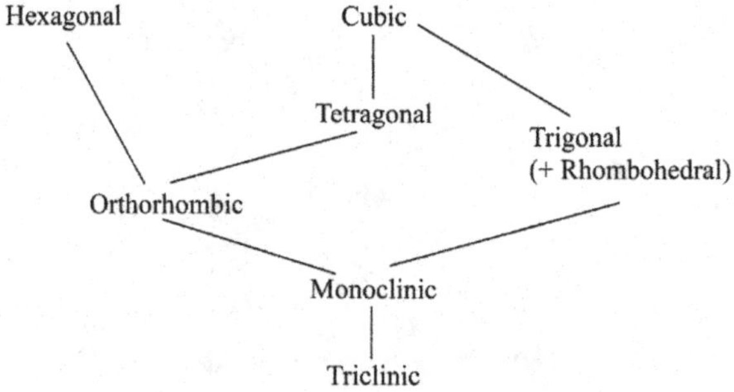

3. Draw the space group diagrams for space groups C_4^1, C_{4v}^1 and C_{4h}^1. Be sure to include the International symbol and all of the symmetry operations that are generated.
4. Draw a two-dimensional lattice and in a general position close to each lattice point draw the letter 'B' to act as a motif. Add further motifs to form a pattern in which the point symmetry at each lattice point is 2. Outline a unit cell of the pattern and indicate in another color any symmetry operations that it displays. Repeat the above procedure to give patterns in which the point symmetry at each lattice point is *m*, *mm*2, 4 and 4*mm*, respectively. In each case, base your pattern on a lattice having **a**, **b** and γ appropriate to the point symmetry.
5. Show that the following pattern belongs to the plane group (wallpaper group) *p*4*gm*.

6. Find the coordinates of the point generated from the point (x, y, z) by the action of the following symmetry operations: (a) a c-glide at $(x, ¼, z)$ followed by a 2_1 screw rotation about an axis through $(0, y, ¼)$. (b) An n-glide at $(x, ¼, z)$ followed by a 4_2 screw about $(0, ½, z)$. What is the result when these operations are applied in reverse order?
7. Draw space group diagrams and give the general equivalent positions for $Cbc2_1$, $Pmba$, $Abc2_1$ and $Pnma$.
8. Consider the space group $P4nc(C_{4v}^6)$ in the *ITA* (No. 104). What is the resulting position produced by the n-glide with a reflection perpendicular to **a** acting on the general position (x, y, z)? What is the result for the n-glide plane parallel to the diagonal [110]? Notice that in the first case, a translation of (0, ½, ½) is involved, whereas in the second case it is (½, ½, ½).
9. With reference to the space groups with a primitive lattice and point group $mm2(C_{2v})$, discussed in Section 5.5, show that the space groups $Pb{-}2$, $Pb{-}2_1$ and $Pc{-}2_1$ are the same as those shown in the complete list in that section but with a redefinition of the **a** and **b**-axes.
10. Find the type and position of the conventional symmetry operation that is involved in each of the following mappings:

$$(x, y, z) \rightarrow (½ - x, ½ - y, ½ + z)$$

$$(x, y, z) \rightarrow (½ + z, ½ - y, z)$$

$$(y - x, y, ⅔ - z) \rightarrow (y - x, \overline{x}, ⅓ + z)$$

CHAPTER 6

Reading the Tables

Contents

The magic is inside you. There ain't no crystal ball.

Dolly Parton

WHAT DOES THE *ITA* TELL US?

In this chapter, we show how a large amount of information of interest to the solid state scientist can be obtained from knowledge of the space group of a crystal. Normally, a solid state scientist works with a crystal whose space

Space Groups for Solid State Scientists
http://dx.doi.org/10.1016/B978-0-12-394400-9.00006-X

group and structure are known. There are a number of questions that could, or should, arise:

(a) What are the positions of the various atoms in the crystal?
(b) What symmetry do these atoms see?
(c) How are they related to other similar atoms?
(d) What is the maximum number of similar atoms allowed by the space group?
(e) What are the restrictions on the macroscopic properties imposed by the space group?
(f) What is the crystal class and Bravais lattice?
(g) What is the point group of the space group and what is its significance?
(h) What are the implications of having glide reflections and screw rotations as symmetry operations instead of point operations?
(i) What is the relationship (the isomorphism or the one-to-one correspondence) between these glides and screws that are symmetry operations, and the simple point operations that are not necessarily symmetry operations?
(j) How can we find the smallest number of symmetry operations needed to specify the space group fully?

These are the kinds of questions that one would normally ask, or be aware of, whether the interest is in the atomic positions *per se*, electronic band theory, lattice vibrations or practically any other solid-state questions. One should remember that, from a quantum mechanical point of view, the space group describes all of the spatial symmetry of the crystal, and so from it one should be able to determine all of the exact solutions of the Hamiltonian that result from the spatial symmetry. This means that we need to describe how all the different eigenfunctions transform under the symmetry operations, the same thing as labelling the eigenfunctions with the irreducible representations of the point group of the space group and **k** (wave vector) values. We shall not deal here with the specialized problems of the irreducible representations of the space groups. However, we shall show how the symmetry operations of the space groups are related to the operations that appear in character tables (see any group theory text) for the 32 point groups.

Clearly, the answers to many of these questions involve simple geometry, while others require involved group theory. In this chapter, we shall start with the simple aspects of what one can learn when the space group and crystal structure are known, and then cover some of the more involved matters towards the end of the chapter. Some of this will be repetition of the important points made in the preceding chapters, and so, to some extent, this chapter can be read on its own and serves as a summary of the important features of space groups.

6.1. CRYSTAL STRUCTURE AND SPACE GROUPS

You will recall from Section 2.3 that it is possible to define a **crystal structure** as simply the repetition of molecules or atoms according to translational symmetry. To do this, we used a convolution between a lattice and a physical object, consisting of an atom or group of atoms (molecules) that is sometimes called a 'basis'. Now, while this is relatively straightforward it does ignore any of the symmetry that the 'basis' itself might possess. This is where the use of space groups becomes important, for, as we shall see, by specifying the positions of atoms according to the symmetry of a space group, it is only then necessary to specify the positions of a few atoms and then let all the space group operations do the work of generating the whole crystal structure.

The structural crystallographer can determine relatively easily the crystal system and unit-cell geometry for a particular crystal, usually by employing X-ray, neutron or electron diffraction techniques. The determination of the point and space groups is often more difficult. It is normally accomplished by looking for the systematic presence or absence of certain diffraction spots (confusingly in the context of symmetry nomenclature, called 'reflections'). It is known that lattice centering and non-symmorphic operations cause certain classes of spots to have zero intensity. Hence it is, in principle, possible to work back and determine what the centering and non-symmorphic operations are. However, it is often found that there are still several possible space groups that satisfy these diffraction conditions, and it can be difficult to make the correct choice. For example, as discussed briefly in dealing with the Laue classes in Section 4.4, to a good approximation the diffraction process appears to add a center of inversion (**Friedel's Law**), so that distinction between centrosymmetric and non-centrosymmetric[1] space groups may have to be made by resorting to examining the crystal properties. That is, it may be necessary to see if the crystal shows chiral properties such as optical activity, or polar properties such as pyroelectricity or piezoelectricity, or optical second harmonic generation. Alternatively, it may be necessary to try all the possibilities and see which one fits best (trial-and-error method). Note, however, Friedel's Law can be seen to be

[1] Some scientists mistakenly use the terms centric and acentric, but these are terms that refer to the statistical distribution of diffraction intensities from which one can determine if a crystal is centrosymmetric or not.

broken when an atom in the crystal shows strong X-ray absorption, in which case small differences in centrosymmetrically related diffracted intensities (i.e. for *hkl* and $\bar{h}\bar{k}\bar{l}$ reflections) can be used to determine the space group (the so-called **anomalous dispersion** effect).

Once the space group and the number of formula units (or molecules) per unit cell are known, the relative positions of the atoms in the unit cell can often be guessed. The number of formula units per cell, Z, is often determined from the measured crystal density ρ and its unit cell volume V. It is given by $Z = \rho NV/M$, where M is the formula weight and N is Avogadro's number. However, with the exception of simple compounds, a complete crystal structure analysis is required to determine with any degree of certainty the exact positions of the atoms, that is whether the atoms are at certain special positions or at general positions, and what the values of their fractional coordinates (x, y, z) are. This is what is called the **crystal structure determination**, that is the determination of the space group and the positions of the atoms in the unit cell. This is commonly carried out by careful measurement of the diffracted intensities, which are proportional to the square of the scattered wave amplitudes. From this, provided all the phase relationships between the scattered waves are known, the complete crystal structure can be worked out. In general, however, these phase relationships cannot be measured directly and their determination constitutes one of the greatest unsolved problems in crystal structure analysis. Several sophisticated techniques have been devised over the years to address this so-called 'phase problem', including the least-squares refinement of trial structures and statistical inference (direct methods). The derivation of most crystal structures is no longer the difficult and time-consuming operation that it used to be, and, to some extent, has become largely automatic. These days macromolecular crystal structures with thousands of atoms in the unit cell are routinely solved. For a full explanation of crystal structure determination, any standard book on X-ray crystallography can be consulted.

An illustration of how crystal structure information can be given succinctly is shown here for the mineral calcite ($CaCO_3$):

$$R\bar{3}c(D_{3d}^{6}) \quad a = 4.990\text{Å} \ \ c = 17.06\text{Å}$$

$$\text{Ca at } 6b: 0, 0, 0$$

$$\text{C at } 6a: 0, 0, \tfrac{1}{4}$$

$$\text{O at } 18e: x, 0, \tfrac{1}{4} \quad \text{with } x = 0.257$$

This contains sufficient information to enable one to draw a diagram of the crystal structure. In the unit cell of calcite there are 30 atoms, and yet, as can be seen, it is not necessary to give the positions of all of them separately; only the three basic positions are given and the space group symmetry determines where the other 27 are, showing how the use of symmetry simplifies the description of something that may be inherently quite complicated. In fact, there is only one free positional parameter in this example that needs to be determined, namely the x-coordinate for the oxygen atom.

6.2. 'TYPICAL' PAGES OF THE *ITA*

There are 230 space groups, any one being 'typical'. We shall show examples of various pages of the *ITA* for several space groups and discuss the meaning of the information given. Note that crystallographers tend to take different origins within the unit cell for different symmetry operations. However, it is possible to take them with respect to *one* origin and we shall show how this is done.

6.3. EXAMPLE PAGES FROM THE *ITA*

6.3.1. Top Lines

Figure 6.1 shows the page from the *ITA* for space group $P4/m(C_{4h}^1)$. Consider the symbols at the top of the page starting at the left and going to the right. First, the symbol for the space group $P4/m$ is given in the International notation followed by the Schoenflies notation, (C_{4h}^1). This is followed by the point group $4/m$ of the space group and then the crystal system, Tetragonal. On the next line is the serial number of the space group in the *ITA*, No. 83. The space groups are numbered sequentially from 1 to 230 starting with the triclinic space group $P1$, number 1, and ending with the cubic space group $Ia\bar{3}d(O_h^{10})$, number 230; this can be used as a rapid means of finding a particular space group provided that it is supplied along with the crystal data. Next, the 'full' International symbol for the space group appears. In this case, it is the same as the short symbol. The last item on this line is the so-called Patterson symmetry: $P4/m$.

6.3.2. Patterson Symmetry

When X-rays, electrons or neutrons are diffracted from a crystal, the scattered waves constructively interfere to give rise to scattered intensity in well-defined directions forming 'Bragg reflections'. These intensities, which arise from the scattering of the radiation by electrons (in the case of X-rays and electrons) or by nuclei (in the case of neutrons) lying in crystallographic planes, are proportional to the square of the modulus of the scattered amplitudes. Because the amplitude is squared, information about the phase relationships between the scattered waves is lost (the so-called 'phase problem'), and it is this information that is needed in order to reconstruct the crystal structure that gave rise to the diffraction pattern in the first place.

The Patterson method, due to Arthur Lindo Patterson in 1935, is an important technique for obtaining a model structure without knowing the phases. A Fourier map is constructed using the intensities (to be more correct, the squares of the moduli of the structure factors) with all contributions given the same phase. The resulting map contains peaks that represent *vectors* between atoms (rather than atoms themselves), and a skilled crystallographer can use this information to suggest a structural model that fits this set of vectors.

To understand how the symmetry of the Patterson map is derived, it is necessary to work out all of the vector distances between general equivalent

$P4/m$	C_{4h}^1	$4/m$	Tetragonal
No. 83	$P4/m$		Patterson symmetry $P4/m$

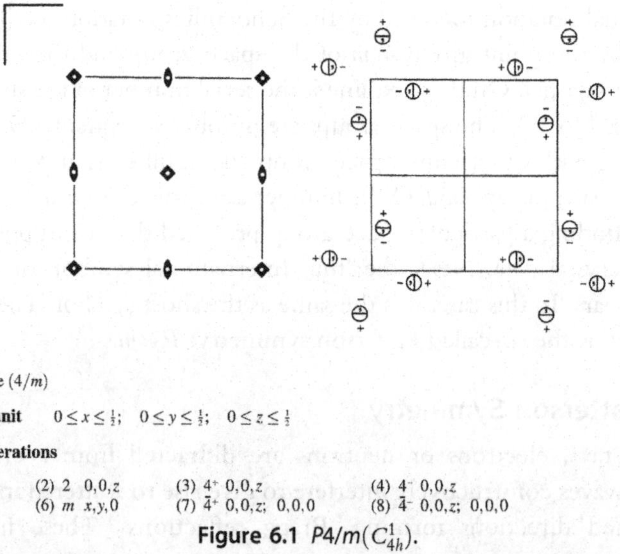

Origin at centre ($4/m$)

Asymmetric unit $0 \le x \le \frac{1}{2}$; $0 \le y \le \frac{1}{2}$; $0 \le z \le \frac{1}{2}$

Symmetry operations

(1) 1	(2) 2 0,0,*z*	(3) 4^+ 0,0,*z*	(4) 4^- 0,0,*z*
(5) $\bar{1}$ 0,0,0	(6) *m* *x*,*y*,0	(7) $\bar{4}^+$ 0,0,*z*; 0,0,0	(8) $\bar{4}^-$ 0,0,*z*; 0,0,0

Figure 6.1 $P4/m(C_{4h}^1)$.

Generators selected (1); $t(1,0,0)$; $t(0,1,0)$; $t(0,0,1)$; (2); (3); (5)

Positions

Multiplicity, Wyckoff letter, Site symmetry			Coordinates				Reflection conditions
							General:
8	l	1	(1) x,y,z	(2) $\bar{x},\bar{y},z$	(3) $\bar{y},x,z$	(4) $y,\bar{x},z$	no conditions
			(5) $\bar{x},\bar{y},\bar{z}$	(6) $x,y,\bar{z}$	(7) $y,\bar{x},\bar{z}$	(8) $\bar{y},x,\bar{z}$	
							Special:
4	k	$m..$	$x,y,\frac{1}{2}$	$\bar{x},\bar{y},\frac{1}{2}$	$\bar{y},x,\frac{1}{2}$	$y,\bar{x},\frac{1}{2}$	no extra conditions
4	j	$m..$	$x,y,0$	$\bar{x},\bar{y},0$	$\bar{y},x,0$	$y,\bar{x},0$	no extra conditions
4	i	$2..$	$0,\frac{1}{2},z$	$\frac{1}{2},0,z$	$0,\frac{1}{2},\bar{z}$	$\frac{1}{2},0,\bar{z}$	hkl : $h+k=2n$
2	h	$4..$	$\frac{1}{2},\frac{1}{2},z$	$\frac{1}{2},\frac{1}{2},\bar{z}$			no extra conditions
2	g	$4..$	$0,0,z$	$0,0,\bar{z}$			no extra conditions
2	f	$2/m..$	$0,\frac{1}{2},\frac{1}{2}$	$\frac{1}{2},0,\frac{1}{2}$			hkl : $h+k=2n$
2	e	$2/m..$	$0,\frac{1}{2},0$	$\frac{1}{2},0,0$			hkl : $h+k=2n$
1	d	$4/m..$	$\frac{1}{2},\frac{1}{2},\frac{1}{2}$				no extra conditions
1	c	$4/m..$	$\frac{1}{2},\frac{1}{2},0$				no extra conditions
1	b	$4/m..$	$0,0,\frac{1}{2}$				no extra conditions
1	a	$4/m..$	$0,0,0$				no extra conditions

Symmetry of special projections

Along [001] $p4$
$\mathbf{a}' = \mathbf{a}$ $\mathbf{b}' = \mathbf{b}$
Origin at $0,0,z$

Along [100] $p2mm$
$\mathbf{a}' = \mathbf{b}$ $\mathbf{b}' = \mathbf{c}$
Origin at $x,0,0$

Along [110] $p2mm$
$\mathbf{a}' = \frac{1}{2}(-\mathbf{a}+\mathbf{b})$ $\mathbf{b}' = \mathbf{c}$
Origin at $x,x,0$

Maximal non-isomorphic subgroups

I [2] $P\bar{4}$ (81) 1; 2; 7; 8
[2] $P4$ (75) 1; 2; 3; 4
[2] $P2/m$ (10) 1; 2; 5; 6

IIa none

IIb [2] $P4_2/m$ ($\mathbf{c}' = 2\mathbf{c}$) (84); [2] $C4/e$ ($\mathbf{a}' = 2\mathbf{a}, \mathbf{b}' = 2\mathbf{b}$) ($P4/n$, 85); [2] $F4/m$ ($\mathbf{a}' = 2\mathbf{a}, \mathbf{b}' = 2\mathbf{b}, \mathbf{c}' = 2\mathbf{c}$) ($I4/m$, 87)

Maximal isomorphic subgroups of lowest index

IIc [2] $P4/m$ ($\mathbf{c}' = 2\mathbf{c}$) (83); [2] $C4/m$ ($\mathbf{a}' = 2\mathbf{a}, \mathbf{b}' = 2\mathbf{b}$) ($P4/m$, 83)

Minimal non-isomorphic supergroups

I [2] $P4/mmm$ (123); [2] $P4/mcc$ (124); [2] $P4/mbm$ (127); [2] $P4/mnc$ (128)

II [2] $I4/m$ (87)

Figure 6.1 ***(Continued).***

positions expressed from a single origin[2]. Table 6.1 illustrates this process for the non-centrosymmetric, non-symmorphic space group $P2_12_12(D_2^3)$. In this space group, there are four general equivalent positions, labelled (1)–(4) here for convenience. We then work out all of the possible vector differences to produce the set of vector distances in the second column. Notice that this forms a centrosymmetric set. In the third column, we have transformed these to a new set of equivalent coordinates.

2 The crystallographer C.H. Carlisle used to say that the 'peasant's' definition of a Patterson was 'All vectors to a common origin'.

Table 6.1 Construction of Patterson symmetry for $P2_12_12$. The general equivalent positions are given by

(1) x, y, z	(2) $\bar{x}, \bar{y}, z$
(3) $\bar{x} + ½, y + ½, \bar{z}$	(4) $x + ½, \bar{y} + ½, \bar{z}$

Each vector between a general position and itself gives a peak at the origin (0,0,0).

Vector pairs	Vector distances	Transformed coordinates
(1)−(2)	$2x, 2y, 0$	$x_1, y_1, 0$
(2)−(1)	$-2x, -2y, 0$	$\bar{x}_1, \bar{y}_1, 0$
(1)−(3)	$2x - ½, -½, 2z$	$x_2, ½, z_2$
(3)−(1)	$-2x + ½, ½, -2z$	$\bar{x}_2, ½, \bar{z}_2$
(1)−(4)	$-½, 2y - ½, 2z$	$½, y_3, z_3$
(4)−(1)	$½, -2y + ½, -2z$	$½, \bar{y}_3, \bar{z}_3$
(2)−(3)	$-½, -2y - ½, 2z$	$½, \bar{y}_3, z_3$
(3)−(2)	$½, 2y + ½, -2z$	$½, y_3, \bar{z}_3$
(2)−(4)	$-2x - ½, -½, 2z$	$\bar{x}_2, ½, z_2$
(4)−(2)	$2x + ½, ½, -2z$	$x_2, ½, \bar{z}_2$
(3)−(4)	$-2x, 2y, 0$	$\bar{x}_1, y_1, 0$
(4)−(3)	$2x, -2y, 0$	$x_1, \bar{y}_1, 0$

There are now three different sets which correspond to special positions labelled 4*y*, 4*x* and 4*v* of space group $Pmmm(D_{2h}^1)$. These are as follows:

4*y*	$x, y, 0$	$\bar{x}, y, 0$	$\bar{x}, \bar{y}, 0$	$x, \bar{y}, 0$
4*x*	$x, 1/2, z$	$\bar{x}, 1/2, z$	$\bar{x}, 1/2, \bar{z}$	$x, 1/2, \bar{z}$
4*v*	$1/2, y, z$	$1/2, \bar{y}, z$	$1/2, \bar{y}, \bar{z}$	$1/2, y, \bar{z}$

The resulting space group is both centrosymmetric and symmorphic.

In general, for any space group the Patterson symmetry is deduced from the space group symbol in two steps:

(1) Replace glide planes by mirror planes, and screw axes by rotation axes. This results in a symmorphic space group.

(2) If this symmorphic space group is not centrosymmetric, add a center of symmetry.

Note that the Patterson symmetry has the same lattice (and lattice symbol) as that of the space group from which it is derived. In a sense, the Patterson symmetry is to space groups what Laue symmetry (Section 4.4) is to point groups. There are 7 and 24 different Patterson symmetries for the two- and three-dimensional space groups, respectively.

For $P4/m$, the Patterson symmetry is also $P4/m$ because the symbol has neither glide planes nor screw axes, and it is centrosymmetric. Here are a few other space groups and their Patterson symmetries:

$Fd\bar{3}m(O_h^7)$	$Fm\bar{3}m$
$F\bar{4}3m(T_d^2)$	$Fm\bar{3}m$
$Cmca(D_{2h}^{18})$	$Cmmm$
$Pmna(D_{2h}^{7})$	$Pmmm$
$P2_1/b(D_{2h}^{5})$	$P112/m$

The second example, ($F\bar{4}3m$), is a non-centrosymmetric symmorphic space group, and so a center of symmetry is added. The others are non-symmorphic space groups. The full symbol for the last example is $P112_1/b$, and hence it has the Patterson symmetry $P112/m$.

6.3.3. Space Group Diagrams

We have already discussed the space group diagrams in some detail in the preceding chapter. As a reminder, the origin of each diagram is chosen to be at the top left-hand corner, with the **a**-axis pointing down, the **b**-axis to the right and the **c**-axis out of the plane of the paper towards the reader, thus making a right-hand set of axes. The diagram to the left shows all the symmetry elements and where they are positioned, while the right-hand diagram shows the effect of symmetry on a general equivalent position, represented by a circle. Plus and minus signs refer to whether the circle lies above or below the plane of projection, and a comma is inserted whenever the effect of the symmetry operators results in a change in chirality.

An important addition to the *ITA* has been the incorporation of diagrams for the symmetry elements with different choices of axes. A good example of this is the space group $Pmna(D_{2h}^{7})$ – see Fig. 6.6. The full space group symbol is given above each diagram, taking care to maintain the convention that **a** is downward, **b** is to the right and **c** is out of the page. This gives six possible symbols for this space group depending on which way the axes are chosen, namely *Pmna, Pman, Pcnm*, and then turning the page through 90°: *Pnmb, Pbmn* and *Pncm*. This ability of the International space group symbols to change for different settings of the same unit cell is logical, although at times it may seem confusing to a solid state scientist. However, one needs to be aware of this since you may find that a crystal structure publication may use any of these settings. At any rate, it can be seen that there is an advantage in using the Schoenflies symbol as well as the serial number of the space group along with the International space group symbol in order to make it clear to which space group one is referring (unfortunately rarely observed).

Another useful feature in the *ITA* is the inclusion of diagrams for the cubic groups. These were previously omitted because of their complexity, but now they have been introduced together in stereo pairs, so that with appropriate stereo-spectacles they can be visualized.

6.3.4. Space Group Origin

Below the diagram the *ITA* states where the **origin** is to be taken. Thus, in Fig. 6.1 the origin is at a center of inversion, corresponding to the intersection of a 4[001] axis and a *m*[001] mirror plane perpendicular. Figure 6.2

$P4/n$	C_{4h}^3	$4/m$	Tetragonal
No. 85	$P4/n$		Patterson symmetry $P4/m$

ORIGIN CHOICE 1

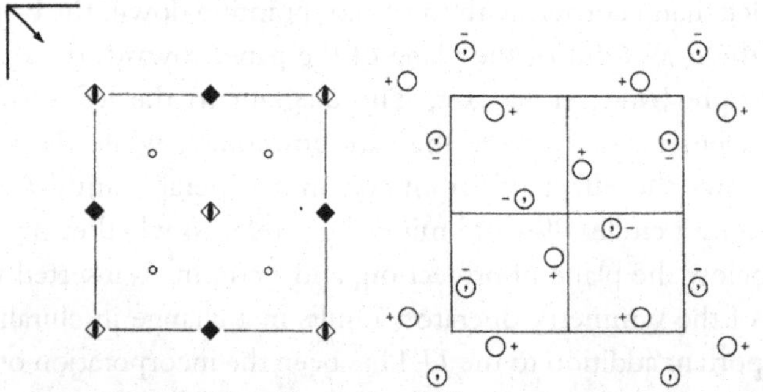

Origin at $\bar{4}$ on n, at $-\frac{1}{4},\frac{1}{4},0$ from $\bar{1}$

Asymmetric unit $0 \le x \le \frac{1}{2};\quad 0 \le y \le \frac{1}{2};\quad 0 \le z \le \frac{1}{2}$

Symmetry operations

(1) 1	(2) 2 $0,0,z$	(3) 4^+ $0,\frac{1}{2},z$	(4) 4^- $\frac{1}{2},0,z$
(5) $\bar{1}$ $\frac{1}{4},\frac{1}{4},0$	(6) $n(\frac{1}{2},\frac{1}{2},0)$ $x,y,0$	(7) $\bar{4}^+$ $0,0,z;$ $0,0,0$	(8) $\bar{4}^-$ $0,0,z;$ $0,0,0$

Figure 6.2 $P4/n(C_{4h}^3)$ Origin choice 1.

Generators selected (1); $t(1,0,0)$; $t(0,1,0)$; $t(0,0,1)$; (2); (3); (5)

Positions

Multiplicity, Wyckoff letter, Site symmetry			Coordinates				Reflection conditions
							General:
8	g	1	(1) x,y,z (5) $\bar{x}+\frac{1}{2},\bar{y}+\frac{1}{2},\bar{z}$	(2) $\bar{x},\bar{y},z$ (6) $x+\frac{1}{2},y+\frac{1}{2},\bar{z}$	(3) $\bar{y}+\frac{1}{2},x+\frac{1}{2},z$ (7) $y,\bar{x},\bar{z}$	(4) $y+\frac{1}{2},\bar{x}+\frac{1}{2},z$ (8) $\bar{y},x,\bar{z}$	$hk0$: $h+k=2n$ $h00$: $h=2n$
							Special: as above, plus
4	f	$2..$	$0,0,z$	$\frac{1}{2},\frac{1}{2},z$	$\frac{1}{2},\frac{1}{2},\bar{z}$	$0,0,\bar{z}$	hkl : $h+k=2n$
4	e	$\bar{1}$	$\frac{1}{4},\frac{1}{4},\frac{1}{2}$	$\frac{3}{4},\frac{3}{4},\frac{1}{2}$	$\frac{1}{4},\frac{3}{4},\frac{1}{2}$	$\frac{3}{4},\frac{1}{4},\frac{1}{2}$	hkl : $h,k=2n$
4	d	$\bar{1}$	$\frac{1}{4},\frac{1}{4},0$	$\frac{3}{4},\frac{3}{4},0$	$\frac{1}{4},\frac{3}{4},0$	$\frac{3}{4},\frac{1}{4},0$	hkl : $h,k=2n$
2	c	$4..$	$0,\frac{1}{2},z$	$\frac{1}{2},0,\bar{z}$			no extra conditions
2	b	$\bar{4}..$	$0,0,\frac{1}{2}$	$\frac{1}{2},\frac{1}{2},\frac{1}{2}$			hkl : $h+k=2n$
2	a	$\bar{4}..$	$0,0,0$	$\frac{1}{2},\frac{1}{2},0$			hkl : $h+k=2n$

Symmetry of special projections

Along [001] $p4$
$\mathbf{a}'=\frac{1}{2}(\mathbf{a}-\mathbf{b})$ $\mathbf{b}'=\frac{1}{2}(\mathbf{a}+\mathbf{b})$
Origin at $0,0,z$

Along [100] $p2mg$
$\mathbf{a}'=\mathbf{b}$ $\mathbf{b}'=\mathbf{c}$
Origin at $x,\frac{1}{4},0$

Along [110] $p2mm$
$\mathbf{a}'=\frac{1}{2}(-\mathbf{a}+\mathbf{b})$ $\mathbf{b}'=\mathbf{c}$
Origin at $x,x,0$

Maximal non-isomorphic subgroups

I	[2] $P\bar{4}$ (81)	1; 2; 7; 8
	[2] $P4$ (75)	1; 2; 3; 4
	[2] $P2/n$ ($P2/c$, 13)	1; 2; 5; 6
IIa	none	
IIb	[2] $P4_2/n$ ($\mathbf{c}'=2\mathbf{c}$) (86)	

Maximal isomorphic subgroups of lowest index

IIc [2] $P4/n$ ($\mathbf{c}'=2\mathbf{c}$) (85); [5] $P4/n$ ($\mathbf{a}'=\mathbf{a}+2\mathbf{b},\mathbf{b}'=-2\mathbf{a}+\mathbf{b}$ or $\mathbf{a}'=\mathbf{a}-2\mathbf{b},\mathbf{b}'=2\mathbf{a}+\mathbf{b}$) (85)

Minimal non-isomorphic supergroups

I [2] $P4/nbm$ (125); [2] $P4/nnc$ (126); [2] $P4/nmm$ (129); [2] $P4/ncc$ (130)
II [2] $C4/m$ ($P4/m$, 83); [2] $I4/m$ (87)

Figure 6.2 ***(Continued).***

shows the non-symmorphic space group $P4/n$ (C_{4h}^3), No.85, in which the origin is at the intersection of the $\bar{4}$ axis and the n-glide plane at a point distant −¼, ¼, 0 from the inversion center. On the other hand, Fig. 6.3 shows the same space group but with the origin on the center of inversion $\bar{1}$, which is on the n-glide plane, at a position ¼, −¼, 0 from the $\bar{4}$ axis.

6.3.5. Asymmetric Unit

The next line gives information on the **asymmetric unit**. An asymmetric unit of a space group is a region (volume V_A) of space which fills all space when all the symmetry operations of the space group are applied. It can be thought of rather in the same way as a unit cell, namely a region of space, which fills all space when the translation operations are applied. The difference is that the asymmetric unit is affected not just by the translation operations but also by the rotation, screw, reflection and glide operations.

The asymmetric unit, therefore, is smaller than or equal to a unit cell, but, nevertheless, it contains all the information that describes the entire crystal. The volume of the asymmetric unit is given simply by $V_A = V_{uc}/nh$, where V_{uc} is the volume of the unit cell, n is the number of lattice points per unit cell (1 for primitive, 2 for one-face-centered, 2 for body-centered, 3 for hexagonal centered, and 4 for all-face-centered), and h is the number of operations in the point group.

For example, in $P4/m$ the asymmetric unit is contained within a region defined by coordinates given by the inequalities $0 \leq x \leq ½$; $0 \leq y \leq ½$; $0 \leq z \leq ½$. For the point group $4/m(C_{4h})$, $h = 8$ and $n = 1$ because of the primitive cell. This means that the asymmetric unit has 1/8th of the volume of the unit cell, in agreement with these relationships.

In Section 6.1, we saw a list of atoms that specified the crystal structure of calcite using the symmetry of the space group $R3c(D_{3d}^6)$. Be aware that crystallographers sometimes (incorrectly) refer to such a list of *atomic* positions as the 'asymmetric unit'.

$P4/n$ C_{4h}^3 $4/m$ Tetragonal

No. 85 $P4/n$ Patterson symmetry $P4/m$

ORIGIN CHOICE 2

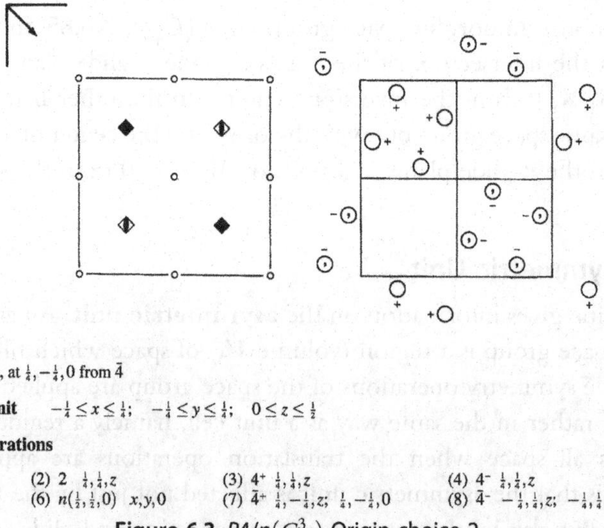

Origin at $\bar{1}$ on n, at $\frac{1}{4},-\frac{1}{4},0$ from $\bar{4}$

Asymmetric unit $-\frac{1}{4} \leq x \leq \frac{1}{4}$; $-\frac{1}{4} \leq y \leq \frac{1}{4}$; $0 \leq z \leq \frac{1}{2}$

Symmetry operations

(1) 1   (2) 2 $\frac{1}{4},\frac{1}{4},z$ (3) 4^+ $\frac{1}{4},\frac{1}{4},z$ (4) 4^- $\frac{1}{4},\frac{1}{4},z$

(5) $\bar{1}$ 0,0,0   (6) $n(\frac{1}{2},\frac{1}{2},0)$ $x,y,0$ (7) $\bar{4}^+$ $\frac{1}{4},-\frac{1}{4},z$; $\frac{1}{4},-\frac{1}{4},0$ (8) $\bar{4}^-$ $-\frac{1}{4},\frac{1}{4},z$; $-\frac{1}{4},\frac{1}{4},0$

Figure 6.3 $P4/n(C_{4h}^3)$ Origin choice 2.

Generators selected (1); $t(1,0,0)$; $t(0,1,0)$; $t(0,0,1)$; (2); (3); (5)

Positions

Multiplicity, Wyckoff letter, Site symmetry	Coordinates				Reflection conditions
					General:
8 g 1	(1) x,y,z (5) $\bar{x},\bar{y},\bar{z}$	(2) $\bar{x}+\frac{1}{2},\bar{y}+\frac{1}{2},z$ (6) $x+\frac{1}{2},y+\frac{1}{2},\bar{z}$	(3) $\bar{y}+\frac{1}{2},x,z$ (7) $y+\frac{1}{2},\bar{x},\bar{z}$	(4) $y,\bar{x}+\frac{1}{2},z$ (8) $\bar{y},x+\frac{1}{2},\bar{z}$	$hk0: h+k=2n$ $h00: h=2n$
					Special: as above, plus
4 f 2..	$\frac{1}{4},\frac{3}{4},z$	$\frac{3}{4},\frac{1}{4},z$	$\frac{3}{4},\frac{1}{4},\bar{z}$	$\frac{1}{4},\frac{3}{4},\bar{z}$	$hkl: h+k=2n$
4 e $\bar{1}$	$0,0,\frac{1}{2}$	$\frac{1}{2},\frac{1}{2},\frac{1}{2}$	$\frac{1}{2},0,\frac{1}{2}$	$0,\frac{1}{2},\frac{1}{2}$	$hkl: h,k=2n$
4 d $\bar{1}$	$0,0,0$	$\frac{1}{2},\frac{1}{2},0$	$\frac{1}{2},0,0$	$0,\frac{1}{2},0$	$hkl: h,k=2n$
2 c 4..	$\frac{1}{4},\frac{1}{4},z$	$\frac{3}{4},\frac{3}{4},\bar{z}$			no extra conditions
2 b $\bar{4}$..	$\frac{1}{4},\frac{3}{4},\frac{1}{2}$	$\frac{3}{4},\frac{1}{4},\frac{1}{2}$			$hkl: h+k=2n$
2 a $\bar{4}$..	$\frac{1}{4},\frac{3}{4},0$	$\frac{3}{4},\frac{1}{4},0$			$hkl: h+k=2n$

Symmetry of special projections

Along [001] $p4$
$\mathbf{a}'=\frac{1}{2}(\mathbf{a}-\mathbf{b})$ $\mathbf{b}'=\frac{1}{2}(\mathbf{a}+\mathbf{b})$
Origin at $\frac{1}{4},\frac{1}{4},z$

Along [100] $p2mg$
$\mathbf{a}'=\mathbf{b}$ $\mathbf{b}'=\mathbf{c}$
Origin at $x,0,0$

Along [110] $p2mm$
$\mathbf{a}'=\frac{1}{2}(-\mathbf{a}+\mathbf{b})$ $\mathbf{b}'=\mathbf{c}$
Origin at $x,x,0$

Maximal non-isomorphic subgroups

I	[2] $P\bar{4}$ (81)	1; 2; 7; 8
	[2] $P4$ (75)	1; 2; 3; 4
	[2] $P2/n$ ($P2/c$, 13)	1; 2; 5; 6
IIa	none	
IIb	[2] $P4_2/n$ ($\mathbf{c}'=2\mathbf{c}$) (86)	

Maximal isomorphic subgroups of lowest index

IIc [2] $P4/n$ ($\mathbf{c}'=2\mathbf{c}$) (85); [5] $P4/n$ ($\mathbf{a}'=\mathbf{a}+2\mathbf{b},\mathbf{b}'=-2\mathbf{a}+\mathbf{b}$ or $\mathbf{a}'=\mathbf{a}-2\mathbf{b},\mathbf{b}'=2\mathbf{a}+\mathbf{b}$) (85)

Minimal non-isomorphic supergroups

I [2] $P4/nbm$ (125); [2] $P4/nnc$ (126); [2] $P4/nmm$ (129); [2] $P4/ncc$ (130)

II [2] $C4/m$ ($P4/m$, 83); [2] $I4/m$ (87)

Figure 6.3 (*Continued*).

6.3.6. Symmetry Operations

This section is a most welcome addition to the tables. In the previous versions of the International Tables the coordinates of the general equivalent positions were listed, but without any indication as to which symmetry operations produced them. In the *ITA*, the symmetry operations are listed (and numbered) in the form of coset representatives of the translation subgroup T in the representative space group G in addition to the general equivalent positions (which are correspondingly numbered).

For $P4/m$, the symmetry operations are given with respect to the translations (0, 0, 0) of the primitive lattice. Each symmetry operation is given a number, in this case from 1 to 8, and these same numbers are also written next to the corresponding general equivalent position that they generate. For example, the coordinate ($\bar{y}$, x, z) is obtained from (x, y, z) by symmetry operation number (3), which is denoted 4^+ 0, 0, z. The 4 means a fourfold rotation and the superscript + denotes an anticlockwise rotation. Thus, 4^+ is equivalent to 4^1, and 4^- is equivalent to 4^3. This symmetry operation acts about an axis specified by the coordinate triplet (0, 0, *z*). For rotation-inversion operations the location of the inversion point is also given; for example, symmetry operator number (8) is seen to denote clockwise rotation about the same axis with its inversion point at (0, 0, 0). In Fig. 6.3 symmetry operator (8) acts about the line −¼, ¼, *z* at a distance of −¼, ¼, 0 from the inversion center. Thus, symmetry operation number (8) takes the point (*x*, *y*, *z*) to the point ($\bar{y}$, *x*+½, $\bar{z}$) and this general equivalent point will then be numbered (8). As another example, consider the reflection operation numbered (6) in space group *P*4/*m* (Fig. 6.1); the location of the mirror plane is given by *x*, *y*, 0 and takes the point at (*x*, *y*, *z*) to one at (*x*, *y*, $\bar{z}$), number (6) in the list of general equivalent positions. In the case of an *n*-glide (Fig. 6.2), operation number (6) is on the plane denoted by *x*, *y*, 0, and takes the point (*x*, *y*, *z*) to one at (*x*+½, *y*+½, $\bar{z}$).

Recently, there has been some discussion about the question of also supplying the symmetry operators in Seitz notation, and so this may be included in a future revision of the *ITA*.

6.3.7. Generators Selected

The generators of any group are a set of symmetry operators of that group which when successively multiplied yield *all* of the operators of the group. In general, the set of generators is arbitrary. At the top of the second page of our examples from the *ITA* can be seen a list under the heading 'Generators selected'. This gives a list of the symmetry operations selected that can generate all of the symmetry operations of the space group.

The first generator in *P*4/*m* is (1), the identity operation, and this is followed by, in this case, translations **t**(1, 0, 0), **t**(0, 1, 0) and **t**(0, 0, 1) along the **a**, **b** and **c**-axes. Following this, we see that symmetry operations (2), (3) and (5) are included in the set of generators, so that when combined with each other and with the translations, all of the symmetry operations of the space group can be obtained. Hence, by successive application of these generators to the position (*x*, *y*, *z*) all 8 general equivalent positions can be obtained. Thus, generators (1) just give the translations. Generator (2) gives position (2) ($\bar{x}$, $\bar{y}$, *z*). Generator (3) gives position (3) ($\bar{y}$, *x*, z) and generator (3) applied three times gives position (4) (*y*, $\bar{x}$, *z*). Generator (5) gives position (5) ($\bar{x}$, $\bar{y}$, $\bar{z}$) and generator (3) applied twice followed by generator (5) gives position (6) (*x*, *y*, $\bar{z}$), and so on.

6.3.8. Positions, Multiplicity, Wyckoff Letter and Site Symmetry

The information given under these four headings is closely related and so we include it in one subsection. Each type of possible site is given a letter, starting with '*a*' for a site of the highest site symmetry (however note, several sites can have equally high site symmetry), and sequentially going through the alphabet until the lowest site symmetry $1(C_1)$ is reached (that of a general position). The letters associated with the sites are called the **Wyckoff[3] notation** and are normally listed down the page in reverse order. The earlier example of calcite given in Section 6.1 shows how crystallographers usually make use of the Wyckoff notation in describing the site information for a structure. The site symmetry of a general position is, of necessity, $1(C_1)$, since by definition it does not lie on any symmetry element. Atoms that are situated at points other than those that have $1(C_1)$ symmetry are said to be at **special positions.**

The infinite set of symmetry-related points is known as a **crystallographic orbit**. If the coordinates of the points are completely fixed by symmetry, for example (0, 0, 0) or (½, ½, ½), then the orbit is the same as the Wyckoff position. On the other hand, if there is a free coordinate, for example (*x*, ¼, 0), the Wyckoff position consists of an infinite number of possible orbits. Thus, in this example one orbit could be, say, (0.15, ¼, 0), while another orbit in the same Wyckoff position could be, for example, (0.24, ¼, 0). Along with the Wyckoff symbol is the **multiplicity**, namely the number of **symmetry-related positions** that the particular site has in each space group.

In the space group $P4/m(C_{4h}^1)$, shown in Fig. 6.1, there is only one possible position for each of the sites labelled *a*, *b*, *c* and *d*; there are two possible positions for sites labelled *e*, *f*, *g* and *h*; four for *i*, *j* and *k* and eight possible positions for sites labelled *l*. For each type of site, the symmetry operations $\{R|\boldsymbol{\tau}\}$ of the space group permute the various coordinates among themselves. All of these coordinates are explicitly written out for each type of site. In $P4/m$, for example, the site 8*l*, has eight coordinates that describe the eight general equivalent positions in the unit cell of this space group. These eight positions are, of course, related to one another by the symmetry operations $\{R|\boldsymbol{\tau}\}$ of the space group, as you should check for yourself using Fig. 6.1. Under all these symmetry operations, the positions described as 1*a*, 1*b*, 1*c* and 1*d* each transform into themselves. For the 2*e* site

[3] After Ralph Walter Graystone Wyckoff, Sr. (1897–1994).

there are two points (0, ½, 0) and (½, 0, 0) that transform between each other, for the 4*i* site, four points, and so on for the other sites.

For each position the **site symmetry** is the point symmetry of the crystal when viewed from that particular position; in other words, with the particular position taken as an origin. It tells us under which operations the crystal will leave this point fixed. This set of point symmetry operations always forms one of the 32 crystallographic point groups and is a subgroup of the point group of the space group. This group is sometimes also known by physicists and mathematicians as the **stabilizer, isotropy group** or **little group**. Thus, for example, any atom placed at the special position 2*e* in *P*4/*m* sits at a place in the crystal that is on a site whose point symmetry is denoted by 2/*m*. Note the dots in this symbol: these define **oriented site symmetries**. Sets of equivalent symmetry directions that do not contribute any operation to the site-symmetry group are represented by a dot. For example, the Wyckoff site 8*j* in *I*4/*mmm* has the oriented site-symmetry denoted by '*m* 2*m* .' (note the spaces and the dot) signifying that this site has point-group symmetry *m*2*m*. To understand the meaning of the dot, we must note its position in the symbol. The order of the symbols is *m*, followed by 2*m*, followed by the dot. Recall that in the tetragonal system the order of the symbols is with respect to (1) the principal direction (**c**-axis), (2) the **a** and **b**-axes, and (3) the diagonal directions **a** + **b** and **a** − **b**, in that order. As the space group here is tetragonal, the oriented site-symmetry symbol shows us how the *m*2*m* operations are oriented with respect to the tetragonal axes. Thus, the first *m* refers to a mirror plane perpendicular to the **c**-axis. Then, the 2*m*, in second place, means that the 2-fold axis lies along **a** and **b**, that is [100] and [010], while the accompanying mirror plane is perpendicular to these directions. The dot in the third position, therefore, signifies that there are no mirrors or rotation axes associated with the diagonal directions **a** + **b** and **a** − **b**, that is [110] and $[1\bar{1}0]$.

In non-symmorphic space groups, operations with non-zero τ are omitted when determining site symmetry. For instance, while the positions (0, 0, 0) and (0, 0, ½) in the space group *P*4/*m* have site symmetry 4/*m*.., in the non-symmorphic space group *P*4/*n* these same positions only have $\bar{4}$.. site symmetry.

For the multiply-primitive unit cells used for the centered *I*, *F*, *C* (*A* or *B*) Bravais lattices, there are, respectively, 2, 4 and 2 times as many positions listed as there are in the primitive unit cell of each of these lattices. In *I*4/*m*, for example, the 'coordinates of equivalent points' are obtained by taking the coordinates as listed in *P*4/*m* and adding to each (½, ½, ½). This operation is indicated on the page in a shorthand fashion by '(0, 0, 0) + (½, ½, ½)', which is precisely the centering condition for a body-centered lattice. Figure 6.4 shows the example of space group $I4_1/a(C_{4h}^6)$. For a space group with an F-centered lattice, the expression is '(0, 0, 0) + (0, ½, ½) + (½, 0, ½) + (½, ½, 0)', with a similar meaning. For a C-centered lattice, '(0, 0, 0) + (½, ½, 0)' appears. This enables fewer coordinates to be written out explicitly on the page.

Note that for a crystal to belong to a specific space group it is not necessary for the atoms to be at all the different Wyckoff positions, *a*, *b*, *c*, ... For example, a crystal may belong to space group $P4/m(C_{4h}^1)$ and yet have atoms only on the 1*a* and 8*l* positions. A different crystal can have the same space group and yet have atoms on all 1*a*, 1*b*, ..., 8*l* positions or just on the 8*l* positions.

Another point to note is that having placed an atom at a certain site, symmetry demands symmetry-related atoms of an identical type. Thus, if an

$I4_1/a$ C_{4h}^6 $4/m$ Tetragonal

No. 88 $I4_1/a$ Patterson symmetry $I4/m$

ORIGIN CHOICE 1

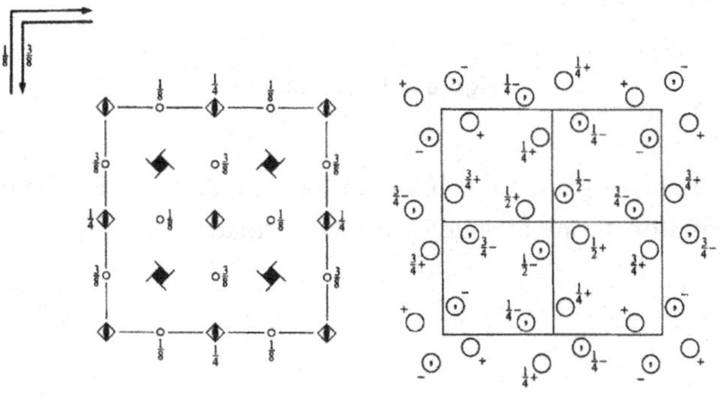

Origin at $\bar{4}$, at $0, -\frac{1}{4}, -\frac{1}{8}$ from $\bar{1}$

Asymmetric unit $0 \le x \le \frac{1}{4}$; $0 \le y \le \frac{1}{4}$; $0 \le z \le 1$

Symmetry operations

For (0,0,0)+ set

(1) 1	(2) $2(0,0,\frac{1}{2})$ $\frac{1}{4},\frac{1}{4},z$	(3) $4^+(0,0,\frac{1}{4})$ $-\frac{1}{4},\frac{1}{4},z$	(4) $4^-(0,0,\frac{3}{4})$ $\frac{1}{4},-\frac{1}{4},z$
(5) $\bar{1}$ $0,\frac{1}{4},\frac{1}{8}$	(6) a $x,y,\frac{3}{8}$	(7) $\bar{4}^+$ $0,0,z$; $0,0,0$	(8) $\bar{4}^-$ $0,\frac{1}{2},z$; $0,\frac{1}{2},\frac{1}{4}$

For $(\frac{1}{2},\frac{1}{2},\frac{1}{2})$+ set

(1) $t(\frac{1}{2},\frac{1}{2},\frac{1}{2})$	(2) 2 $0,0,z$	(3) $4^+(0,0,\frac{3}{4})$ $\frac{1}{4},\frac{1}{4},z$	(4) $4^-(0,0,\frac{1}{4})$ $\frac{1}{4},\frac{1}{4},z$
(5) $\bar{1}$ $\frac{1}{4},0,\frac{3}{8}$	(6) b $x,y,\frac{1}{8}$	(7) $\bar{4}^+$ $\frac{1}{2},0,z$; $\frac{1}{2},0,\frac{1}{4}$	(8) $\bar{4}^-$ $0,0,z$; $0,0,0$

Figure 6.4 $I4_1/a(C_{4h}^6)$.

Generators selected (1); $t(1,0,0)$; $t(0,1,0)$; $t(0,0,1)$; $t(\frac{1}{2},\frac{1}{2},\frac{1}{2})$; (2); (3); (5)

Positions

Multiplicity, Wyckoff letter, Site symmetry			Coordinates				Reflection conditions
			$(0,0,0)+\ \ (\frac{1}{2},\frac{1}{2},\frac{1}{2})+$				General:
16	*f*	1	(1) x,y,z	(2) $\bar{x}+\frac{1}{2},\bar{y}+\frac{1}{2},z+\frac{1}{2}$	(3) $\bar{y},x+\frac{1}{2},z+\frac{1}{4}$	(4) $y+\frac{1}{2},\bar{x},z+\frac{3}{4}$	hkl : $h+k+l=2n$
			(5) $\bar{x},\bar{y}+\frac{1}{2},\bar{z}+\frac{1}{4}$	(6) $x+\frac{1}{2},y,\bar{z}+\frac{3}{4}$	(7) $y,\bar{x},\bar{z}$	(8) $\bar{y}+\frac{1}{2},x+\frac{1}{2},\bar{z}+\frac{1}{2}$	$hk0$: $h,k=2n$
							$0kl$: $k+l=2n$
							hhl : $l=2n$
							$00l$: $l=4n$
							$h00$: $h=2n$
							$h\bar{h}0$: $h=2n$
							Special: as above, plus
8	*e*	2..	$0,0,z$	$0,\frac{1}{2},z+\frac{1}{4}$	$0,\frac{1}{2},\bar{z}+\frac{1}{4}$	$0,0,\bar{z}$	hkl : $l=2n+1$ or $2h+l=4n$
8	*d*	$\bar{1}$	$0,\frac{1}{4},\frac{5}{8}$	$\frac{1}{2},\frac{1}{4},\frac{1}{8}$	$\frac{3}{4},\frac{1}{2},\frac{7}{8}$	$\frac{3}{4},0,\frac{3}{8}$ }	hkl : $l=2n+1$ or $h,k=2n$, $h+k+l=4n$
8	*c*	$\bar{1}$	$0,\frac{1}{4},\frac{1}{8}$	$\frac{1}{2},\frac{1}{4},\frac{5}{8}$	$\frac{3}{4},\frac{1}{2},\frac{3}{8}$	$\frac{3}{4},0,\frac{7}{8}$ }	
4	*b*	$\bar{4}$..	$0,0,\frac{1}{2}$	$0,\frac{1}{2},\frac{3}{4}$ }			hkl : $l=2n+1$ or $2h+l=4n$
4	*a*	$\bar{4}$..	$0,0,0$	$0,\frac{1}{2},\frac{1}{4}$ }			

Symmetry of special projections

Along [001] $p4$	Along [100] $c2mm$	Along [110] $p2mg$
$\mathbf{a}'=\frac{1}{2}\mathbf{a}$ $\mathbf{b}'=\frac{1}{2}\mathbf{b}$	$\mathbf{a}'=\mathbf{b}$ $\mathbf{b}'=\mathbf{c}$	$\mathbf{a}'=\frac{1}{2}(-\mathbf{a}+\mathbf{b})$ $\mathbf{b}'=\frac{1}{2}\mathbf{c}$
Origin at $0,0,z$	Origin at $x,0,\frac{3}{8}$	Origin at $x,x+\frac{1}{4},\frac{1}{8}$

Maximal non-isomorphic subgroups

I [2] $I\bar{4}$ (82) (1; 2; 7; 8)+
[2] $I4_1$ (80) (1; 2; 3; 4)+
[2] $I2/a$ ($C2/c$, 15) (1; 2; 5; 6)+

IIa none

IIb none

Maximal isomorphic subgroups of lowest index

IIc [3] $I4_1/a$ ($\mathbf{c}'=3\mathbf{c}$) (88); [5] $I4_1/a$ ($\mathbf{a}'=\mathbf{a}+2\mathbf{b},\mathbf{b}'=-2\mathbf{a}+\mathbf{b}$ or $\mathbf{a}'=\mathbf{a}-2\mathbf{b},\mathbf{b}'=2\mathbf{a}+\mathbf{b}$) (88)

Minimal non-isomorphic supergroups

I [2] $I4_1/amd$ (141); [2] $I4_1/acd$ (142)

II [2] $C4_2/a$ ($\mathbf{c}'=\frac{1}{2}\mathbf{c}$) ($P4_2/n$, 86)

Figure 6.4 (*Continued*).

oxygen atom, say, is placed at a $4k$ site in $P4/m$, then there will be necessarily four oxygen atoms, one at each of the coordinate positions

$$x,\ y,\ \tfrac{1}{2} \qquad \bar{x},\ \bar{y},\ \tfrac{1}{2} \qquad \bar{y},\ x,\ \tfrac{1}{2} \qquad y,\ \bar{x},\ \tfrac{1}{2}$$

related by the mirror symmetry. Perhaps we should stress this by saying that under all the symmetry operations of the space group, the eight positions, labelled, say, $8l$, transform among themselves, and so *all the positions must be occupied by the same kind of atom.* However, it is possible that there could be eight atoms of one kind in the $8l$ positions and eight atoms of another kind also in the $8l$ positions. The values of (x, y, z) will differ for the two sets of atoms since they cannot physically be at the same places. There may, in fact, be many sets of atoms with this same type of position. It is only possible to have more than one kind of atom in the same type of position when there is at least one freely adjustable parameter in the coordinates, that is a parameter not fixed by symmetry. For example, in $P4/m(C_{4h}^1)$ positions *a–f* are completely fixed by the symmetry, whereas positions *g–l* have adjustable parameters.

The list of positions and Wyckoff symbols can be used to determine whether a space group is symmorphic or not. We have already seen that the International standard short symbol, the symbol that appears on the top right or left of the page in the *ITA*, has all the information necessary to determine if the space group is symmorphic or non-symmorphic. In this symbol, if there are no screw axes or glide planes marked, then the space group is symmorphic. If any of these appear, then the space group is non-symmorphic. However, care must be taken in using this criterion since the International notation changes according to the setting chosen for the unit-cell axes. There is another simple method to detect if the space group is symmorphic from the special positions. *A symmorphic space group must have at least one point in the unit cell that has a site symmetry that is the same as the point group of the space group.* By looking at the positions in the space groups in Figs 6.1, 6.2, and 6.4, we immediately see that for space group $P4/m(C_{4h}^1)$ there are several positions that have site symmetry $4/m$. However, for space groups $P4/n(C_{4h}^3)$ and $I4_1/a(C_{4h}^6)$ there are no positions with site symmetry $4/m$. Thus, these latter space groups must be non-symmorphic.

Consider primitive unit cells for the moment. In a symmorphic space group, the 'number of positions' for positions that have site symmetry equal to the point group of the space group must always be equal to one, for example, $1a$, $1b$, $1c$ or $1d$ for $P4/m$. Clearly, this is because this position must always transform into itself under all the point group and space group operations (remember we can always move points by a primitive lattice translation $\{1|\mathbf{t}_n\}$ and all points related by this translation we call the same or equivalent points). For a non-symmorphic space group, all positions must have at least two for the 'number of positions' because there is no position in the unit cell that remains stationary under all the symmetry operations of the space group. Thus, in $P4/n$ we see that for the site with highest symmetry ($\bar{4}$) there are two equivalent positions. Generally, for a given set of equivalent positions, the number of equivalent positions times the order of the site-symmetry point group is equal to the order of the point group of the space group. All these statements clearly apply to any Bravais lattice when working with a primitive cell instead of the centered cell. Alternatively, the above statements apply if for an *I*, *F* or *C* (or *A* or *B*) Bravais lattice, the 'number of positions' listed in the *ITA* is divided by 2, 4 or 2. This can be seen in Fig. 6.4 for space group $I4_1/a$.

6.3.9. Reflection Conditions

On the right-hand side of each page in the *ITA* a list of conditions that limit the possible *hkl* (diffraction) reflections is given. These conditions are of interest to crystallographers, since they help in the determination of the space group from the diffraction pattern. They are also relevant in lattice-potential calculations of electronic band structures. For instance, in Fig. 6.4 the statement *hkl:* $h+k+l=2n$ means that any reflections whose indices disobey this equation must be absent: this particular condition arises from the body-centering of the lattice. Similarly, *00l:* $l=4n$ means that reflections of the form *00l* are allowed provided that $l=4n$: this is a result of having a 4_1 axis.

6.3.10. Symmetry of Special Projections

If a three-dimensional space group is projected onto a plane, the result is one of the 17 two-dimensional plane groups (Section 5.7). The *ITA* gives the resulting plane group for certain special projections. For example, if space group $I4_1/a(C_{4h}^6)$ is projected down [001], that is down the **c**-axis, the result (Fig. 6.4) is the plane group *p*4. This redefines the axes to new axes $\mathbf{a}'$, $\mathbf{b}'$ so that

$$\mathbf{a}' = \tfrac{1}{2}\mathbf{a} \quad \mathbf{b}' = \tfrac{1}{2}\mathbf{b} \qquad [6.1]$$

with the origin set at (0, 0, *z*). When projected along [100], the **a**-axis, the plane group becomes *c*2*mm* with the origin at (*x*, 0, ⅜) and the axes defined by

$$\mathbf{a}' = \mathbf{b} \quad \mathbf{b}' = \mathbf{c} \qquad [6.2]$$

Similarly, when projected along [110] the projection has the symmetry of plane group *p*2*mg* and

$$\mathbf{a}' = \tfrac{1}{2}(-\mathbf{a}+\mathbf{b}) \quad \mathbf{b}' = \tfrac{1}{2}\mathbf{c} \qquad [6.3]$$

with the origin at $(x, x+\tfrac{1}{4}, \tfrac{1}{8})$.

6.4. SUBGROUPS AND SUPERGROUPS[4]

The *ITA* contains a wealth of important information on subgroups and supergroups of space groups. This feature will be of significant use for physicists and chemists, especially those interested in phase transformations. The *ITA* classifies subgroups and supergroups according to specific rules.

[4] In future editions of the *ITA*, this information will be removed and included in International Tables for Crystallography Volume *A*1 (*ITA1*) on 'Symmetry relations between space groups'.

6.4.1. Maximal Subgroups

Suppose G is a subgroup of a parent group G_0, that is the elements of G are all contained within G_0. G is called a **proper subgroup** if G_0 contains elements that are not in G. If there is no other proper subgroup, say H, such that G is a proper subgroup of H, G is then termed a **maximal subgroup**. In other words, G is a maximal subgroup if there is no proper subgroup of G_0 which 'lies between' G and G_0. There are two types of maximal subgroups defined in crystallographic space groups:

I translationengleiche or **t-subgroups**

II klassengleiche or **k-subgroups**

If the maximal subgroup is formed from the parent group by retaining the translational elements, but the order of the point group is reduced, the subgroup is a t-subgroup or **equi-translational** subgroup. If it is formed by preserving the point group, but with a loss of translations, it is a k-subgroup or **equi-class** subgroup. The k-subgroups are further subdivided as follows:

IIa in which the conventional unit cells of G and G_0 are the same.

IIb in which the conventional unit cell of G is larger than that of G_0.

IIc in which G and G_0 are isomorphic (i.e. they denote the same space group).

Note that subgroups of types I, IIa and IIb are maximal **non-isomorphic** subgroups. Consider the *I*4/*mmm* space group (Table 6.2).

6.4.1.1. t-Subgroups

Just below the list of special projections, we see seven maximal non-isomorphic subgroups of type I. For example, the fourth one to appear is written '[2] *I*422'. This is a t-subgroup since it is formed from *I*4/*mmm* by a reduction from point group 4/*mmm* to 422. The latter has half the symmetry operations of 4/*mmm*, and this is indicated by [2] placed before the symbol. Note that the *I*-centering is retained so that there is no loss of translational symmetry in going from *I*4/*mmm* to *I*422, and therefore *I*422 is a t-subgroup of *I*4/*mmm*. Since there is no subgroup of *I*4/*mmm* for which *I*422 is a subgroup, then *I*422 is a **maximal subgroup**. The *ITA* also lists the symmetry operations of the parent group that are retained: in this case they are numbers 1, 2, 3, 4, 5, 6, 7 and 8 (plus the body-centering operations) taken from the list of symmetry operations. Figures 6.5a and b are from the *ITA* and they show the symmetry operations for *I*4/*mmm* and for *I*422. It is clear that all of the symmetry operations of *I*422 are contained within *I*4/*mmm*, the loss of the mirror planes in *I*422 being particularly obvious.

Table 6.2 Maximal subgroups and minimal non-isomorphic supergroups for space group $I4/mmm(D_{4h}^{17})$

Maximal non-isomorphic subgroups

I

[2] $I\bar{4}2m$ (121)	(1; 2; 5; 6; 11; 12; 15; 16)+
[2] $I\bar{4}m2$ (119)	(1; 2; 7; 8; 11; 12; 13; 14)+
[2] $I4mm$ (107)	(1; 2; 3; 4; 13; 14; 15; 16)+
[2] $I422$ (97)	(1; 2; 3; 4; 5; 6; 7; 8)+
[2] $I4/m11$ ($I4/m$, 87)	(1; 2; 3; 4; 9; 10; 11; 12)+
[2] $I2/m2/m1$($Immm$, 71)	(1; 2; 5; 6; 9; 10; 13; 14)+
[2] $I2/m12/m$($Fmmm$, 69)	(1; 2; 7; 8; 9; 10; 15; 16)+

IIa

[2] $P4_2/nmc$ (137)	1;2;7;8;11;12;13;14;(3;4;5;6;9;10;15;16)+(½, ½,½)
[2] $P4_2/mnm$ (136)	1;2;7;8;9;10;15;16;(3;4;5;6;11;12;13;14)+(½, ½,½)
[2] $P4_2/nnm$ (134)	1;2;5;6;11;12;15;16;(3;4;7;8;9;10;13;14)+(½, ½,½)
[2] $P4_2/mmc$ (131)	1;2;5;6;9;10;13;14;(3;4;7;8;11;12;15;16)+(½, ½,½)
[2] $P4_2/nnm$ (129)	1;2;3;4;13;14;15;16;(5;6;7;8;9;10;11;12)+(½, ½,½)
[2] $P4_2/mnc$ (128)	1;2;3;4;9;10;11;12;(5;6;7;8;13;14;15;16)+(½, ½,½)
[2] $P4_2/nnc$ (126)	1;2;3;4;5;6;7;8;(9;10;11;12;13;14;15;16)+(½, ½,½)
[2] $P4/mmm$ (123)	1;2;3;4;5;6;7;8;9;10;11;12;13;14;15;16

IIb

none

Maximal isomorphic subgroups of lowest index

IIc

[3] $I4/mmm$($\mathbf{c}'=3\mathbf{c}$) (139); [9] $I4/mmm$($\mathbf{a}'=3\mathbf{a}$, $\mathbf{b}'=3\mathbf{b}$) (139)

Minimal non-isomorphic supergroups

I

[3]$Fm\bar{3}\,m$ (225); [3]$Im\bar{3}\,m$ (229)

II

[2]$C4/mmm$($\mathbf{c}'=½\mathbf{c}$) ($P4/mmm$, 123)

Consider another type I subgroup. The symbol '[2]$I\ 2/m\ 2/m\ 1$' is written with reference to the coordinate system of G_0 (i.e. $I4/mmm$). Once again, in the tetragonal system, the first symbol after the lattice type refers to the **c**-axis direction, the second to the **a** and **b**-axes directions, and the third to the diagonal directions $\mathbf{a}+\mathbf{b}$ and $\mathbf{a}-\mathbf{b}$ (Appendix 4). Thus, the first $2/m$ means that the subgroup has its 2-fold axis along [001] of the $I4/mmm$ group and its mirror plane perpendicular. The second $2/m$,

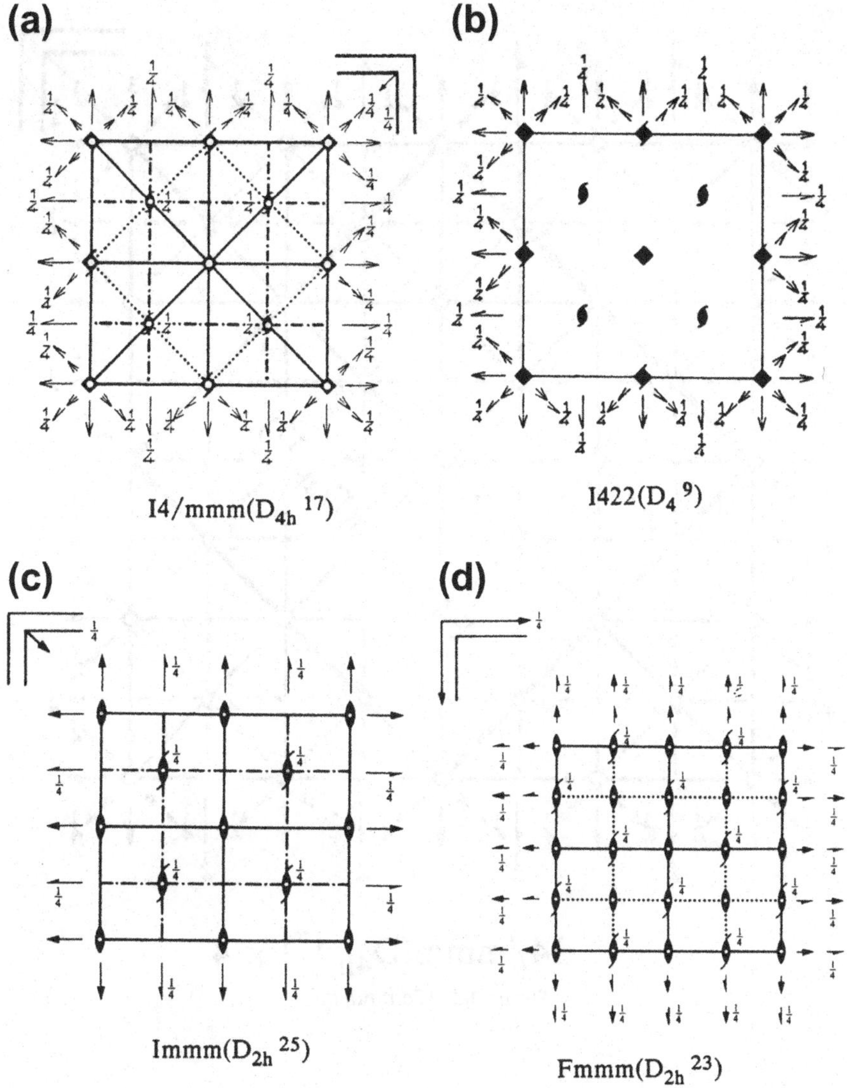

Figure 6.5 (a)–(d) The symmetry elements of *I4/mmm* and some of its subgroups. (e) The same as in (a), but here four cells are shown; see the text for further discussion.

(e)

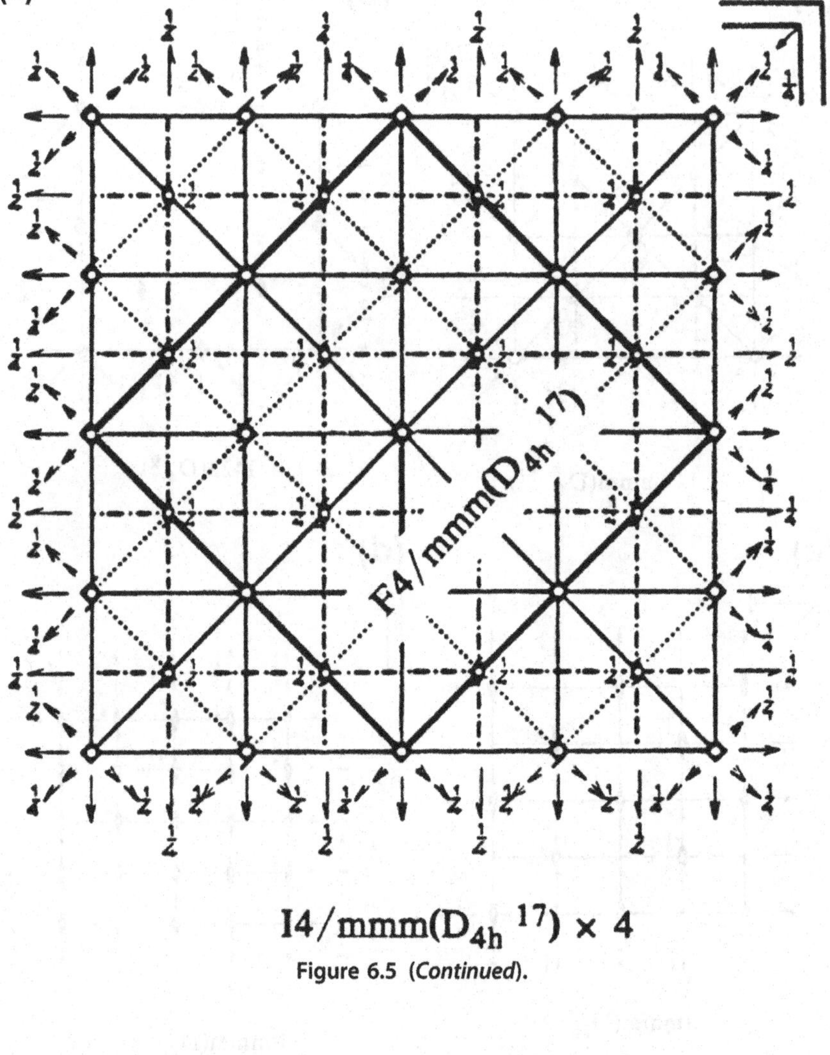

$I4/mmm(D_{4h}{}^{17}) \times 4$

Figure 6.5 (*Continued*).

because it is in second place in a symbol referring to a tetragonal parent group, means that 2-fold axes are directed along [100] and [010] of the *I*4/*mmm* group with associated mirror planes perpendicular. Finally, the one in the third place indicates that there are no further symmetry operations referring to the tetragonal diagonal directions. The symbol *I* 2/*m* 2/*m* 1 is not a conventional space group symbol, since it is specified with respect to the tetragonal system. Because there are two perpendicular mirror planes plus a center of symmetry (operation 9) retained, this can then be called by the conventional orthorhombic space group symbol *Immm* (recall that in the orthorhombic system the order of the symbols refers to the **a**, **b** and **c**-axes, and so '*mmm*' means mirror-planes perpendicular to these axes). Again, the translational operations have been retained and so this is a t-subgroup. Figure 6.5c shows the symmetry operations for *Immm* and those of the parent group *I*4/*mmm*. From Fig. 6.5c, the meaning of the '1' in the third place should be obvious, since the diagonal mirror planes are now missing.

On the other hand, the last subgroup on this page has the symbol '[2] *I* 2/*m* 1 2/*m*'. There are important differences between this subgroup and the one discussed before. First, there are no longer 2-fold operations about the **a** and **b**-axes, because the '1' is in the second position. Second, there are 2-fold operations about axes directed along [110] and $[1\bar{1}0]$ of the *I*4/*mmm* unit cell (with mirror planes perpendicular), as indicated by the '2/*m*' in the third position. This corresponds to another orthorhombic space group *Fmmm* (Fig. 6.5d) whose axes are directed along [001], [110] and $[1\bar{1}0]$ of the tetragonal cell. Although the resulting lattice is *F*-centered, it has not been formed by loss of translational operations of the tetragonal space group, but rather through a redefinition of the orientation of unit cell axes. Therefore, this too is a t-subgroup. Figure 6.5e shows the symmetry operations for $I4/mmm(D_{4h}^{17})$: but here 4 unit cells are drawn, and the heavy lines at 45° outline the face-centered cell which would be called $F4/mmm(D_{4h}^{17})$. The symmetry operations for $F4/mmm(D_{4h}^{17})$, shown in Fig. 6.5e, can be compared with those for $Fmmm(D_{2h}^{23})$, shown in Fig. 6.5d (turned through 45°). These two diagrams indicate which symmetry operations are lost (one half of them) and which are retained. The latter are also given in the *ITA*.

6.4.1.2. k-Subgroups

Turning now to the type IIa subgroups, we see that the last one in the list is *P*4/*mmm*. This is obtained from *I*4/*mmm* by losing the body-centering symmetry operations. Note that the list of operations specified after the symbol contains all of the first 16 operations before centering is added. Thus, the point group is still 4/*mmm*, and so this subgroup is a k-subgroup. The unit cell is the same size as in the *I*4/*mmm* group and so this is type IIa klassengleiche. Notice that another IIa subgroup, $P4_2/nnc$, is obtained by keeping operations 1, 2, 3, 4, 5, 6, 7 and 8 plus the operations formed by adding (½, ½, ½) to operations 9, 10, 11, 12, 13, 14, 15 and 16. Thus, operations formed by adding (½, ½, ½) to operations 1, 2, 3, 4, 5, 6, 7 and 8 have been eliminated.

Pmna	D_{2h}^{7}	*mmm*	Orthorhombic
No. 53	$P\ 2/m\ 2/n\ 2_1/a$		Patterson symmetry *Pmmm*

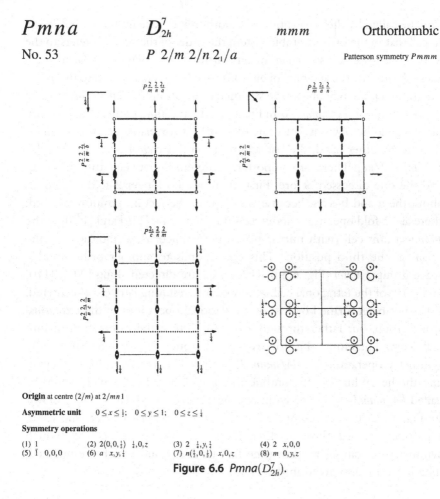

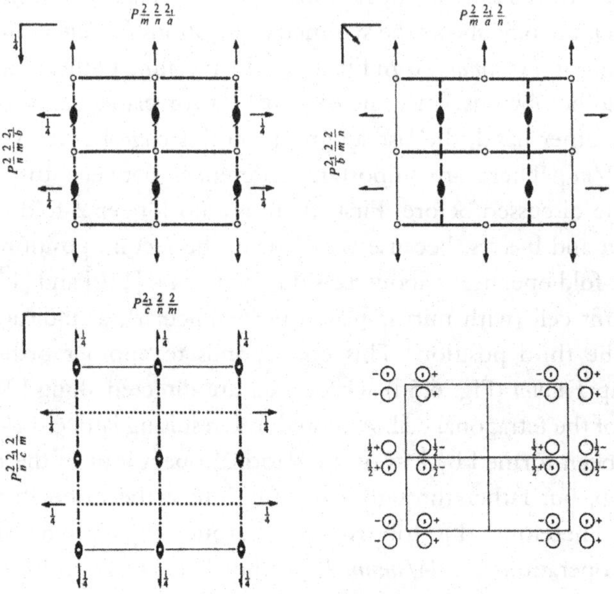

Origin at centre $(2/m)$ at $2/mn1$

Asymmetric unit $0 \le x \le \frac{1}{2}$; $0 \le y \le 1$; $0 \le z \le \frac{1}{4}$

Symmetry operations

(1) 1	(2) $2(0,0,\frac{1}{2})$ $\frac{1}{4},0,z$	(3) 2 $\frac{1}{4},y,\frac{1}{4}$	(4) 2 $x,0,0$
(5) $\bar{1}$ 0,0,0	(6) a $x,y,\frac{1}{4}$	(7) $n(\frac{1}{2},0,\frac{1}{2})$ $x,0,z$	(8) m $0,y,z$

Figure 6.6 *Pmna*(D_{2h}^{7}).

For the *I4/mmm* space group, there are no examples of type IIb subgroups. An example of one can be found in *Pmna*(D_{2h}^{7}), No. 53. For this case (Fig. 6.6), there are no type IIa subgroups. However, there are three type IIb k-subgroups. For instance, we see [2]*Pmnn* ($\mathbf{b}' = 2\mathbf{b}$)(*Pnnm*). The meaning of this is that by eliminating every other translational operation along the **b**-axis repeat of the unit cell of *Pmna*, the space group *Pmnn* is obtained. Clearly, the unit cell repeat along **b** is now twice as large as in the parent cell, although the point group (*mmm*) is unchanged. This is therefore a k-subgroup of type IIb. The symbol in parentheses, *Pnnm*, is the conventional short symbol for this space group.

Generators selected (1); $t(1,0,0)$; $t(0,1,0)$; $t(0,0,1)$; (2); (3); (5)

Positions

Multiplicity, Wyckoff letter, Site symmetry			Coordinates				Reflection conditions
							General:
8	i	1	(1) x,y,z (5) $\bar{x},\bar{y},\bar{z}$	(2) $\bar{x}+\frac{1}{2},\bar{y},z+\frac{1}{2}$ (6) $x+\frac{1}{2},y,\bar{z}+\frac{1}{2}$	(3) $\bar{x}+\frac{1}{2},y,\bar{z}+\frac{1}{2}$ (7) $x+\frac{1}{2},\bar{y},z+\frac{1}{2}$	(4) $x,\bar{y},\bar{z}$ (8) $\bar{x},y,z$	$h0l$: $h+l=2n$ $hk0$: $h=2n$ $h00$: $h=2n$ $00l$: $l=2n$
							Special: as above, plus
4	h	$m\,.\,.$	$0,y,z$	$\frac{1}{2},\bar{y},z+\frac{1}{2}$	$\frac{1}{2},y,\bar{z}+\frac{1}{2}$	$0,\bar{y},\bar{z}$	no extra conditions
4	g	$.\,2\,.$	$\frac{1}{4},y,\frac{1}{4}$	$\frac{1}{4},\bar{y},\frac{3}{4}$	$\frac{3}{4},\bar{y},\frac{3}{4}$	$\frac{3}{4},y,\frac{1}{4}$	hkl : $h=2n$
4	f	$2\,.\,.$	$x,\frac{1}{2},0$	$\bar{x}+\frac{1}{2},\frac{1}{2},\frac{1}{2}$	$\bar{x},\frac{1}{2},0$	$x+\frac{1}{2},\frac{1}{2},\frac{1}{2}$	hkl : $h+l=2n$
4	e	$2\,.\,.$	$x,0,0$	$\bar{x}+\frac{1}{2},0,\frac{1}{2}$	$\bar{x},0,0$	$x+\frac{1}{2},0,\frac{1}{2}$	hkl : $h+l=2n$
2	d	$2/m\,.\,.$	$0,\frac{1}{2},0$	$\frac{1}{2},\frac{1}{2},\frac{1}{2}$			hkl : $h+l=2n$
2	c	$2/m\,.\,.$	$\frac{1}{2},\frac{1}{2},0$	$0,\frac{1}{2},\frac{1}{2}$			hkl : $h+l=2n$
2	b	$2/m\,.\,.$	$\frac{1}{2},0,0$	$0,0,\frac{1}{2}$			hkl : $h+l=2n$
2	a	$2/m\,.\,.$	$0,0,0$	$\frac{1}{2},0,\frac{1}{2}$			hkl : $h+l=2n$

Symmetry of special projections

Along [001] $p2mm$
$\mathbf{a}'=\frac{1}{2}\mathbf{a}$ $\mathbf{b}'=\mathbf{b}$
Origin at $0,0,z$

Along [100] $p2gm$
$\mathbf{a}'=\mathbf{b}$ $\mathbf{b}'=\mathbf{c}$
Origin at $x,0,0$

Along [010] $c2mm$
$\mathbf{a}'=\mathbf{c}$ $\mathbf{b}'=\mathbf{a}$
Origin at $0,y,0$

Maximal non-isomorphic subgroups

I
[2] $Pmn2_1$ (31) 1; 2; 7; 8
[2] $P2na$ ($Pnc2$, 30) 1; 4; 6; 7
[2] $Pm2a$ ($Pma2$, 28) 1; 3; 6; 8
[2] $P222_1$ (17) 1; 2; 3; 4
[2] $P112_1/a$ ($P2_1/c$, 14) 1; 2; 5; 6
[2] $P12/n1$ ($P2/c$, 13) 1; 3; 5; 7
[2] $P2/m11$ ($P2/m$, 10) 1; 4; 5; 8

IIa none

IIb [2] $Pbna$ ($\mathbf{b}'=2\mathbf{b}$) ($Pbcn$, 60); [2] $Pmnn$ ($\mathbf{b}'=2\mathbf{b}$) ($Pnnm$, 58); [2] $Pbnn$ ($\mathbf{b}'=2\mathbf{b}$) ($Pnna$, 52)

Maximal isomorphic subgroups of lowest index

IIc [2] $Pmna$ ($\mathbf{b}'=2\mathbf{b}$) (53); [3] $Pmna$ ($\mathbf{a}'=3\mathbf{a}$) (53); [3] $Pmna$ ($\mathbf{c}'=3\mathbf{c}$) (53)

Minimal non-isomorphic supergroups

I none

II [2] $Cmce$ (64); [2] $Bmmm$ ($Cmmm$, 65); [2] $Amaa$ ($Cccm$, 66); [2] $Imma$ (74); [2] $Pmaa$ ($\mathbf{c}'=\frac{1}{2}\mathbf{c}$) ($Pccm$, 49); [2] $Pmcm$ ($\mathbf{a}'=\frac{1}{2}\mathbf{a}$) ($Pmma$, 51)

Figure 6.6 (*Continued*).

6.4.1.3. Maximal isomorphic subgroups of lowest index

These are type IIc k-subgroups. Thus, referring to *Pmna*(D_{2h}^7), we see that if the **a** repeat is multiplied by 3 (by removal of two-thirds of the translational operations along **a**), the Tables tell us that the resulting subgroup is *Pmna* still. As seen on the same line, *Pmna* is obtained also when the **b**-axis is doubled or when the **c**-axis is tripled after removal of the appropriate translations. Similarly, in *I*4/*mmm*, we see two type IIc maximal isomorphic subgroups of lowest index, one involving a tripling of the **c** repeat and the other involving tripling of **a** and **b**.

6.4.2. Minimal Non-Isomorphic Supergroups

Just as we could form maximal subgroups from any space group, it is possible to determine the nearest supergroups as well. A group G is a **minimal supergroup** of G_0 if G_0 is a maximal subgroup of G. The minimal supergroups are also divided into two distinct types, **translationengleiche** or **t-supergroups** and **klassengleiche** or **k-supergroups**.

Referring to the tables for $I4/mmm$, we see that there are two minimal non-isomorphic t-supergroups. Both are cubic with the [3] indicating 3 times the number of symmetry operations as in $I4/mmm$. To understand how these supergroups are related to the tetragonal cell, it is necessary to refer to the page in the *ITA* in which these supergroups are shown. Thus, by consulting the page for space group $Fm\bar{3}m(O_h^5)$, we find $I4/mmm$ as a type I maximal k-subgroup, and the operations of the cubic group that form $I4/mmm$ are listed.

Also note that there is a type II minimal k-supergroup of $I4/mmm$. This is $C4/mmm$, which is obtained by adding twice as many translations along the **c**-axis. Its conventional short symbol is $P4/mmm$ with the unit cell redefined in the usual way, as in Fig. 3.6.

6.5. SPACE GROUP SYMMETRY OPERATIONS

We have discussed at length the symmetry operations of space groups. However, we have not explicitly shown how, in general, the appropriate number of symmetry operations for the space groups may be determined. As mentioned before, there is for every space group an infinite number of primitive lattice translations $\{1|\mathbf{t}_n\}$. Now, we must determine the h symmetry operations ($h =$ order of the point group of the space group) of the form $\{R|\boldsymbol{\tau}\}$. The fundamental theorem, proved in Section 5.3.1, tells us that the h symmetry operations R, belonging to the symmetry operations $\{R|\boldsymbol{\tau}\}$, are the same as the h operations in the point group of the space group. Thus, only the values of $\boldsymbol{\tau}$ need to be determined for each value of R. In this section, we shall determine all the h $\{R|\boldsymbol{\tau}\}$ symmetry operations with respect to a *single* origin. This is not absolutely necessary, but it is certainly a convenience (recall from Section 5.4 that the space groups were described using only the International symbol as a guide. We found that, in general, different origins were required for the different symmetry operations). It is important to remember that if attention is focussed on a specific atom in a specific unit cell and one of the $\{R|\boldsymbol{\tau}\}$ symmetry operations takes the atom out of this unit cell, then it can always be translated back into the original cell by one of the translational symmetry operations $\{1|\mathbf{t}_n\}$.

6.5.1. Symmorphic Space Group Operations

For symmorphic space groups it is easy to find the h symmetry operations of the form $\{R|\mathbf{0}\}$. The h-values of R from the point group of the space group are clearly the symmetry operations of the space group. The origin, the point that is fixed for these symmetry operations, is taken at any position with site symmetry the same as the point group. For example, consider the space group $P4/m(C_{4h}^1)$ given in Fig. 6.1. The eight space group symmetry operations in the International and Schoenflies notations are:

$$\{1|\mathbf{0}\},\ \{4|\mathbf{0}\},\ \{2|\mathbf{0}\},\ \{4^3|\mathbf{0}\},\ \{\bar{1}|\mathbf{0}\},\ \{\bar{4}|\mathbf{0}\},\ \{m|\mathbf{0}\},\ \{\bar{4}^3|\mathbf{0}\}$$
$$\{E|\mathbf{0}\},\ \{C_4|\mathbf{0}\},\ \{C_2|\mathbf{0}\},\ \{C_4^3|\mathbf{0}\},\ \{i|\mathbf{0}\},\ \{S_4^3|\mathbf{0}\},\ \{\sigma_h|\mathbf{0}\},\ \{S_4|\mathbf{0}\}$$

By taking the origin at Wyckoff sites a, b, c or d, we see that these eight symmetry operations reproduce the eight general positions (circles) in the unit cell. The general positions outside the unit cell are obtained by applying a primitive translation $\{1|\mathbf{t}_n\}$ to those inside the primitive cell. The general positions outside the unit cell are included in the diagram for the convenience of showing the symmetry operations clearly, but it must be remembered that they belong to neighbouring unit cells.

These same eight symmetry operations also apply to the primitive cell of the space group $I4/m(C_{4h}^5)$. However, as usual, the diagrams (Fig. 6.7) in the *ITA* show the conventional Bravais body-centered unit cell. Therefore, in this diagram, we see that the eight circles grouped at the corner of the unit cell transform among themselves under all of the eight symmetry operations of the point group with respect to an origin at (0, 0, 0), which is one of the Wyckoff 2*a* positions. The other eight circles in this centered unit cell are obtained from these first eight circles by adding the centering condition. For the conventional *I*-unit cell one adds the quantity (½, ½, ½) remembering that (1, 0, 0), (0, 1, 0), etc., can always be added or subtracted for any position because of the translational symmetry.

Thus, the symmorphic space groups only require a knowledge of point groups to understand them. Now, we go to the non-symmorphic space groups where it is slightly more difficult to determine the h-symmetry operations.

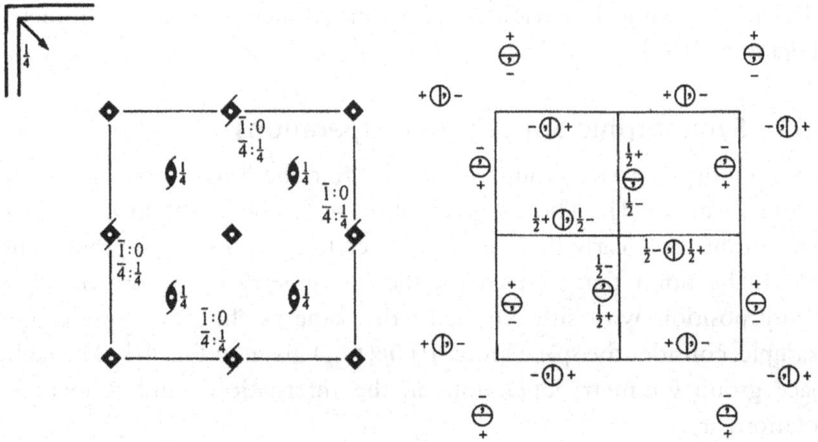

Figure 6.7 $I4/m(C_{4h}^5)$.

6.5.2. Non-Symmorphic Space Group Operations

If the lattice is primitive, the space group diagrams in the *ITA* can be used to determine the *h*-space group symmetry operations for the non-symmorphic space groups. Consider the space group $P4/n(C_{4h}^3)$ in Fig. 6.2. We wish to determine the eight symmetry operations $\{R|\boldsymbol{\tau}\}$, where $\boldsymbol{\tau}$ must be non-zero for at least one R, because the space group is non-symmorphic. For convenience, take the origin to be at the top left-hand corner of the unit cell (the normal convention in the *ITA*) and start with the circle just below it and to the right. The four symmorphic operations, $\{R_1|\mathbf{0}\}$, namely $\{1|\mathbf{0}\}$, $\{\overline{4}^3|\mathbf{0}\}$, $\{2|\mathbf{0}\}$ and $\{\overline{4}|\mathbf{0}\}$, generate from it the four circles grouped around this origin, of course, leaving the original circle unchanged. How do we obtain the other four circles close to the position (½, ½, 0) and yet keep the origin fixed? This can be done in the following way. Operate on the starting circle with operation $\overline{1}$ and follow this by a translation (½, ½, 0). The result of this is the circle, with a comma, above and to the left of the (½, ½, 0) position. Operating on the starting circle with {4|½, ½, 0} one gets the circle above and to the right of the (½, ½, 0) position. The remaining two circles are obtained by $\{4^3|$½, ½, 0} and {m[001]|½, ½, 0}. Notice, in this case, that the new set of operations $\{R_2|\tau\}$ is produced from the first set $\{R_1|\mathbf{0}\}$ by

$$\{\overline{1}|(½, ½, 0)\}\ \{R_1|\mathbf{0}\} = \{R_2|\boldsymbol{\tau}\} \qquad [6.4]$$

Thus, the eight symmetry operations of this non-symmorphic space group taken with respect to one origin are (in two notations):

$$\{1|\mathbf{0}\} \quad \{4|\boldsymbol{\tau}\} \quad \{2|\mathbf{0}\} \quad \{4^3|\boldsymbol{\tau}\} \qquad \{\bar{1}|\boldsymbol{\tau}\} \quad \{\bar{4}|\mathbf{0}\} \quad \{m[001]|\boldsymbol{\tau}\} \quad \{\bar{4}^3|\mathbf{0}\}$$

$$\{E|\mathbf{0}\} \quad \{C_4|\boldsymbol{\tau}\} \quad \{C_2|\mathbf{0}\} \quad \{C_4^3|\boldsymbol{\tau}\} \qquad \{i|\boldsymbol{\tau}\} \quad \{S_4^3|\mathbf{0}\} \quad \{\sigma_h|\boldsymbol{\tau}\} \quad \{S_4|\mathbf{0}\}$$

with $\boldsymbol{\tau} = \mathbf{a}/2 + \mathbf{b}/2$. Note that in this case the same $\boldsymbol{\tau}$ appears in all of the non-symmorphic symmetry operations.

Clearly, with some practice, this procedure can be repeated for any space group, even if a space group diagram is not available. The procedure is straightforward. We may pick an origin at any convenient place. For instance, the center of the unit cell might have been picked in the above example. We start with a general position to represent the identity symmetry operation. We then write down $(h-1)$ other symmetry operations that relate this original general position to the $(h-1)$ *different* general positions in the unit cell, always remembering that we may also use translational symmetry to move any general position back into the unit cell.

We use this occasion to bring up an interesting point about the symmetry operations given above. In Chapter 5, we talked about glide reflections and screw rotations. From the symbol $P4/n$ and the diagram we see that there is a diagonal glide but no screw axes. However, some of the symmetry operations we have obtained are neither glide reflections nor screw rotations. $\{4^3|½, ½, 0\}$ is not a screw rotation because $\boldsymbol{\tau}$ is not in the direction of the 4-fold axis, and $\{\bar{1}\,|½, ½, 0\}$ is not even remotely like a glide or screw operation. Here again, we run into the different approaches taken by crystallographers and other solid state scientists. By allowing rotation axes and reflection planes to be at various positions in the unit cell, the crystallographer has pure glides and screws. However, by causing all of the point operators to be taken with respect to a single origin, we find $\{R|\boldsymbol{\tau}\}$ operators that are not known as glides or screws. The approach of using one origin can, however, be a great convenience.

6.5.3. Symmetry Operations from the General Equivalent Positions

The *ITA* lists the coordinates of the general positions for all the 230 space groups. From these coordinates one can determine the symmetry operations of the space group with respect to a single origin, in much the same way as from the diagram. Take the general equivalent position (x, y, z) as the starting point (identity operation). Then, the other $(h-1)$ symmetry operations of the form $\{R|\boldsymbol{\tau}\}$, taken with respect to one origin, that give the other $(h-1)$ positions, may be written down almost by inspection in the same way as in Section 6.5.2. As above, the h symmetry operations R are known from the point group.

A procedure for determining the h symmetry operations $\{R|\boldsymbol{\tau}\}$, but with respect to different origins, as in the *ITA*, has been given by Wondratschek and Neubüser, where tables giving the geometric characterizations of all space-group symmetry operations are provided. The same results can be obtained by using the methods discussed in Chapter 5, where the only tables that need to be consulted are those for the matrix operators in Appendix 1. For example, to determine which symmetry operation maps the point (x, y, z) to $(y-x, y, \frac{1}{2}-z)$, we write

$$\{R|\boldsymbol{\tau}\}\,\mathbf{r} = \begin{bmatrix} -1 & 1 & 0 \\ 0 & 1 & 0 \\ 0 & 0 & -1 \end{bmatrix} \mathbf{r}' + \begin{bmatrix} 0 \\ 0 \\ \frac{1}{2} \end{bmatrix} \qquad [6.5]$$

which from Appendix 1 is simply $\{2[120]|\mathbf{0}\}$ passing through the point $\frac{1}{2}(0, 0, \frac{1}{2}) = (0, 0, \frac{1}{4})$.

6.6. HALL SPACE GROUP SYMBOLS

In 1981, S.R. Hall[5] used the idea of specifying all the space group operations from a common origin to derive a complete set of space group symbols based on the use of Seitz operators. The result is a concise and unambiguous notation that is well suited to handling symmetry in computing and database applications. The notation follows the format

$$\mathrm{L}[\mathrm{N}_t^{\mathrm{A}}]_1[\mathrm{N}_t^{\mathrm{A}}]_2[\mathrm{N}_t^{\mathrm{A}}]_3$$

L is the lattice symbol, N^{A} is the symbol for the rotation matrix and t is the symbol for the translation vector. For a full understanding of the notation we refer the reader to the original paper by Hall and to the International Tables Volume B. The following are a few examples of the Hall symbols as specified for computer usage

5 S.R. Hall, *Acta Cryst.*, **A37**, 517–525 (1981) and International Tables for Crystallography, Volume B, Section 1.4.

$Pm11$	P -2x
$P1m1$	P -2y
$P11m$	P -2
$P2_1/c$	-P 2ybc
$Pnma$	-P 2ac 2n
$R3m$ (hexagonal axes)	R 3 -2″
$R3m$ (rhombohedral axes)	P 3* -2
$Fd\bar{3}m$ (origin at $\bar{4}3m$)	F 4D 2 3 -1D
$Fd\bar{3}$m (origin at $\bar{3}m$)	-F 4VW 2VW 3

Problems

1. Consider the space groups $P4_122(D_4^3)$ and $P4_322(D_4^7)$. Write out the coordinates of equivalent positions for all of the special and general points as well as the site symmetry for each point. These two space groups are an enantiomorphic pair. Which symmetry operation relates these two space groups?
2. Show that by using the short symbol $Fd\bar{3}m$ alone it is possible to deduce the presence of a 4_1 axis along [100] and its position with respect to a chosen origin. Hint: take the combination of a d-glide perpendicular to [001] with a mirror perpendicular to $[0\bar{1}\bar{1}]$ and use the matrix multiplication method of Chapter 5.
3. What are the point groups of the following space groups and what is the symmetry of the special position with the highest site symmetry in each case: $P\bar{4}m2$, $P\bar{4}c2$, $P3$m1, $R\bar{3}$c and $I23$?
4. Consider the space group $Pban(D_{2h}^4)$ given in the *ITA*. Write out the essential symmetry operations with respect to a fixed origin taken to be at (a) the intersection of the 2-fold axes and (b) a center of symmetry. Hint: the *ITA* has separate pages for these two choices of origin. Notice that in (a) a single $\boldsymbol{\tau} = \mathbf{a}/2 + \mathbf{b}/2$ is all that is needed, while in (b) three different values of $\boldsymbol{\tau}$ are required: $\mathbf{a}/2 + \mathbf{b}/2$, $\mathbf{a}/2$ and $\mathbf{b}/2$. Careful observation of the coordinates of the general equivalent positions for these two choices of origin will indicate that these two different sets of $\boldsymbol{\tau}$'s will occur.
5. The general equivalent positions for space group $Pmmm(D_{2h}^1)$ are:

(1) x, y, z	(2) $\bar{x}, \bar{y}, z$	(3) $\bar{x}, y, \bar{z}$	(4) $x, \bar{y}, \bar{z}$
(5) $\bar{x}, \bar{y}, \bar{z}$	(6) $x, y, \bar{z}$	(7) $x, \bar{y}, z$	(8) $\bar{x}, y, z$

Write down the corresponding symmetry operations that give these positions.

6. The following space groups are maximal subgroups of $R3m(C_{3v}^5)$ specified on hexagonal axes: $P3m1$, $R31$, $R3c$ ($\mathbf{a}' = -\mathbf{a}$, $\mathbf{b}' = -\mathbf{b}$, $\mathbf{c}' = 2\mathbf{c}$) and $R3m$ ($\mathbf{a}' = -2\mathbf{a}$, $\mathbf{b}' = -2\mathbf{b}$). Classify each of these according to types I, IIa, IIb and IIc. (You will find it useful first to sketch the symmetry operator diagram for $R3m$).
7. Using the *ITA*, start with space group $Fd\bar{3}m(O_h^7)$ and derive a sequence of maximal subgroups that ends with $C2(C_2^3)$, noting at each step the type of subgroup. Similarly, starting with $P4(C_4^1)$, use the minimal supergroup information to find a sequence that gets to $Im\bar{3}m(O_h^9)$.

CHAPTER 7

Space Group Applications

Contents

Space Groups for Solid State Scientists
http://dx.doi.org/10.1016/B978-0-12-394400-9.00007-1

We looked at the books about crystals but they are so dreadful.

J. Ruskin, The Ethics of the Dust, Ten Lectures to Little Housewives on the Elements of Crystallisation (1866)

AND NOW ATOMS

In order to make more concrete the points discussed in the preceding sections, we consider the way in which crystal structures can be described using the concept of space groups. We deliberately start with some rather simple, but important, structures that belong to the cubic system.

7.1. FACE-CENTERED CUBIC STRUCTURES

Figure 7.1 shows the crystal structure of the metal copper (Cu). The structure is very simple with one copper atom located at each lattice point of a face-centered cubic (fcc) lattice. Examination of the diagram shows that, in addition to the 3-fold axes along <111>, there are mirror planes perpendicular to <100> and <110>. The space group is $Fm\overline{3}m(O_h^5)$. We can therefore specify this structure thus:

Space group	$Fm\overline{3}m(O_h^5)$	
Cu	4*a*	0, 0, 0

The Wyckoff site 4*a* (Fig. 7.3) means that the copper atoms will be located at the positions (0, 0, 0), (0, ½, ½), (½, 0, ½) and (½, ½, 0). In this case, the coordinates are just the same as those of the lattice points. As always, there is one lattice point, and therefore in this structure one atom in the primitive unit cell, as was discussed in Chapter 3, and four lattice points with four atoms in the *F*-unit cell as shown in Fig. 7.1. The basis consists of a single Cu atom, and, because of this, the structure diagram looks at first sight like a unit cell of the lattice itself (the difference is that the structure consists of *atoms*, whereas the lattice of consists of *points* – do not confuse these two concepts!).

Figure 7.1 shows different forms of the same crystal structure. Even in this very simple crystal structure, different views emphasize different aspects of the structure, but in all of them you should be able to discern the elements of symmetry of space group $Fm\overline{3}m(O_h^5)$.

Figure 7.2 shows this crystal structure again, along with other structures, such as NaCl and CaF_2 that have the same space group.

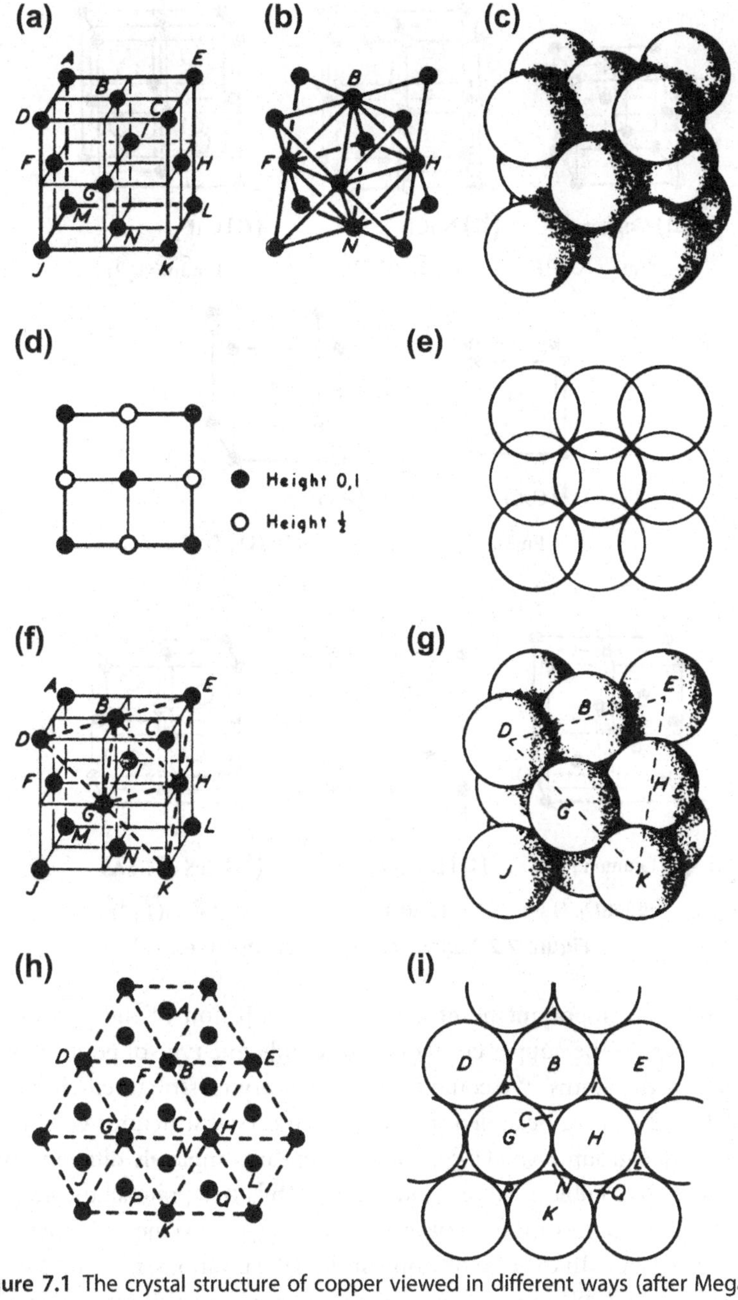

Figure 7.1 The crystal structure of copper viewed in different ways (after Megaw).

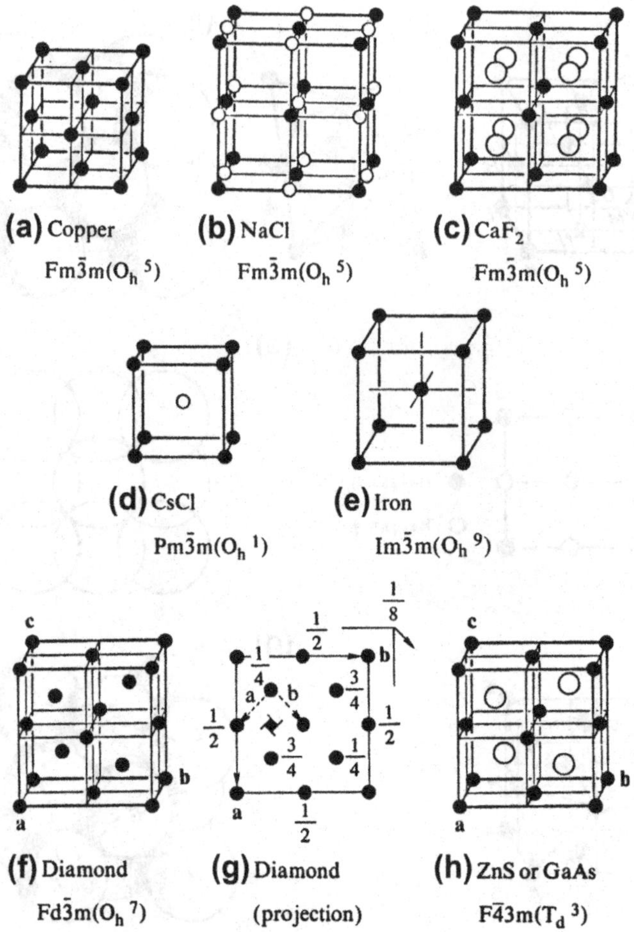

Figure 7.2 Various cubic crystal structures.

There is an important point about the occupation of the positions of a space group. In the copper crystal structure only one type of position, $4a$, is occupied by the atoms. We could have another crystal structure, where only the $192l$ positions are occupied. Both of these crystal structures would have the same space group $Fm\bar{3}m(O_h^5)$, but they appear completely different. After all, the lattice is a cubic F-lattice, and about each lattice point all the 48 point symmetry operations of the point group $m\bar{3}m(O_h)$ are symmetry operations of the space group. In the case of atoms at the $192l$ positions, the atoms would be located at the 48 positions around the chosen space group origin with coordinates given by the list in Fig. 7.3. The fact that there are four lattice points in the conventional cell means that there would be $4 \times 48 = 192$

Positions

Multiplicity, Wyckoff letter, Site symmetry			Coordinates					
			(0,0,0)+	(0,$\frac{1}{2}$,$\frac{1}{2}$)+	($\frac{1}{2}$,0,$\frac{1}{2}$)+	($\frac{1}{2}$,$\frac{1}{2}$,0)+		
192	*l*	1	(1) x,y,z	(2) $\bar{x},\bar{y},z$	(3) $\bar{x},y,\bar{z}$	(4) $x,\bar{y},\bar{z}$		
			(5) z,x,y	(6) $z,\bar{x},\bar{y}$	(7) $\bar{z},\bar{x},y$	(8) $\bar{z},x,\bar{y}$		
			(9) y,z,x	(10) $\bar{y},z,\bar{x}$	(11) $y,\bar{z},\bar{x}$	(12) $\bar{y},\bar{z},x$		
			(13) $y,x,\bar{z}$	(14) $\bar{y},\bar{x},\bar{z}$	(15) $y,\bar{x},z$	(16) $\bar{y},x,z$		
			(17) $x,z,\bar{y}$	(18) $\bar{x},z,y$	(19) $\bar{x},\bar{z},\bar{y}$	(20) $x,\bar{z},y$		
			(21) $z,y,\bar{x}$	(22) $z,\bar{y},x$	(23) $\bar{z},y,x$	(24) $\bar{z},\bar{y},\bar{x}$		
			(25) $\bar{x},\bar{y},\bar{z}$	(26) $x,y,\bar{z}$	(27) $x,\bar{y},z$	(28) $\bar{x},y,z$		
			(29) $\bar{z},\bar{x},\bar{y}$	(30) $\bar{z},x,y$	(31) $z,x,\bar{y}$	(32) $z,\bar{x},y$		
			(33) $\bar{y},\bar{z},\bar{x}$	(34) $y,\bar{z},x$	(35) $\bar{y},z,x$	(36) $y,z,\bar{x}$		
			(37) $\bar{y},\bar{x},z$	(38) y,x,z	(39) $\bar{y},x,\bar{z}$	(40) $y,\bar{x},\bar{z}$		
			(41) $\bar{x},\bar{z},y$	(42) $x,\bar{z},\bar{y}$	(43) x,z,y	(44) $\bar{x},z,\bar{y}$		
			(45) $\bar{z},\bar{y},x$	(46) $\bar{z},y,\bar{x}$	(47) $z,\bar{y},\bar{x}$	(48) z,y,x		
96	*k*	..*m*	x,x,z	$\bar{x},\bar{x},z$	$\bar{x},x,\bar{z}$	$x,\bar{x},\bar{z}$	z,x,x	$z,\bar{x},\bar{x}$
			$\bar{z},\bar{x},x$	$\bar{z},x,\bar{x}$	x,z,x	$\bar{x},z,\bar{x}$	$x,\bar{z},\bar{x}$	$\bar{x},\bar{z},x$
			$x,x,\bar{z}$	$\bar{x},\bar{x},\bar{z}$	$x,\bar{x},z$	$\bar{x},x,z$	$x,z,\bar{x}$	$\bar{x},z,x$
			$\bar{x},\bar{z},\bar{x}$	$x,\bar{z},x$	$z,x,\bar{x}$	$z,\bar{x},x$	$\bar{z},x,x$	$\bar{z},\bar{x},\bar{x}$
96	*j*	*m*..	$0,y,z$	$0,\bar{y},z$	$0,y,\bar{z}$	$0,\bar{y},\bar{z}$	$z,0,y$	$z,0,\bar{y}$
			$\bar{z},0,y$	$\bar{z},0,\bar{y}$	$y,z,0$	$\bar{y},z,0$	$y,\bar{z},0$	$\bar{y},\bar{z},0$
			$y,0,\bar{z}$	$\bar{y},0,\bar{z}$	$y,0,z$	$\bar{y},0,z$	$0,z,\bar{y}$	$0,z,y$
			$0,\bar{z},\bar{y}$	$0,\bar{z},y$	$z,y,0$	$z,\bar{y},0$	$\bar{z},y,0$	$\bar{z},\bar{y},0$
48	*i*	*m*.*m*2	$\frac{1}{2},y,y$	$\frac{1}{2},\bar{y},y$	$\frac{1}{2},y,\bar{y}$	$\frac{1}{2},\bar{y},\bar{y}$	$y,\frac{1}{2},y$	$y,\frac{1}{2},\bar{y}$
			$\bar{y},\frac{1}{2},y$	$\bar{y},\frac{1}{2},\bar{y}$	$y,y,\frac{1}{2}$	$\bar{y},y,\frac{1}{2}$	$y,\bar{y},\frac{1}{2}$	$\bar{y},\bar{y},\frac{1}{2}$
48	*h*	*m*.*m*2	$0,y,y$	$0,\bar{y},y$	$0,y,\bar{y}$	$0,\bar{y},\bar{y}$	$y,0,y$	$y,0,\bar{y}$
			$\bar{y},0,y$	$\bar{y},0,\bar{y}$	$y,y,0$	$\bar{y},y,0$	$y,\bar{y},0$	$\bar{y},\bar{y},0$
48	*g*	2.*mm*	$x,\frac{1}{4},\frac{1}{4}$	$\bar{x},\frac{3}{4},\frac{1}{4}$	$\frac{1}{4},x,\frac{1}{4}$	$\frac{1}{4},\bar{x},\frac{3}{4}$	$\frac{1}{4},\frac{1}{4},x$	$\frac{3}{4},\frac{1}{4},\bar{x}$
			$\frac{1}{4},x,\frac{3}{4}$	$\frac{3}{4},\bar{x},\frac{3}{4}$	$x,\frac{1}{4},\frac{3}{4}$	$\bar{x},\frac{1}{4},\frac{1}{4}$	$\frac{1}{4},\frac{1}{4},\bar{x}$	$\frac{1}{4},\frac{3}{4},x$
32	*f*	.3*m*	x,x,x	$\bar{x},\bar{x},x$	$\bar{x},x,\bar{x}$	$x,\bar{x},\bar{x}$		
			$x,x,\bar{x}$	$\bar{x},\bar{x},\bar{x}$	$x,\bar{x},x$	$\bar{x},x,x$		
24	*e*	4*m*.*m*	$x,0,0$	$\bar{x},0,0$	$0,x,0$	$0,\bar{x},0$	$0,0,x$	$0,0,\bar{x}$
24	*d*	*m*.*mm*	$0,\frac{1}{4},\frac{1}{4}$	$0,\frac{3}{4},\frac{1}{4}$	$\frac{1}{4},0,\frac{1}{4}$	$\frac{1}{4},0,\frac{3}{4}$	$\frac{1}{4},\frac{1}{4},0$	$\frac{3}{4},\frac{1}{4},0$
8	*c*	$\bar{4}3m$	$\frac{1}{4},\frac{1}{4},\frac{1}{4}$	$\frac{1}{4},\frac{1}{4},\frac{3}{4}$				
4	*b*	$m\bar{3}m$	$\frac{1}{2},\frac{1}{2},\frac{1}{2}$					
4	*a*	$m\bar{3}m$	$0,0,0$					

Figure 7.3 $Fm\overline{3}m(O_h^5)$.

atoms. Naturally the coordinates of atoms about the origin transform among themselves under all the symmetry operations of the point group $m\bar{3}m(O_h)$. In general, each of the various positions for this space group may or may not be occupied with different types or the same types of atoms. Complicated chemical compounds may crystallize into a crystal structure with this space group, in which case a number of different types of position may be occupied.

Figure 7.2b shows the crystal structure of sodium chloride (NaCl), sometimes called the rock-salt structure. The space group is again $Fm\bar{3}m(O_h^5)$. For this crystal structure, Na atoms are located at the 4*a* positions, and Cl atoms at the 4*b* positions. As can be seen in the diagram, the site symmetry of both of these positions is the full point symmetry $m\bar{3}m(O_h)$, and therefore we may equally well take the Cl atoms at 4*a* positions and the Na atoms at 4*b* positions. In order to describe this crystal structure, we can write

Space group	$Fm\bar{3}m(O_h^5)$	
Na	4*a*	0, 0, 0
Cl	4*b*	½, ½,½

We may also say that the space group is $Fm\bar{3}m(O_h^5)$ and the basis is Na at (0, 0, 0) and Cl at (½, ½, ½). Because this structure has a basis of two atoms, there are two atoms per lattice point and the resulting crystal structure should not be confused with the diagram for the lattice. Yet another way of describing this crystal structure is in terms of a Na atom surrounded by six Cl atoms at the corners of an octahedron. This is consistent with the fact that an octahedron has $m\bar{3}m(O_h)$ symmetry itself. It is interesting, and even amusing, to note that when this structure was first derived by Lawrence Bragg there was considerable disbelief within the chemistry community, since till then it had been expected that Na and Cl would form molecular units rather than distribute themselves in this alternating pattern. The following letter from Professor H.E. Armstrong sent to Nature in 1927 is an example of the depth of feeling:

Poor Common Salt

Some books are lies frae end to end' says Burns. Scientific (save the mark) speculation would seem to be on the way to this state! ... Professor W.L. Bragg asserts that in 'Sodium Chloride there appear to be no molecules represented by NaCl. The equality in number of sodium and chlorine atoms is arrived at by a chess-board pattern of these atoms: it is a result of geometry and not of a pairing of the atoms'. This statement is more than 'repugnant to common sense'. It is absurd to the nth degree, not chemical cricket. Chemistry is neither chess nor geometry, whatever X-ray physics might be. Such unjustified aspersion of the molecular character of our most necessary condiment must not be allowed any longer to pass *unchallenged. It were time that chemists took charge of chemistry once more and protected neophytes against the worship of false gods; at least taught them to ask for something more than chess-board evidence.*

In Fig. 7.2, we show the crystal structure of calcium fluoride (CaF_2), sometimes called the fluorite structure. The space group again is $Fm\bar{3}m(O_h^5)$, but this time it has Ca atoms at 4*a* positions and F atoms at the 8*c* positions (clearly, from the chemical formula, there must be twice as many positions for the F atoms as for the Ca atoms). We may also describe this structure by saying that the space group is $Fm\bar{3}m(O_h^5)$ and that the basis is Ca at (0, 0, 0) and F at ±(¼, ¼, ¼), that is three atoms in the basis. Here again, we may describe this structure in terms of a Ca atom surrounded by eight F atoms at the corners of a cube, so that around each lattice point the point symmetry is $m\bar{3}m(O_h)$ (the symmetry of a cube).

We can find many more complicated examples of structures with this space group, for example, Na_3FeF_6. Fe atoms are on the 4*a* positions, Na(1) on the 4*b* positions, Na(2) on the 8*c* positions, and Fe atoms on the 24*e* positions with $x \approx 0.23$. With this small amount of information you should be able to draw a diagram of the crystal structure for yourself. Note how the same types of atoms may be on different positions in the unit cell. It should therefore be obvious that there is no limit to the number of different crystals that can adopt structures with the same space group.

7.2. PRIMITIVE CUBIC STRUCTURES

The simplest structure possible in this category is where the space group is $Pm\bar{3}m(O_h^1)$ with one atom per lattice point. This structure is known to physicists and chemists as **simple cubic**: it consists of a single atom at each corner of the cubic unit cell. However, despite its inclusion in standard solid state texts, this structure is extremely rare, and as far as we know it only occurs under normal conditions in the metal polonium! More interestingly, consider Fig. 7.2d which shows the caesium chloride (CsCl) crystal structure. The space group is $Pm\bar{3}m(O_h^1)$, which is a primitive cubic symmorphic space group. The relevant page of the *ITA* is shown in Fig. 7.4. The structure is given by

Space group	$Pm\bar{3}m(O_h^1)$	
Cs	1*a*	0, 0, 0
Cl	1*b*	½, ½,½

Positions

Multiplicity,	Wyckoff letter,	Site symmetry	Coordinates					
48	*n*	1	(1) x,y,z	(2) $\bar{x},\bar{y},z$	(3) $\bar{x},y,\bar{z}$	(4) $x,\bar{y},\bar{z}$		
			(5) z,x,y	(6) $z,\bar{x},\bar{y}$	(7) $\bar{z},\bar{x},y$	(8) $\bar{z},x,\bar{y}$		
			(9) y,z,x	(10) $\bar{y},z,\bar{x}$	(11) $y,\bar{z},\bar{x}$	(12) $\bar{y},\bar{z},x$		
			(13) $y,x,\bar{z}$	(14) $\bar{y},\bar{x},\bar{z}$	(15) $y,\bar{x},z$	(16) $\bar{y},x,z$		
			(17) $x,z,\bar{y}$	(18) $\bar{x},z,y$	(19) $\bar{x},\bar{z},\bar{y}$	(20) $x,\bar{z},y$		
			(21) $z,y,\bar{x}$	(22) $z,\bar{y},x$	(23) $\bar{z},y,x$	(24) $\bar{z},\bar{y},\bar{x}$		
			(25) $\bar{x},\bar{y},\bar{z}$	(26) $x,y,\bar{z}$	(27) $x,\bar{y},z$	(28) $\bar{x},y,z$		
			(29) $\bar{z},\bar{x},\bar{y}$	(30) $\bar{z},x,y$	(31) $z,x,\bar{y}$	(32) $z,\bar{x},y$		
			(33) $\bar{y},\bar{z},\bar{x}$	(34) $y,\bar{z},x$	(35) $\bar{y},z,x$	(36) $y,z,\bar{x}$		
			(37) $\bar{y},\bar{x},z$	(38) y,x,z	(39) $\bar{y},x,\bar{z}$	(40) $y,\bar{x},\bar{z}$		
			(41) $\bar{x},\bar{z},y$	(42) $x,\bar{z},\bar{y}$	(43) x,z,y	(44) $\bar{x},z,\bar{y}$		
			(45) $\bar{z},\bar{y},x$	(46) $\bar{z},y,\bar{x}$	(47) $z,\bar{y},\bar{x}$	(48) z,y,x		
24	*m*	. . *m*	x,x,z	$\bar{x},\bar{x},z$	$\bar{x},x,\bar{z}$	$x,\bar{x},\bar{z}$	z,x,x	$z,\bar{x},\bar{x}$
			$\bar{z},\bar{x},x$	$\bar{z},x,\bar{x}$	x,z,x	$\bar{x},z,\bar{x}$	$x,\bar{z},\bar{x}$	$\bar{x},\bar{z},x$
			$x,x,\bar{z}$	$\bar{x},\bar{x},\bar{z}$	$x,\bar{x},z$	$\bar{x},x,z$	$x,z,\bar{x}$	$\bar{x},z,x$
			$\bar{x},\bar{z},\bar{x}$	$x,\bar{z},x$	$z,x,\bar{x}$	$z,\bar{x},x$	$\bar{z},x,x$	$\bar{z},\bar{x},\bar{x}$
24	*l*	*m* . .	$\frac{1}{2},y,z$	$\frac{1}{2},\bar{y},z$	$\frac{1}{2},y,\bar{z}$	$\frac{1}{2},\bar{y},\bar{z}$	$z,\frac{1}{2},y$	$z,\frac{1}{2},\bar{y}$
			$\bar{z},\frac{1}{2},y$	$\bar{z},\frac{1}{2},\bar{y}$	$y,z,\frac{1}{2}$	$\bar{y},z,\frac{1}{2}$	$y,\bar{z},\frac{1}{2}$	$\bar{y},\bar{z},\frac{1}{2}$
			$y,\frac{1}{2},\bar{z}$	$\bar{y},\frac{1}{2},\bar{z}$	$y,\frac{1}{2},z$	$\bar{y},\frac{1}{2},z$	$\frac{1}{2},z,\bar{y}$	$\frac{1}{2},z,y$
			$\frac{1}{2},\bar{z},\bar{y}$	$\frac{1}{2},\bar{z},y$	$z,y,\frac{1}{2}$	$z,\bar{y},\frac{1}{2}$	$\bar{z},y,\frac{1}{2}$	$\bar{z},\bar{y},\frac{1}{2}$
24	*k*	*m* . .	$0,y,z$	$0,\bar{y},z$	$0,y,\bar{z}$	$0,\bar{y},\bar{z}$	$z,0,y$	$z,0,\bar{y}$
			$\bar{z},0,y$	$\bar{z},0,\bar{y}$	$y,z,0$	$\bar{y},z,0$	$y,\bar{z},0$	$\bar{y},\bar{z},0$
			$y,0,\bar{z}$	$\bar{y},0,\bar{z}$	$y,0,z$	$\bar{y},0,z$	$0,z,\bar{y}$	$0,z,y$
			$0,\bar{z},\bar{y}$	$0,\bar{z},y$	$z,y,0$	$z,\bar{y},0$	$\bar{z},y,0$	$\bar{z},\bar{y},0$
12	*j*	*m* . *m*2	$\frac{1}{2},y,y$	$\frac{1}{2},\bar{y},y$	$\frac{1}{2},y,\bar{y}$	$\frac{1}{2},\bar{y},\bar{y}$	$y,\frac{1}{2},y$	$y,\frac{1}{2},\bar{y}$
			$\bar{y},\frac{1}{2},y$	$\bar{y},\frac{1}{2},\bar{y}$	$y,y,\frac{1}{2}$	$\bar{y},y,\frac{1}{2}$	$y,\bar{y},\frac{1}{2}$	$\bar{y},\bar{y},\frac{1}{2}$
12	*i*	*m* . *m*2	$0,y,y$	$0,\bar{y},y$	$0,y,\bar{y}$	$0,\bar{y},\bar{y}$	$y,0,y$	$y,0,\bar{y}$
			$\bar{y},0,y$	$\bar{y},0,\bar{y}$	$y,y,0$	$\bar{y},y,0$	$y,\bar{y},0$	$\bar{y},\bar{y},0$
12	*h*	*m m*2 . .	$x,\frac{1}{2},0$	$\bar{x},\frac{1}{2},0$	$0,x,\frac{1}{2}$	$0,\bar{x},\frac{1}{2}$	$\frac{1}{2},0,x$	$\frac{1}{2},0,\bar{x}$
			$\frac{1}{2},x,0$	$\frac{1}{2},\bar{x},0$	$x,0,\frac{1}{2}$	$\bar{x},0,\frac{1}{2}$	$0,\frac{1}{2},\bar{x}$	$0,\frac{1}{2},x$
8	*g*	. 3 *m*	x,x,x	$\bar{x},\bar{x},x$	$\bar{x},x,\bar{x}$	$x,\bar{x},\bar{x}$		
			$x,x,\bar{x}$	$\bar{x},\bar{x},\bar{x}$	$x,\bar{x},x$	$\bar{x},x,x$		
6	*f*	4 *m* . *m*	$x,\frac{1}{2},\frac{1}{2}$	$\bar{x},\frac{1}{2},\frac{1}{2}$	$\frac{1}{2},x,\frac{1}{2}$	$\frac{1}{2},\bar{x},\frac{1}{2}$	$\frac{1}{2},\frac{1}{2},x$	$\frac{1}{2},\frac{1}{2},\bar{x}$
6	*e*	4 *m* . *m*	$x,0,0$	$\bar{x},0,0$	$0,x,0$	$0,\bar{x},0$	$0,0,x$	$0,0,\bar{x}$
3	*d*	4/*m m* . *m*	$\frac{1}{2},0,0$	$0,\frac{1}{2},0$	$0,0,\frac{1}{2}$			
3	*c*	4/*m m* . *m*	$0,\frac{1}{2},\frac{1}{2}$	$\frac{1}{2},0,\frac{1}{2}$	$\frac{1}{2},\frac{1}{2},0$			
1	*b*	$m\bar{3}m$	$\frac{1}{2},\frac{1}{2},\frac{1}{2}$					
1	*a*	$m\bar{3}m$	$0,0,0$					

Figure 7.4 $Pm\bar{3}m(O_h^1)$.

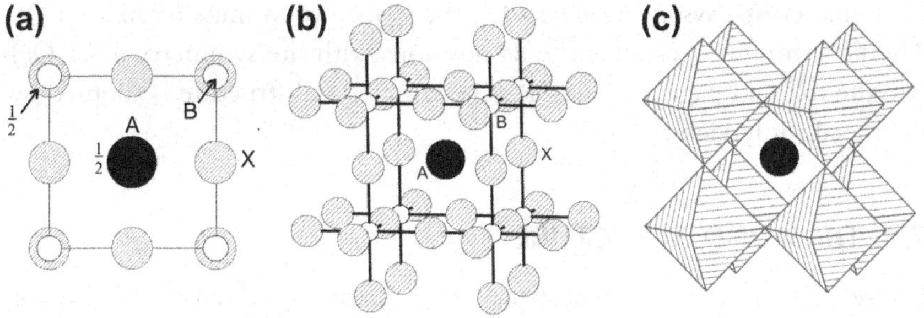

Figure 7.5 The ideal perovskite crystal structure: (a) plan view, (b) three-dimensional view and (c) anion octahedra.

Clearly from the equality of site symmetry, these two positions can be interchanged. Another way of describing this crystal structure is to say that the lattice is primitive cubic and the basis is Cs at (0, 0, 0) and Cl at (½, ½, ½). Otherwise, we may describe this structure by saying that at each lattice point of a *cP* lattice, there is a Cs atom surrounded by eight Cl atoms at the corners of a cube. Thus, surrounding each lattice point, one has an object with point symmetry $m\overline{3}m(O_h)$ immediately leading to space group $Pm\overline{3}m(O_h^1)$. It is a common mistake to refer to this structure as body-centered. Remember that a body-centered structure means that if an atom is placed at (0, 0, 0), the same type of atom is at (½, ½, ½). In CsCl we have different atoms at each site, and hence the unit cell is primitive.

Another important structure with this space group is that of perovskite. A number of compounds with the formula ABX_3 have this crystal structure which in its ideal form belongs to space group $Pm\overline{3}m(O_h^1)$. The A cation is at the 1*a* position, the B cation at the 1*b* position and the X anions (usually oxygen) are at the 3*c* positions. Thus one may see (Fig. 7.5) that surrounding each B atom there are 6 O atoms at the corners of an octahedron and surrounding each A atom are 12 O atoms, each at the mid-point of a cube edge. In Fig. 7.5a, the fractional heights of the A and X atoms are marked, all other atoms being at zero height. The compound WO_3 also has the perovskite structure and space group $Pm\overline{3}m(O_h^1)$, with the W atom at the 1*b* position and the O atoms again at 3*c* positions.

7.3. BODY-CENTERED CUBIC STRUCTURES

Figure 7.2e shows the structure of iron (Fe). The Fe atoms are located at the lattice points of a *cI* lattice, and because the symmetry at a lattice point of the cubic crystal system is $m\bar{3}m(O_h)$, the space group must be $Im\bar{3}m(O_h^9)$. The Fe atoms are located at the 2*a* positions with site symmetry $m\bar{3}m(O_h)$ (see the relevant page of the *ITA*). The same crystal structure is adopted by the metal molybdenum.

7.4. DIAMOND STRUCTURE

Figure 7.2f shows the crystal structure of diamond, which serves also for silicon (Si) and germanium (Ge), and Fig. 7.2g is a projection with some space-group operations indicated. The space group is $Fd\bar{3}m(O_h^7)$, which means that the lattice is all-face-centered, nonsymmorphic, and has point group $m\bar{3}m(O_h)$. Figure 7.6 shows the relevant page of the *ITA*. The atoms are at the 8*a* positions, with point symmetry $\bar{4}3m(T_d)$, which, is of lower symmetry than the point group of the space group, as indeed it must always be for nonsymmorphic space groups. From the full space group symbol, we see that there is a 4_1-axis at ½, ¼, *z*, as shown in Fig. 7.2g. This symmetry operation takes the atom located at (¼, ¼, ¼) to the atom at the face-center (½, 0, ½) (path *a*, marked on the Fig. 7.2g). This operation applied twice moves the original atom to the one at (¾, ¼, ¾), and applied three times moves it to the atom at (½, ½, 1), which is the same as (½, ½, 0). This symmetry operation applied four times yields the original atom. Similarly, the diamond glide plane perpendicular to this 4_1-axis is located at $z = ⅛$.

The atom at (¼, ¼, ¼) lies ⅛ above this glide plane, and hence it is reflected to the $z = 0$ plane and subsequently translated through (¼, ¼, 0) to the atom at (½, ½, 0) (path b, marked on the Fig. 7.2g). If the operation is applied twice, then the atom at (¾, ¾, ¼) is obtained. Applying three and four times, the atom at (1, 1, 0), which is the same as (0, 0, 0), plus the original atom are obtained. The atoms do not lie at the center of inversion, that is the center of inversion is not a symmetry operation of the site group $\bar{4}3m(T_d)$. However, the space group does have $\bar{1}(i)$ as a symmetry operation. Normally this would be shown in the space group diagram, but we can also determine this fact by noting that the site symmetry of the 16*c* position is $\bar{3}m(D_{3d})$, which does have a center of symmetry. From the *ITA*, we see that there are centers at (⅛, ⅛ ⅛) and (⅛, ⅜, ⅜). Clearly the center of symmetry at (⅛, ⅛, ⅛) takes the atom at (¼, ¼, ¼) to the atom at (0, 0, 0), while the center at (⅛, ⅜, ⅜) takes it into the one at (0, ½, ½).

Multiplicity, Wyckoff letter, Site symmetry			Coordinates					
			$(0,0,0)+$	$(0,\frac{1}{2},\frac{1}{2})+$	$(\frac{1}{2},0,\frac{1}{2})+$	$(\frac{1}{2},\frac{1}{2},0)+$		
192	i	1	(1) x,y,z	(2) $\bar{x},\bar{y}+\frac{1}{2},z+\frac{1}{2}$	(3) $\bar{x}+\frac{1}{2},y+\frac{1}{2},\bar{z}$	(4) $x+\frac{1}{2},\bar{y},\bar{z}+\frac{1}{2}$		
			(5) z,x,y	(6) $z+\frac{1}{2},\bar{x},\bar{y}+\frac{1}{2}$	(7) $\bar{z},\bar{x}+\frac{1}{2},y+\frac{1}{2}$	(8) $\bar{z}+\frac{1}{2},x+\frac{1}{2},\bar{y}$		
			(9) y,z,x	(10) $\bar{y}+\frac{1}{2},z+\frac{1}{2},\bar{x}$	(11) $y+\frac{1}{2},\bar{z},\bar{x}+\frac{1}{2}$	(12) $\bar{y},\bar{z}+\frac{1}{2},x+\frac{1}{2}$		
			(13) $y+\frac{3}{4},x+\frac{1}{4},\bar{z}+\frac{3}{4}$	(14) $\bar{y}+\frac{1}{4},\bar{x}+\frac{1}{4},\bar{z}+\frac{1}{4}$	(15) $y+\frac{1}{4},\bar{x}+\frac{3}{4},z+\frac{3}{4}$	(16) $\bar{y}+\frac{3}{4},x+\frac{3}{4},z+\frac{1}{4}$		
			(17) $x+\frac{3}{4},z+\frac{1}{4},\bar{y}+\frac{3}{4}$	(18) $\bar{x}+\frac{3}{4},z+\frac{3}{4},y+\frac{1}{4}$	(19) $\bar{x}+\frac{1}{4},\bar{z}+\frac{1}{4},\bar{y}+\frac{1}{4}$	(20) $x+\frac{1}{4},\bar{z}+\frac{3}{4},y+\frac{3}{4}$		
			(21) $z+\frac{3}{4},y+\frac{1}{4},\bar{x}+\frac{3}{4}$	(22) $z+\frac{1}{4},\bar{y}+\frac{3}{4},x+\frac{3}{4}$	(23) $\bar{z}+\frac{3}{4},y+\frac{3}{4},x+\frac{1}{4}$	(24) $\bar{z}+\frac{1}{4},\bar{y}+\frac{1}{4},\bar{x}+\frac{1}{4}$		
			(25) $\bar{x}+\frac{1}{4},\bar{y}+\frac{1}{4},\bar{z}+\frac{1}{4}$	(26) $x+\frac{1}{4},y+\frac{3}{4},\bar{z}+\frac{3}{4}$	(27) $x+\frac{3}{4},\bar{y}+\frac{3}{4},z+\frac{1}{4}$	(28) $\bar{x}+\frac{3}{4},y+\frac{1}{4},z+\frac{3}{4}$		
			(29) $\bar{z}+\frac{1}{4},\bar{x}+\frac{1}{4},\bar{y}+\frac{1}{4}$	(30) $\bar{z}+\frac{3}{4},x+\frac{1}{4},y+\frac{3}{4}$	(31) $z+\frac{1}{4},x+\frac{3}{4},\bar{y}+\frac{3}{4}$	(32) $z+\frac{3}{4},\bar{x}+\frac{3}{4},y+\frac{1}{4}$		
			(33) $\bar{y}+\frac{1}{4},\bar{z}+\frac{1}{4},\bar{x}+\frac{1}{4}$	(34) $y+\frac{3}{4},\bar{z}+\frac{3}{4},x+\frac{1}{4}$	(35) $\bar{y}+\frac{3}{4},z+\frac{1}{4},x+\frac{3}{4}$	(36) $y+\frac{1}{4},z+\frac{3}{4},\bar{x}+\frac{3}{4}$		
			(37) $\bar{y}+\frac{1}{2},\bar{x},z+\frac{1}{2}$	(38) y,x,z	(39) $\bar{y},x+\frac{1}{2},\bar{z}+\frac{1}{2}$	(40) $y+\frac{1}{2},\bar{x}+\frac{1}{2},\bar{z}$		
			(41) $\bar{x}+\frac{1}{2},\bar{z},y+\frac{1}{2}$	(42) $x+\frac{1}{2},\bar{z}+\frac{1}{2},\bar{y}$	(43) x,z,y	(44) $\bar{x},z+\frac{1}{2},\bar{y}+\frac{1}{2}$		
			(45) $\bar{z}+\frac{1}{2},\bar{y},x+\frac{1}{2}$	(46) $\bar{z},y+\frac{1}{2},\bar{x}+\frac{1}{2}$	(47) $z+\frac{1}{2},\bar{y}+\frac{1}{2},\bar{x}$	(48) z,y,x		
96	h	$..2$	$\frac{1}{8},y,\bar{y}+\frac{1}{4}$	$\frac{7}{8},\bar{y}+\frac{1}{2},\bar{y}+\frac{3}{4}$	$\frac{3}{8},y+\frac{1}{2},y+\frac{3}{4}$	$\frac{5}{8},\bar{y},y+\frac{1}{4}$		
			$\bar{y}+\frac{1}{4},\frac{1}{8},y$	$\bar{y}+\frac{3}{4},\frac{7}{8},\bar{y}+\frac{1}{2}$	$y+\frac{3}{4},\frac{3}{8},y+\frac{1}{2}$	$y+\frac{1}{4},\frac{5}{8},\bar{y}$		
			$y,\bar{y}+\frac{1}{4},\frac{1}{8}$	$\bar{y}+\frac{1}{2},\bar{y}+\frac{3}{4},\frac{7}{8}$	$y+\frac{1}{2},y+\frac{3}{4},\frac{3}{8}$	$\bar{y},y+\frac{1}{4},\frac{5}{8}$		
			$\frac{1}{8},\bar{y}+\frac{1}{4},y$	$\frac{3}{8},y+\frac{3}{4},y+\frac{1}{2}$	$\frac{7}{8},\bar{y}+\frac{3}{4},\bar{y}+\frac{1}{2}$	$\frac{5}{8},y+\frac{1}{4},\bar{y}$		
			$y,\frac{1}{8},\bar{y}+\frac{1}{4}$	$y+\frac{1}{2},\frac{3}{8},y+\frac{3}{4}$	$\bar{y}+\frac{1}{2},\frac{7}{8},\bar{y}+\frac{3}{4}$	$\bar{y},\frac{5}{8},y+\frac{1}{4}$		
			$\bar{y}+\frac{1}{4},y,\frac{1}{8}$	$y+\frac{3}{4},y+\frac{1}{2},\frac{3}{8}$	$\bar{y}+\frac{3}{4},\bar{y}+\frac{1}{2},\frac{7}{8}$	$y+\frac{1}{4},\bar{y},\frac{5}{8}$		
96	g	$..m$	x,x,z	$\bar{x},\bar{x}+\frac{1}{2},z+\frac{1}{2}$	$\bar{x}+\frac{1}{2},x+\frac{1}{2},\bar{z}$	$x+\frac{1}{2},\bar{x},\bar{z}+\frac{1}{2}$		
			z,x,x	$z+\frac{1}{2},\bar{x},\bar{x}+\frac{1}{2}$	$\bar{z},\bar{x}+\frac{1}{2},x+\frac{1}{2}$	$\bar{z}+\frac{1}{2},x+\frac{1}{2},\bar{x}$		
			x,z,x	$\bar{x}+\frac{1}{2},z+\frac{1}{2},\bar{x}$	$x+\frac{1}{2},\bar{z},\bar{x}+\frac{1}{2}$	$\bar{x},\bar{z}+\frac{1}{2},x+\frac{1}{2}$		
			$x+\frac{3}{4},x+\frac{1}{4},\bar{z}+\frac{3}{4}$	$\bar{x}+\frac{1}{4},\bar{x}+\frac{1}{4},\bar{z}+\frac{1}{4}$	$x+\frac{1}{4},\bar{x}+\frac{3}{4},z+\frac{3}{4}$	$\bar{x}+\frac{3}{4},x+\frac{3}{4},z+\frac{1}{4}$		
			$x+\frac{3}{4},z+\frac{1}{4},\bar{x}+\frac{3}{4}$	$\bar{x}+\frac{3}{4},z+\frac{3}{4},x+\frac{1}{4}$	$\bar{x}+\frac{1}{4},\bar{z}+\frac{1}{4},\bar{x}+\frac{1}{4}$	$x+\frac{1}{4},\bar{z}+\frac{3}{4},x+\frac{3}{4}$		
			$z+\frac{3}{4},x+\frac{1}{4},\bar{x}+\frac{3}{4}$	$z+\frac{1}{4},\bar{x}+\frac{3}{4},x+\frac{3}{4}$	$\bar{z}+\frac{3}{4},x+\frac{3}{4},x+\frac{1}{4}$	$\bar{z}+\frac{1}{4},\bar{x}+\frac{1}{4},\bar{x}+\frac{1}{4}$		
48	f	$2.mm$	$x,0,0$	$\bar{x},\frac{1}{2},\frac{1}{2}$	$0,x,0$	$\frac{1}{2},\bar{x},\frac{1}{2}$	$0,0,x$	$\frac{1}{2},\frac{1}{2},\bar{x}$
			$\frac{3}{4},x+\frac{1}{4},\frac{3}{4}$	$\frac{1}{4},\bar{x}+\frac{1}{4},\frac{1}{4}$	$x+\frac{3}{4},\frac{1}{4},\frac{3}{4}$	$\bar{x}+\frac{3}{4},\frac{3}{4},\frac{1}{4}$	$\frac{3}{4},\frac{1}{4},\bar{x}+\frac{3}{4}$	$\frac{1}{4},\frac{3}{4},x+\frac{3}{4}$
32	e	$.3m$	x,x,x	$\bar{x},\bar{x}+\frac{1}{2},x+\frac{1}{2}$				
			$\bar{x}+\frac{1}{2},x+\frac{1}{2},\bar{x}$	$x+\frac{1}{2},\bar{x},\bar{x}+\frac{1}{2}$				
			$x+\frac{3}{4},x+\frac{1}{4},\bar{x}+\frac{3}{4}$	$\bar{x}+\frac{1}{4},\bar{x}+\frac{1}{4},\bar{x}+\frac{1}{4}$				
			$x+\frac{1}{4},\bar{x}+\frac{3}{4},x+\frac{3}{4}$	$\bar{x}+\frac{3}{4},x+\frac{3}{4},x+\frac{1}{4}$				
16	d	$.\bar{3}m$	$\frac{5}{8},\frac{5}{8},\frac{5}{8}$	$\frac{3}{8},\frac{7}{8},\frac{1}{8}$	$\frac{7}{8},\frac{1}{8},\frac{3}{8}$	$\frac{1}{8},\frac{3}{8},\frac{7}{8}$		
16	c	$.\bar{3}m$	$\frac{1}{8},\frac{1}{8},\frac{1}{8}$	$\frac{7}{8},\frac{3}{8},\frac{5}{8}$	$\frac{3}{8},\frac{5}{8},\frac{7}{8}$	$\frac{5}{8},\frac{7}{8},\frac{3}{8}$		
8	b	$\bar{4}3m$	$\frac{1}{2},\frac{1}{2},\frac{1}{2}$	$\frac{1}{4},\frac{3}{4},\frac{1}{4}$				
8	a	$\bar{4}3m$	$0,0,0$	$\frac{3}{4},\frac{1}{4},\frac{3}{4}$				

Figure 7.6 $Fd\bar{3}m(O_h^7)$.

Another way of describing this crystal structure is to say that the lattice is fcc and that the basis consists of atoms at (0, 0, 0) and (¼, ¼, ¼). Note that, at times, one may want to take the origin at the center of inversion instead of at the position with site symmetry $\bar{4}3m(T_d)$. The next page in the *ITA*, which is not shown here, does exactly that. Thus the 8*a* position with $\bar{4}3m$ site symmetry has coordinates (⅛, ⅛, ⅛), (⅞, ⅞, ⅞), and the 16*c* position with site symmetry $\bar{3}m$ has coordinates (0, 0, 0), and so on.

7.5. SPINEL STRUCTURE

Spinel is an important mineral, with a formula of the type $MgAl_2O_4$. Like diamond it adopts the cubic space group $Fd\overline{3}m(O_h^7)$ with the atoms, Wyckoff symbols and the site symmetries (Fig. 7.6):

$$
\begin{array}{lll}
\mathrm{Mg}: & 8a & \overline{4}3m(T_d) \\
\mathrm{Al}: & 16d & \overline{3}m(D_{3d}) \\
\mathrm{O}: & 32e & 3m(C_{3v})
\end{array}
$$

For most materials with this crystal structure the free parameter for the 32*e* position is close to ⅜. From the structure, it can be determined that atoms on the 8*a* sites are tetrahedrally surrounded by O atoms, while those on the 16*d* sites are surrounded by six O atoms in the form of a distorted octahedron.

The spinel structure has an interesting variant, $MgFe_2O_4$. Here half of the Fe atoms are on the 8*a* sites, and the remaining Fe plus the Mg atoms are on the 16*d* sites. This structure is then called an **inverted spinel.** There are many materials that have the spinel structure, but with distributions of metal atoms somewhere between a normal and inverted arrangement.

7.6. ZINC SULPHIDE STRUCTURE

Figure 7.2h shows the zinc-blende (ZnS) or gallium arsenide (GaAs) structure, with space group $F\overline{4}3m(T_d^2)$. It can be seen that there is a close relationship between this structure, the diamond structure and the fluorite structure. Figure 7.7 shows the relevant page of the *ITA* for this space group. The Zn or Ga atoms are located at 4*a* positions, and the S or As atoms at 4*c* positions (or vice versa). The relationship between the ZnS or GaAs structure and the diamond structure is particularly interesting. For $Fd\overline{3}m$, the diamond glide reflections and the screw rotations would take the atoms at positions totally within the unit cell, such as (¼, ¼, ¼), to positions at the face-centers of the cell as in Fig. 7.2h. Such operations cannot be symmetric for $F\overline{4}3m$, because these two positions are not occupied by the same types of atoms. We see from the symbol that $F\overline{4}3m$ is a symmorphic space group. The point group is $\overline{4}3m(T_d)$, which has half as many symmetry operations as for the point group of the $Fd\overline{3}m$ space group. In fact one may show that the

Multiplicity, Wyckoff letter, Site symmetry	Coordinates					
	(0, 0, 0)+ (0, ½, ½)+ (½, 0, ½)+ (½, ½, 0)+					
96 *i* 1	(1) x, y, z	(2) $\bar{x}, \bar{y}, z$	(3) $\bar{x}, y, \bar{z}$	(4) $x, \bar{y}, \bar{z}$		
	(5) z, x, y	(6) $z, \bar{x}, \bar{y}$	(7) $\bar{z}, \bar{x}, y$	(8) $\bar{z}, x, \bar{y}$		
	(9) y, z, x	(10) $\bar{y}, z, \bar{x}$	(11) $y, \bar{z}, \bar{x}$	(12) $\bar{y}, \bar{z}, x$		
	(13) y, x, z	(14) $\bar{y}, \bar{x}, z$	(15) $y, \bar{x}, \bar{z}$	(16) $\bar{y}, x, \bar{z}$		
	(17) x, z, y	(18) $\bar{x}, z, \bar{y}$	(19) $\bar{x}, \bar{z}, y$	(20) $x, \bar{z}, \bar{y}$		
	(21) z, y, x	(22) $z, \bar{y}, \bar{x}$	(23) $\bar{z}, y, \bar{x}$	(24) $\bar{z}, \bar{y}, x$		
48 *h* . . *m*	x, x, z	$\bar{x}, \bar{x}, z$	$\bar{x}, x, \bar{z}$	$x, \bar{x}, \bar{z}$	z, x, x	$z, \bar{x}, \bar{x}$
	$\bar{z}, \bar{x}, x$	$\bar{z}, x, \bar{x}$	x, z, x	$\bar{x}, z, \bar{x}$	$x, \bar{z}, \bar{x}$	$\bar{x}, \bar{z}, x$
24 *g* 2 . *m m*	x, ¼, ¼	$\bar{x}$, ¾, ¼	¼, x, ¼	¼, $\bar{x}$, ¾	¼, ¼, x	¾, ¼, $\bar{x}$
24 *f* 2 . *m m*	x, 0, 0	$\bar{x}$, 0, 0	0, x, 0	0, $\bar{x}$, 0	0, 0, x	0, 0, $\bar{x}$
16 *e* . 3 *m*	x, x, x	$\bar{x}, \bar{x}, x$	$\bar{x}, x, \bar{x}$	$x, \bar{x}, \bar{x}$		
4 *d* $\bar{4}$ 3 *m*	¾, ¾, ¾					
4 *c* $\bar{4}$ 3 *m*	¼, ¼, ¼					
4 *b* $\bar{4}$ 3 *m*	½, ½, ½					
4 *a* $\bar{4}$ 3 *m*	0, 0, 0					

Figure 7.7 $F\bar{4}3m(T_d^2)$.

space group $F\bar{4}3m(T_d^2)$ is a subgroup of $Fd\bar{3}m(O_h^7)$. We also see in Fig. 7.6 that for space group $F\bar{4}3m(T_d^2)$, there is no position that has a center of inversion since this would result in taking a Zn or Ga atom at (0, 0, 0) into a S or As atom at (¼, ¼, ¼).

As an example of a more complicated material that has this space group we note that the mineral zunyite $Al_{13}Si_5O_{20}(OH,F)Cl$ has Cl – 4*b*; Si(1) – 4*c*; Si(2) – 16*e*; Al(1) – 4*d*; Al(2) – 48*h*; O(1) – 16*e*; O(2) – 16*e*; O(3) – 48*h*; (OH,F)(1) – 24*f*; (OH,F)(2) – 48*h*. We quote this result only to remind the reader that complicated as well as simple crystal structures can still have the same space group.

The cubic zinc-blende structure, considered here, is closely related to the hexagonal wurtzite structure. In fact several compounds can be found to occur in both structures. Problem 1, at the end of this chapter, discusses the wurtzite as well as the simpler hexagonal close-packed (hcp) structures.

7.7. CHALCOPYRITE

The chalcopyrite structure, which derives its name from copper pyrite ($CuFeS_2$), is closely related to the zinc-blende structure. We consider it for two reasons. First, there have been many studies of materials with this or closely related structures. These include $AgGaS_2$, $AgGaSe_2$ and $AgGaTe_2$, as well as $CdGeP_2$ and $CdGeAs_2$, and similar compounds. Second, this structure, which is based on a superstructure of zinc-blende, serves as an excellent introduction to artificial superlattices, the subject of the next section.

It is instructive to consider the chalcopyrite structure by starting with zinc-blende (Fig. 7.2h). When viewed perpendicular to the **c**-axis, ZnS appears to consist of layers of Zn on planes at height 0 and ½ with S atoms on planes at height ¼ and ¾. Now stack two zinc-blende unit cells upon each other as in Fig. 7.8 and replace the Zn atoms with Cu and Fe atoms as shown. Note that on the planes at 0, ¼, ½ and ¾ of the *new* unit cell there are as many Cu as there are Fe atoms (two on each plane within the unit cell). The resulting arrangement is the chalcopyrite structure with a unit-cell repeat twice as long as in zinc-blende. However, just like in zinc-blende and diamond, all of the atoms are tetrahedrally coordinated to four other atoms.

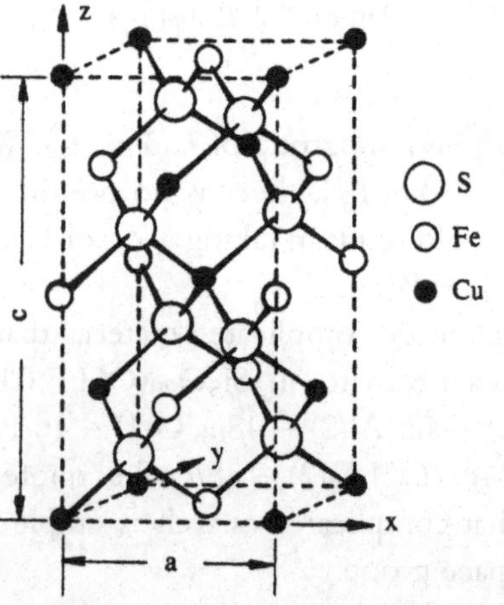

Figure 7.8 The conventional body-centered unit cell of $CuFeS_2$ in the chalcopyrite structure.

The structure can be described by space group $I\bar{4}2d(D_{2d}^{12})$ with the atoms at positions:

$$\begin{array}{lll} \text{Cu}: & 4a & (0,0,0);\ (\tfrac{1}{2},0,\tfrac{3}{4});+ \\ \text{Fe}: & 4b & (0,0,\tfrac{1}{2});\ (\tfrac{1}{2},0,\tfrac{1}{4});+ \\ \text{S}: & 8d & (u,\tfrac{1}{4},\tfrac{1}{8});\ (\bar{u},\tfrac{3}{4},\tfrac{1}{8});\ (\tfrac{1}{4},u,\tfrac{7}{8});\ (\tfrac{1}{4},\bar{u},\tfrac{7}{8});+ \end{array}$$

where + signifies that the body-centering condition (½, ½, ½) must be added and $u \approx$ ¼.

There are several reasons that materials with this structure are of interest. First, if magnetic impurity atoms (e.g. Mn) are added to $AgGaS_2$, for example, they tend to occupy the Ga sites preferentially. Thus they are considerably farther apart from each other than when substituted into ZnS. This increased separation can advantageously reduce their mutual interaction. Second, the chalcopyrite structure is a superstructure of zinc-blende (Appendix 9). New diffraction spots occur in one direction at half the distance expected for the zinc-blende structure, as required by the $I\bar{4}2d(D_{2d}^{12})$ space group. Thus the **c**-axis length in the chalcopyrite structure is about twice that for the corresponding zinc-blende compound (Fig. 7.8). Then, to a reasonable approximation, a material such as $CdGeAs_2$ is similar to GaAs, except that one dimension of the unit cell is doubled. In reciprocal space, this halves the Brillouin zone in that direction, which, in terms of electronic band structures or phonon dispersion curves, can be described as a folding back of the energy versus wave-vector curves in that direction.

There are other materials whose chemical formulae resemble that of chalcopyrite, but that can be legitimately described as having the zinc-blende crystal structure, with space group $F4\bar{3}m(T_d^2)$. Typical examples are $ZnSnAs_2$ and $MgGeP_2$. For these materials the two metal atoms randomly occupy the Zn positions of the ZnS structure. Of course, it is often of interest to see how a small amount of randomness of the metal atoms in a true chalcopyrite can affect some of its properties. Some degree of randomness can usually be obtained by quenching the sample from a high temperature.

7.8. SEMICONDUCTOR SUPERLATTICES

For many systems, it is possible to grow, in an ultrahigh vacuum (UHV) environment by molecular-beam epitaxy (MBE), extremely thin, very perfect films of one substance on a crystalline surface of another (the latter is called the **substrate**). With the correct materials and growth conditions, the film can maintain perfect atomic registry with the substrate. The resulting structure is an **artificial structure** or **artificial heterostructure**.

New structures of a given material can be formed; for example, if a thin film of CsCl is grown on NaCl, the CsCl can adopt the NaCl structure. Metal artificial structures can also be made. For example, Fe normally has a body-centered cubic (bcc) structure but when grown on fcc Cu, Fe adopts the fcc structure, at least up 15 or 20 Fe layers. In this epitaxial form, the Fe layers have unusual magnetic ordering properties, some layers being ferromagnetic and some apparently coupling antiferromagnetically. However, at present, most epitaxial work is concentrated on semiconductors and this is the area that we discuss here.

In addition to a heterostructure made with materials A and B, a **superstructure** or **superlattice** can be grown by constantly repeating the process. Thus a real (periodic) crystal structure can be obtained in the form ABAB.... Since a third and fourth material can be used (ABCDAB...), the varieties of superlattices can become very large. These superstructures can be detected by the appearance of new diffraction spots.

We shall discuss just two important semiconductor superlattice systems. Consider GaAs and AlAs both of which have the zinc-blende structure (Fig. 7.2h), with space group $F\bar{4}3m(T_d^2)$ and lattice constant **a**. A superlattice of these materials[1] can be grown along the [001] direction by the periodic growth of *n* layers of GaAs, followed by *m* layers of AlAs, and then *n* layers of GaAs, and so on. This superlattice is designated $(GaAs)_n/(AlAs)_m$, and we assume that all of the atoms lie on the normally occupied sites of the zinc-blende structure. For example, the metal atoms in this structure lie on sites with $z = 0$, ½, 1, 1½, ..., with respect to the cubic lattice constant **a**. Thus, for $(GaAs)_2/(AlAs)_1$, Ga is at 0, ½, 1½, 2, ... and Al is at 1, 2½, ...

For $(GaAs)_n/(AlAs)_m$ superlattices grown along the [001] direction, there are an infinite number of structures but only two possible space groups depending on whether $(n + m)$ is an even or odd number:

$$n + m = even \rightarrow P\bar{4}m2(D_{2d}^5)$$

$$n + m = odd \rightarrow I\bar{4}m2(D_{2d}^9)$$

[1] See, for example, J. Sapriel, J. C. Michel, J. C. Toledano, R. Vacher, J. Kervarec & A. Regreny, *Phys. Rev.* B **28**, 2007 (1983).

The first space group is primitive, while the second is body-centered. When $(n + m)$ is even, the basis vectors of the primitive tetragonal Bravais lattice are given by

$$[110]\tfrac{1}{2}\mathbf{a} \quad [\bar{1}10]\tfrac{1}{2}\mathbf{a} \quad [001](n+m)\tfrac{1}{2}\mathbf{a}$$

Both of the space groups are symmorphic, as is their parent $F\bar{4}3m$, and they are both subgroups of $F\bar{4}3m$.

For a $(GaAs)_n/(AlAs)_m$ superlattice grown along the [111] direction, there are also only two possible space groups:

$$(n+m)/3 = k \rightarrow P3m1(C^1_{3v})$$

$$(n+m)/3 \neq k \rightarrow R3m(C^5_{3v})$$

where k is an integer. For this superlattice, but grown along the [110] direction, there are four possible space groups depending on whether n, m and $(n + m)$ are even or odd.

The other superlattice system that we consider is that composed of Si and Ge[2], both of which normally have the diamond structure (Fig. 7.2f), with space group $Fd\bar{3}m(O^7_h)$. Take for instance the superlattice Si_n/Ge_m, grown along [001]. The formula Si_n/Ge_m means that in the [001] direction there are n layers of Si atoms, followed by m layers of Ge atoms, and then the pattern repeats with both the Si and Ge atoms on the normal positions of the diamond structure. For example, for Si_3/Ge_2 the same types of atoms extend throughout the crystal along the **a** and **b**-directions, but along the **c**-axis there are three layers of Si atoms at $z = 0$, ¼, ½, followed by two layers of Ge atoms at ¾ and 1. The pattern then repeats itself with Si at 1¼, 1½, 1¾, and so on. The result is a superstructure of the diamond structure, whose periodicity depends on n and m as well as on the stacking (growth) direction, which we have taken to be the [001] direction. Note that for Si_1/Ge_1, the zinc-blende structure is obtained, which is interesting in that it is obtained using two group IV elements.

Thus we can obtain the diamond and zinc-blende structures if $n = 0 \neq m$, or $n = 1 = m$, respectively. In addition to the space groups of these structures, only five other space groups are obtained, depending on whether n and m are both even (two space groups), both odd (two space groups), or if one is even and the other odd (one space group). Letting k be any integer, the results can be summarized in the following table.

[2] M. I. Alonso, M. Cardona & G. Kanellis. *Solid State Commun.* **69**, 479 (1989).

n, m even:

$$n + m = even = 4k \rightarrow Pmma(D_{2h}^{5})$$

$$n + m = even = 4k + 2 \rightarrow Imma(D_{2h}^{28})$$

n, m odd:

$$n + m = even = 4k \rightarrow P\bar{4}m2(D_{2d}^{5})$$

$$n + m = even = 4k + 2 \rightarrow I\bar{4}m2(D_{2d}^{9})$$

n, even and m odd, or vice versa:

$$n + m = 2k + 1 \rightarrow I4_1/amd(D_{4h}^{19})$$

For each n and m pair, the superstructure is different, and yet all of these structures can be classified according to these five space groups plus the zinc-blende space group. For example, for $n = 8$ and $m = 4$, we have $n + m = 12 = 4(3)$ and so the structure has the *Pmma* space group; but if $n = 8$ and $m = 6$ we have $n + m = 14 = 4(3) + 2$ and therefore the *Imma* space group.

Some generalizations can be made about the important properties of these structures. In particular, their electronic and vibrational properties can be understood in terms of these space groups. For example, when both n and m are odd the structure has either the $P\bar{4}m2(D_{2d}^{5})$ or the $I\bar{4}m2(D_{2d}^{9})$ space group, which in turn is a type I t-subgroup of the zinc-blende $F\bar{4}3m(T_{d}^{2})$ space group. Because these space groups are not centrosymmetric, these materials could in principle show bulk optical second harmonic generation and their phonons should be simultaneously infrared and Raman active, even though they are composed of just Si and Ge atoms.

Note that even for the relatively simple $(GaAs)_n/(AlAs)_m$ and Si_n/Ge_m systems, we have only discussed growth along a few of the important directions. With MBE growth, it is practical to grow superstructures of these materials on many low-index crystallographic planes such as (012), (310), (211). Clearly, this is a rich field of study.

7.9. STRUCTURAL PHASE TRANSITIONS IN CRYSTALS

A phase transition means an abrupt change of thermodynamic variables characterizing a material. Typically this occurs at a particular temperature T_c or pressure p_c, although, for some materials, it is possible to induce a phase transition by other means, such as through an applied electric, magnetic or mechanical (stress) field. In a **structural phase transition (SPT)**,[3] the crystal structure changes with an accompanying symmetry change. The change in structure may be quite small, and sometimes it is difficult to detect; on the other hand, dramatic changes can also occur. Recall from Chapter 1 that at the phase transition the symmetry cannot change continuously: there is no 'half-way house' for symmetry – it is either one or the other. Thus symmetry is discrete and the phase transition usually requires breaking of the symmetry found in one phase to form a new phase with its particular symmetry.

Broadly speaking, SPTs can be classified into three types, reconstructive, order–disorder and displacive. **A reconstructive phase transition** is one in which transition from one phase to the other results in breaking and appropriate rejoining of bonds in the other phase such that the orientations of the bonds in the two phases may be distinctly different. That is, the topological linkage pattern of the bonds in the two phases may be drastically altered. Reconstructive phase transitions tend to be slow (seconds to 10^{-9} s), because atoms must diffuse from one set of positions in one structure to different positions in the other so that a new set of bonds may be formed. If one starts with a single crystal then in the other phase it may actually fragment into smaller crystallites because the bonds in the new phase have few systematic orientations (relationships) with the old phase. Thus, in different parts of the crystal, the reformed bond directions are incompatible with those in the other phase.

Order–disorder type SPTs can be divided into two types. In a **substitutional order–disorder phase transition**, both phases have close (or the same) orientational relations of the bonds, even though atoms may diffuse in order to go from one phase to the other. The time required to complete the phase transition is similar to that found in a reconstructive SPT. In an **orientational order–disorder phase transition**, small groups of atoms change their orientation by small amounts, and so they do not disturb the fundamental bonding in the material. In both of these order–disorder SPTs, the high-temperature phase involves disorder to some extent, either positional (substitutional) or orientational. Since, in the latter, no long-range

[3] H. T. Stokes & D. M. Hatch, 'Isotropy Subgroups of the 230 Crystallographic Space Groups' (World Scientific, Singapore, 1988) give an exhaustive list of possible symmetry changes that can occur in phase transitions where the space-group symmetry of one phase is a subgroup of the space-group symmetry of the other phase. Each symmetry change is associated with an irreducible representation of the space group of the higher-symmetry phase.

diffusion is required, these phase transformations usually occur more rapidly than do reconstructive SPTs.

In a **displacive phase transition** bonds are not broken; the atoms are simply displaced with respect to one another by small distances (small compared with the internuclear distances). Thus, in this phase transition, the topology of the linkage pattern in the two phases is unaltered. Since the atomic movements are small, the time required to complete the motion is of the order of the period of a phonon (a lattice vibration), which is fast (10^{-11} to 10^{-15} s).

There are few general comments relating to space groups that can be made about reconstructive and order–disorder phase transitions. This is because the two phases often have such different structures (and space groups) that each must be described separately. For example, at room temperature and pressure RbI has the NaCl structure (Fig. 7.2b) and each atom has 6-coordination. Then, at a fairly low pressure ($p_c = 4.3$ kbar), there is a phase transition to the CsCl structure (Fig. 7.2d), so that the atoms have 8-coordination. To go from the NaCl to the CsCl structure, bonds must be broken and new bonds must be made. This increase in coordination number (and density) is fairly typical for the high-pressure phases of materials. Materials undergoing an orientational order–disorder phase transition typically have a high-temperature structure with multi-well potentials for a single set of atoms. Thus, in each unit cell, an atom will randomly occupy one of several possible sites at the minima of the potential energy function. For example, at room temperature, NH_4Cl has a disordered CsCl structure (Fig. 7.2d), where the NH_4 group can have one of two possible orientations, so that the tetrahedrally arranged N–H bonds point to one set of four of the eight nearest-neighbouring Cl atoms. Since neither of the two possible sets is favoured over the other, on going from one unit cell to the next there is no periodic relationship between the sets chosen, and such a structure has no true translational periodicity. In other words, it is disordered, although if we average over the whole crystal, we can treat the structure as one in which the H atoms fractionally occupy *all* of the eight possible sites equally. The space group of this average structure is $Pm\overline{3}m(O_h^1)$. Below the phase transition, the NH_4 groups are aligned parallel to each other and so true translational symmetry is obtained to give space group $P\overline{4}3m(T_d^1)$. Other examples of order–disorder phase transitions are found in the reorientation of molecular groups such as SO_4, BeF_4, …, and perhaps changes in hydrogen bonds such as in K_2HPO_4 (KDP), or in triglycine sulphate (TGS).

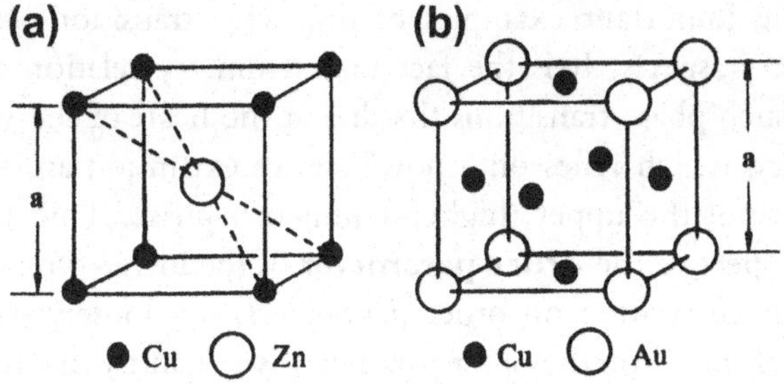

Figure 7.9 The low-temperature ordered structures of (a) CuZn and (b) Cu_3Au.

Two examples of substitutional order–disorder phase transitions, in which space group relationships are useful to consider, occur in β-brass (CuZn) and Cu_3Au. Figure 7.9 shows the room-temperature structures. Their symmetries can both be described by the space group $Pm\overline{3}m(O_h^1)$ (Fig. 7.4) with atoms at

$$\text{CuZn}: \quad \text{Cu at } 1a; \quad \text{Zn at } 1b$$

$$\text{Cu}_3\text{Au}: \quad \text{Au at } 1a; \quad \text{Cu at } 3c$$

Above a certain temperature ($T_c = 460$ and 350 °C), the ordered arrangement of atoms disappears and each site shown in the figures is randomly occupied by either of the two atoms in the material. Thus, at high temperatures, the average symmetries are different. The high-temperature structure of CuZn is bcc, with space group $Im\overline{3}m(O_h^9)$, while for Cu_3Au it is fcc, with space group $Fm\overline{3}m(O_h^5)$.

Many materials have these ordered structures. Of course the ordered CuZn structure is similar to the CsCl structure. MPt_3 alloys (M = Ti, V, Cr, Mn, Fe and Co) have the ordered Cu_3Au structure with interesting magnetic properties; they display all three classical types of magnetism: ferromagnetism, antiferromagnetism and ferrimagnetism.

7.10. DISPLACIVE SPTS

Displacive phase transitions usually involve relatively subtle changes in the crystal structure, which results in interesting symmetry relationships between the two phases. A displacive transition may take place continuously through a gradual change in the atomic displacements (although note that the *symmetry* still changes abruptly at the transition point), known as **continuous** or **second order.** On the other hand, there may be a discontinuity in the volume of the crystal, called **first order.** Here we shall describe some important examples of displacive transitions emphasizing their symmetry aspects. It is the fact that symmetry relationships can be expected in such phase transitions that lies at the heart of the well-known Landau theory, which relies on a power series expansion around the free energy density of the upper (higher-symmetry) phase. This expansion is taken with respect to the **order parameter** of the lower-symmetry phase. Books have been written on order parameters, but loosely speaking, for a second-order transition the order parameter is a quantity that measures the deviation of the structure from the high-symmetry (high-temperature) phase. Thus it changes from zero to non-zero in going from the high to the low-symmetry structure.

7.10.1. Quartz

The mineral quartz (SiO_2) crystallizes in a structure with space group $P3_121(D_3^4)$ or $P3_221(D_3^6)$. These two space groups are enantiomorphically related and so the structures have opposite chirality (or hand). An examination of quartz crystals shows that equal numbers of crystals grow in each space group. By observation of the optical rotation of polarized light propagating along the 3-fold axis of the crystal, the two types of crystals can be separated. Crystals with space group $P3_121(D_3^4)$ rotate the plane of polarization anti-clockwise for the light coming towards the observer (they are said to be **laevorotatory**), while quartz crystals with space group $P3_221(D_3^5)$ are **dextrorotatory.** The relationship between the sense of rotation of the light and the chirality of the structure is outside the scope of this book,[4] but it can be appreciated that the two forms must be related. Both the 'hands' of the structure (and space group) and of the optical properties change together.

Figure 7.10 shows the laevorotatory structure together with the symmetry operations. The small circles are Si and the large circles are O. Heights of oxygen atoms are marked in 1/100th of the **c**-axis repeat. The atoms are at the following positions:

Si : at $3a$ sites $(½-u_1,\ 0,\ 0)$ with $u_1 = 0.03$

O : at $6c$ sites $(x,\ y,\ z)$

with $x = x_0 - u_0$, $y = \bar{x}_0 - u_0$, and $z = ⅙ - w_0$. Here, $x_0 = 0.21$, $u_0 = 0.06$ and $w_0 = 0.05$. Our choice of origin is not that of the *ITA*, which is at

4 See A. M. Glazer & K. Stadnicka. *J. Appl. Cryst.*, **19**, 108 (1986) for a discussion on this topic.

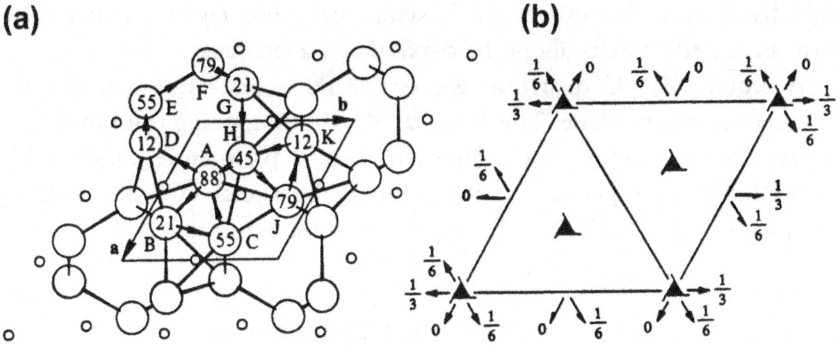

Figure 7.10 (a) A projection on (001) of the laevorotatory structure of α-SiO_2 (low quartz). (b) Symmetry elements of $P3_121(D_3^4)$; the heights have been altered with respect to the *ITA* by displacing the origin through 0, 0, −1/3 to be consistent with the chosen origin for the structure in (a).

height **c**/3 with respect to our choice of axes. Instead, we have moved the origin in order to make the structure and space group diagrams consistent for both the low and high-temperature phases. This allows for a more straightforward comparison. In this description, the atomic positions can be described in terms of small displacements u_0, u_1 and w_0 with respect to 'special' positions in the unit cell (they are not special in this phase, but become special positions in the space group of the high-temperature phase of quartz, discussed as follows).[5]

The oxygen atoms form three different types of helices about the **c**-axis. There are two helices ABC and HJK, each consisting of three oxygen atoms in the **c**-axis repeat, each helix with its center on the 3_1 axis along $\frac{2}{3}$, $\frac{1}{3}$, z and $\frac{1}{3}$, $\frac{2}{3}$, z, as can been seen in Fig. 7.10. All the oxygen atoms on each of these helices are related by the 3_1 screw operation, and so the helices are perfect right-hand screws. However, there is a larger imperfect, so-called **structural helix** DEFGHA of six oxygen atoms around the 3_1-axis at 0, 0, z (i.e. through the origin). In this helix, there are two sets of three oxygen atoms alternating. The oxygen atoms D, F and H are related to each other by the 3_1 operation, as are the oxygen atoms E, G and A. The oxygen atoms of one set are related to those of the other set by the twofold operations about [100], [010] and $[\bar{1}\bar{1}0]$, which are symmetry operations of this space group (Fig. 7.10b). Thus, while the separate helices DFH and EGA are each

5 The description of the structure of quartz is wrong in many publications! For a review see J. D. H. Donnay & Y. Le Page. 'The vicissitudes of the low-quartz crystal setting or the pitfalls of enantiomorphism', *Acta Cryst.* **A34**, 584 (1978).

right-hand screws because of the 3_1 symmetry, when taken together they form an imperfect helix about the **c**-axis that is a left-hand screw!

At around 550 °C, quartz undergoes a nearly second-order transition to a higher-symmetry phase. This is the $\alpha-\beta$ transition, going from the low-temperature α-structure to the high-temperature β-structure. Figure 7.11 shows the high-temperature structure for the laevorotatory crystal, which has space group $P6_422(D_6^5)$. The atoms are at the positions

$$\text{Si}: \quad \text{at } 3c \text{ sites} \quad (\tfrac{1}{2}, 0, 0)$$

$$\text{O}: \quad \text{at } 6j \text{ sites} \quad (x, \bar{x}, \tfrac{1}{6}) \text{ at } x \approx 0.20$$

The only changes in the structure are small movements of the silicon and oxygen atoms to special positions of the high-temperature space group making the structure more symmetric. For instance, the imperfect structural helix DEFGHA of the low-temperature phase is now a perfect sixfold helix in which all the oxygen atoms are related by a 6_4-axis. Here both the screw symmetry as well as the actual helix of atoms have the same hand (i.e. left hand).

It is important to realize here that the chirality of the crystal structure is not determined simply by the space group operations alone, but by the relationships between the atoms in the structure; that is, it is the *structural* helix that is important, not the *symmetry* helix. Thus, in laevorotatory quartz, the space group change is $P3_121$ to $P6_422$. The screw operations 3_1 and 6_4 form helices of opposite hand (right hand and left hand, respectively),

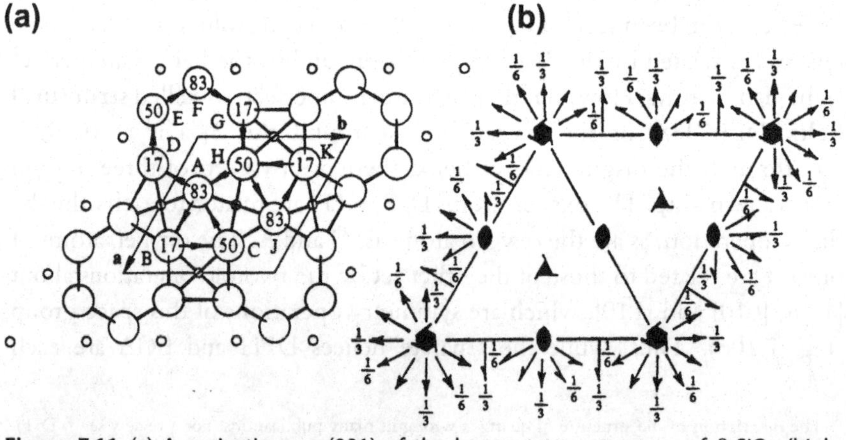

Figure 7.11 (a) A projection on (001) of the laevorotatory structure of β-SiO_2 (high quartz). (b) Symmetry elements of $P6_422(D_6^5)$.

and yet it is possible to transform from one form to the other easily since the hand of the structural helix is not changed (this will become clearer in the $AlPO_4$ example to follow). Indeed, the *ITA* lists $P6_422$ as a Type I minimal non-isomorphic supergroup of $P3_121$ (and $P3_121$ as a Type I maximal non-isomorphic subgroup of $P6_422$). Thus this apparent change in hand of symmetry is given by a group–subgroup relationship, and therefore can be achieved through a displacive transition. In deciding on questions of chirality and its relationship to physical properties, one should not only consider the space-group symmetry in isolation but also the disposition of the atoms in the structure.

Another example of this kind of problem can be found in the **berlinite structure** ($AlPO_4$). The laevorotatory compound is shown in Fig. 7.12. This structure is similar to quartz, except that Al and P replace the silicon atoms *alternately* along the helices. The effect of the alternating substitution is to make the repeat distance along the **c**-axis approximately twice that found in quartz as can be seen by comparing quartz and berlinite in Fig. 7.13. Surprisingly perhaps, we now note that the space group is $P3_221(D_3^6)$, while in laevorotorary quartz it is $P3_121(D_3^4)$. The two structures look superficially similar, and rotate light in the same sense, and yet the space groups are enantiomorphically opposite (Appendix 7).

How can this be possible, when the structures are so closely related? The reason for this comes from the alternating Al and P atoms. First, consider a right-hand helix consisting of three identical Si atoms in a repeat

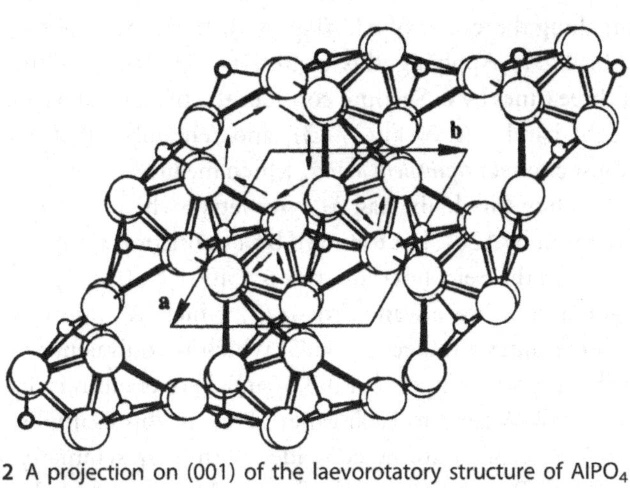

Figure 7.12 A projection on (001) of the laevorotatory structure of $AlPO_4$ (berlinite). Small circles Al and P, and large circles, O. Arrows indicate helical arrangements directed out of the page.

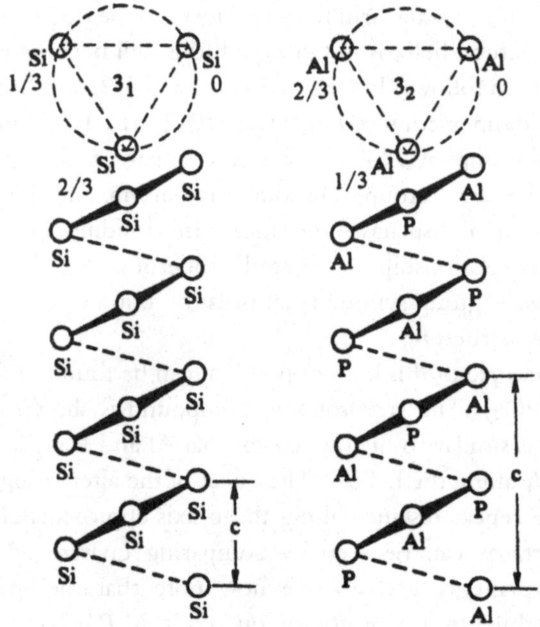

Figure 7.13 Left: A helix of Si atoms in laevo-quartz seen perpendicular to **c** (top view shown above). The atoms are related by a 3_1-axis to form a right-hand helix. Right: On replacing the Si atoms by alternating P and Al atoms, the resulting helix is still right hand, but the screw symmetry is now 3_2 (the opposite hand).

(left-hand side of Fig. 7.13). This helix has 3_1 symmetry, which means that the symmetry operation is a counter-clockwise rotation by 120° followed by a translation along the **c**-axis of **c**/3 (Fig. 5.4). In the Seitz notation, this is {3[001]|0, 0, 1/3}. Applying this symmetry operation to the Si helix advances it three times by **c**. Second, consider the helix of alternating Al and P atoms (right-hand side of Fig. 7.13), and remember that a symmetry operation must connect *equivalent atoms*, which means Al to Al, or P to P, but not Al to P. Thus this helix has 3_2 symmetry, which means a counter clockwise rotation by 120°, but followed by a translation along the **c**-axis of 2**c**/3 (Fig. 5.4). In the Seitz notation, this is {3[001]|0, 0, 2/3}. By applying this operation twice, we advance to 4**c**/3, which is equivalent to **c**/3; applying it three times advances to 6**c**/3, which is equivalent to **c**.

One of the important aspects of the 3_1 and 3_2 operations is that they are enantiomorphically related to each other. Thus in this example of quartz, the structural helix of Si atoms coincides with a 3_1 symmetry helix. In $AlPO_4$, the Al or P atoms separately form a 3_2 symmetry helix, whereas taken together they form a structural helix whose chirality is the same as in quartz. Hence, despite the change in symmetry, the effect is to rotate the polarization of light in the same sense.

Again, note that in the *ITA*, space group $P3_221$ is a Type IIc maximal isomorphic subgroup of lowest index of $P3_121$ in which the **c**-axis is doubled, so that our observation is consistent with a subgroup–supergroup relationship. The phase transitions in quartz and in berlinite are of second order, in agreement with this group–subgroup relationship.

7.10.2. Perovskites

The perovskite structure is quite simple.[6] Perovskites are compounds with general formula ABX_3, where A and B are cations and X is an anion, typically oxygen or fluorine. The basic structure was shown in Fig. 7.5 and has space group $Pm\bar{3}m(O_h^1)$ with atoms at

A	at $1a$ sites	(½, ½, ½)
B	at $1b$ sites	(0, 0, 0)
X	at $3d$ sites	(½, 0, 0); (0, 0, ½); (0, ½, 0)

However, sometimes this structure is described with the A cation placed at the origin, hence the atomic positions are:

A	at $1b$ sites	(0, 0, 0)
B	at $1a$ sites	(½, ½, ½)
X	at $3c$ sites	(½, ½, 0); (½, 0, ½); (0, ½, ½)

Either description is acceptable, depending on which features one wishes to stress. The structure consists of corner-linked X-anion octahedra arranged in a regular fashion, with the A cation at the center of the space between the octahedra and the B cation located at the center of each octahedron.

Many materials that have this crystal structure undergo various kinds of SPTs. Often the low-symmetry (low-temperature) structures have physical properties that are rather different from those found in the (high-symmetry) cubic phase. Thus many of these materials are exploited in technological applications, and so a fundamental understanding of their structures and symmetries is important. The changes that occur are summarized here.

1. Displacements of the cations can be along certain directions. These may all be parallel to each other, in which case the crystal is polar; such crystals are pyroelectric and piezoelectric. Otherwise the displacements

[6] See the book *Perovskites: Modern and Ancient* by R.H. Mitchell (Almaz, 2002) which contains a wealth of information about perovskite structures.

can be anti-parallel leaving the crystal centrosymmetric, but with a larger unit cell.

2. Tilting of the anion octahedra can occur. Imagine slightly rotating one of the octahedra. Then you can see that because of the linkage to neighbouring octahedra, the other octahedra in the crystal must also rotate in some way. It turns out that for a perovskite there are many possible structures that can be formed by tilting alone (Appendix 9).
3. The octahedra can become distorted, especially when Jahn–Teller effects become important, for example, in $KCuF_3$.
4. Combinations of cation displacements, octahedral tilts and distortions can occur.

The archetypal example of the first type is given by $BaTiO_3$ (Fig. 7.14). Above $T_c \approx 135$ °C, $BaTiO_3$ has the cubic $Pm\overline{3}m(O_h^1)$ perovskite structure. However, on cooling below this temperature, the structure undergoes an SPT to a tetragonal phase (Fig. 7.14b), in which the cations are displaced along one of the cube axes with respect to the oxygen octahedra framework. The resulting space group is $P4mm(C_{4v}^1)$, a subgroup of $Pm\overline{3}m(O_h^1)$, although not a *maximal* subgroup (it is a maximal subgroup of $P4/mmm$, which is a Type I maximal non-isomorphic subgroup of $Pm\overline{3}m$). This

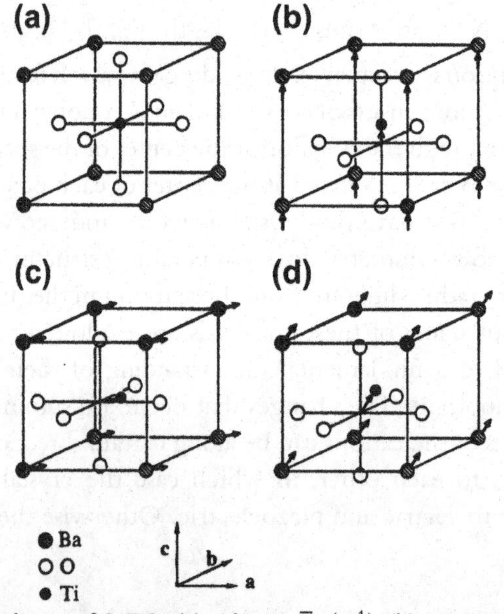

Figure 7.14 The phases of $BaTiO_3$ (a) cubic $Pm\overline{3}m(O_h^1)$, (b) tetragonal $P4mm(C_{4v}^1)$, (c) orthorhombic $C2mm(C_{2v}^{14})$ and (d) rhombohedral $R3m(C_{3v}^5)$.

structure is polar and is the room-temperature structure. Because of the large dielectric anomalies associated with this phase transition, $BaTiO_3$, and closely related materials, are used as high-dielectric constant materials in capacitors in the electronics industry.

The cubic-tetragonal transition could in principle be second order, because the low-temperature phase has a space group that is a subgroup of the high-temperature phase. In fact, it is first order for reasons that lie outside the scope of this book, but have to do with the coupling to elastic strains. Interestingly, $BaTiO_3$ undergoes other displacive phase transitions on further cooling. Thus, on cooling the tetragonal phase, the cations become displaced along the [110] direction (Fig. 7.14c). Examination of this diagram reveals that there are 2-fold axes along [110] with mirror planes perpendicular to [001] and parallel to (110). That is, the point group symmetry is $mm2(C_{2v})$, and the crystal system is orthorhombic. However, because the 2-fold-axis is along a face-diagonal, in order to specify an orthorhombic space group, a new cell has to be constructed with axes containing the 2-fold-axis and a mirror plane. This new cell is outlined in Fig. 7.15, where we see that it is *C*-centered with the 2-fold-axis along the new $\mathbf{a}_0$-axis and mirror planes perpendicular to the new $\mathbf{b}_0$ and $\mathbf{c}_0$-axes. Note that the new $\mathbf{c}_0$-axis is equivalent to the original $\mathbf{c}$-axis. The space group is therefore $C2mm(C_{2v}^{14})$. This space group is not a subgroup of that of the tetragonal phase above ($P4mm$), but it is a subgroup of the parent phase $Pm\overline{3}m(O_h^1)$ according to

$$C2mm \xrightarrow{t\text{-subgroup}} Cmmm \xrightarrow{t\text{-subgroup}} P4/mmm \xrightarrow{t\text{-subgroup}} Pm\overline{3}m$$

Thus, it is not possible to go from the tetragonal $P4mm$ phase to the orthorhombic $C2mm$ phase by a continuous change in the cation displacements, and so this transition must be first order. Finally, at still lower

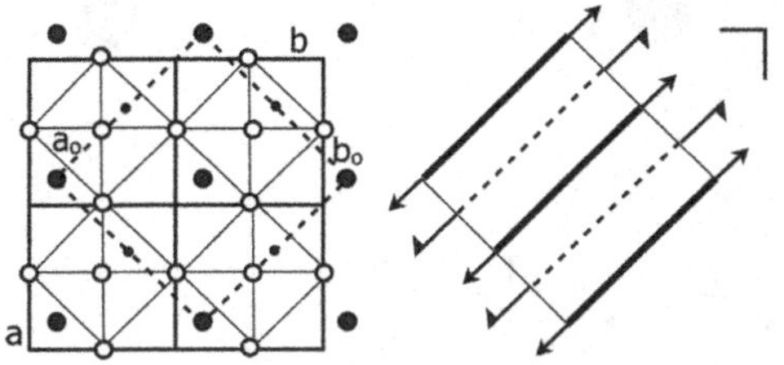

Figure 7.15 *C*-centered cell of the orthorhombic $BaTiO_3$ structure. Large black circles: Ba; small black circles: Ti; open circles: O. Cation displacements exaggerated for display purposes.

temperatures, the displacements are directed along [111] to form space group $R3m(C_{3v}^5)$ (Fig. 7.14d). This too is a subgroup of $Pm\bar{3}m(O_h^1)$ (it is a maximal subgroup of $R\bar{3}m(D_{3d}^5)$, which in turn is a maximal subgroup of $Pm\bar{3}m(O_h^1)$, but it is not a subgroup of the orthorhombic phase space group *C2mm* and so again this transition must be first order.

The best known example of a perovskite with tilted octahedra is $SrTiO_3$. At ≈ 105 K, there is a phase transition from the ideal cubic structure to one where there is a tilt or rotation of the octahedra about the **c**-axis. The sense of rotation of the octahedra along the **c**-axis alternates from one octahedron to the next (Fig. 7.16), resulting in a doubling of the **c**-axis repeat. Because of the corner-linking between the octahedra, this rotation forces the neighbouring octahedra in the **ab** planes to also rotate about **c** in the opposite sense, thus doubling the repeat along **a** and **b** as well. Since the axis of rotation is along **c**, this direction becomes unique and the structure is tetragonal with space group $F4/mmc(D_{4h}^{18})$. Note that by slowly untilting the octahedra, we approach the ideal cubic structure, and so the tetragonal space group must be a subgroup of $Pm\bar{3}m(O_h^1)$. In fact, a glance at the *ITA* shows that *F4/mmc* is a Type IIb maximal non-isomorphic k-subgroup of *P4/mmm*, in which $\mathbf{a}' = 2\mathbf{a}$, $\mathbf{b}' = 2\mathbf{b}$ and $\mathbf{c}' = 2\mathbf{c}$, and of course *P4/mmm* is a maximal non-isomorphic t-subgroup of $Pm\bar{3}m$, again showing the use of the *ITA* in helping to understand phase

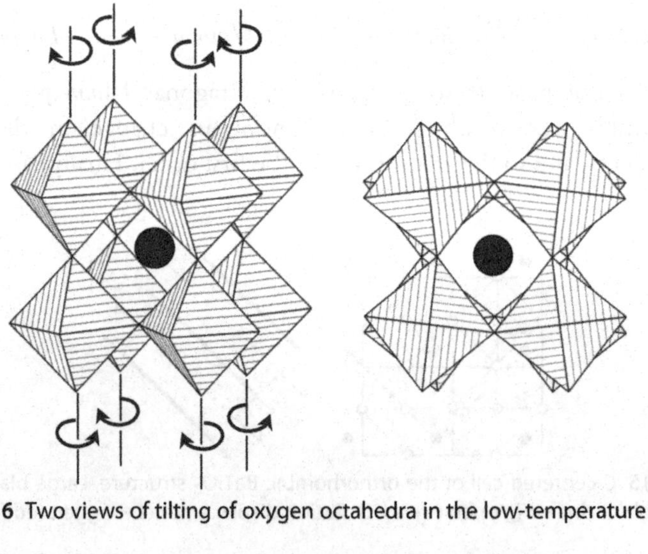

Figure 7.16 Two views of tilting of oxygen octahedra in the low-temperature phase of $SrTiO_3$.

transitions. Tilts, and other aspects of perovskites, are further discussed in Appendix 9.

7.10.3. Bismuth Vanadate

$BiVO_4$ shows an interesting displacive phase transition, which is an example of a ferroelastic transition (see Section 7.13). The high and low-temperature structures are shown in Fig. 7.17a and b, respectively. In the high-temperature structure (above 250 °C), alternate layers of VO_4 tetrahedra along the **c**-axis are rotated through 90° with respect to each other. The Bi atoms occupy sites between the tetrahedra, and the space group is $I4_1/a(C_{4h}^6)$. The structure is given by

Bi :	at 4*b* sites	(0, ¾, ⅜); (0, ¼, ⅝); (½, ¼, ⅞); (½, ¾, ⅛)
V :	at 4*a* sites	(0, ¾, ⅞); (0, ¼, ⅜); (½, ¼, ⅜); (½, ¾, ⅝)
O :	at 16*f* sites	$x = 0.249,\ y = -0.112,\ z = -0.047$
		$x = -0.138, y = -0.001, z = -0.203$
		etc.

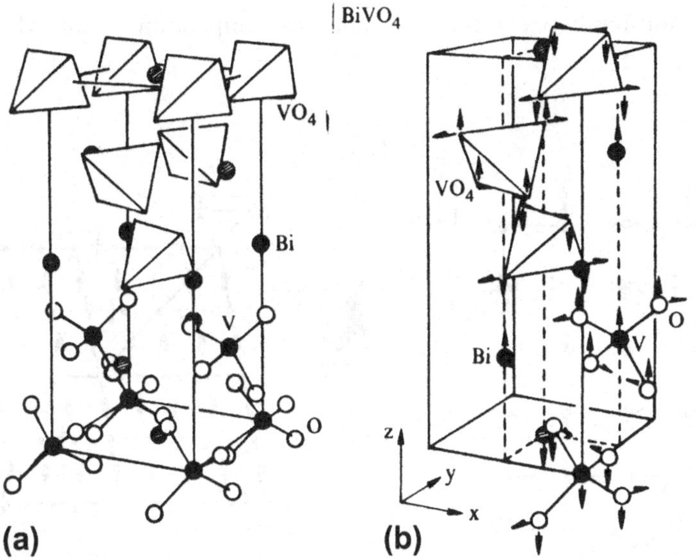

Figure 7.17 (a) Perspective drawing of tetragonal $BiVO_4$. (b) Structural distortion from tetragonal high-temperature structure. Vertical arrows along ±**c** indicate displacements of tetrahedra and Bi. Horizontal arrows denote distortion of the tetrahedra.

Below 250 °C, there is a second-order phase transition to a monoclinic phase, space group $I2/a(C_{2h}^6)$. Note that this space group symbol uses a nonstandard set of axes of $C2/c(C_{2h}^6)$ in first setting (i.e. the **c**-axis is chosen to be unique). This choice maintains the unit-cell relationship between the high- and low-temperature phases. This space group is a Type I maximal non-isomorphic t-subgroup of $I4_1/a(C_{4h}^6)$, which is consistent with the second-order nature of the transition.

The structural changes can be described by alternating displacements of the Bi atoms and VO_4 tetrahedra along the **c**-axis (vertical arrows in Fig. 7.17b) together with alternating distortions of the tetrahedra in the **ab**-planes (horizontal arrows in Fig. 7.17b) caused by the displacement of the oxygen atoms in opposing directions in these planes. It can be seen that by reducing these displacements progressively the high-temperature phase can be reached.

Figure 7.18b shows part of the relevant page from the *ITA* for space group $C2/c(C_{2h}^6)$, first setting, which we use to illustrate a useful aspect of the tables. Four unit cells are shown with the corresponding symmetry operations marked. Several other choices of unit cell are also outlined and in particular the one shaded in the accompanying inset diagram is described by $I112/b$ in the tables. This is equivalent to the choice of $I112/a$, but with **a** and **b** interchanged, and so we see how helpful the *ITA* is in understanding the relationship between the high- and low-temperature phases when it is

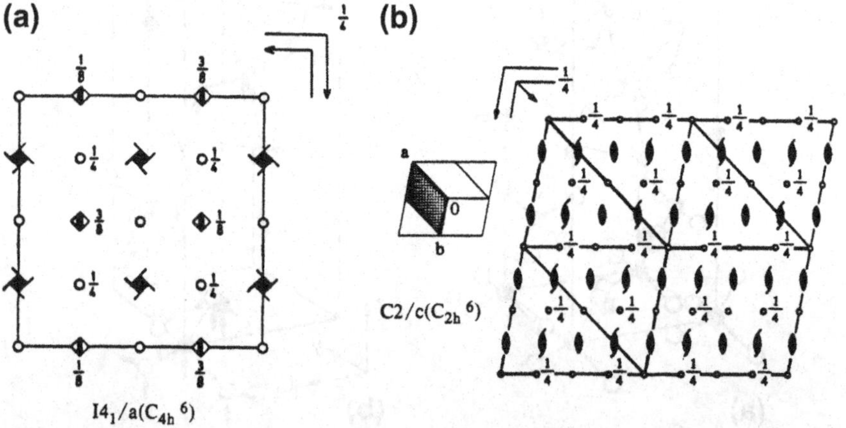

Figure 7.18 (a) Space group operations of $I4_1/a$. (b) The same for $C2/c$. The inset shows the choice of unit cell required to obtain $I112/b$, which is equivalent in turn to $I112/a$.

necessary to use non-standard settings. Returning to the room-temperature structure of $BiVO_4$, the atomic positions are:

Bi	at 4*e*	$(0, \frac{3}{4}, \frac{3}{8}+z)$; $(0, \frac{1}{4}, \frac{5}{8}-z)$;
		$(\frac{1}{2}, \frac{1}{4}, \frac{7}{8}+z)$; $(\frac{1}{2}, \frac{3}{4}, \frac{1}{8}-z)$ with $z = 0.009$
V	at 4*e*	$(0, \frac{3}{4}, \frac{7}{8}+z)$; $(0, \frac{1}{4}, \frac{3}{8}-z)$;
		$(\frac{1}{2}, \frac{1}{4}, \frac{3}{8}+z)$; $(\frac{1}{2}, \frac{3}{4}, \frac{5}{8}-z)$ with $z = -0.003$
O1	at 8*f*	$(x = 0.240, y = -0.119, z = -0.051)$; etc.
O2	at 8*f*	$(x = -0.147, y = 0.009, z = -0.209)$; etc.

Thus the atoms lie close to the positions occupied in the high-temperature phase, as expected. The cation fractional displacements are directed along the **c**-axis only, and are very small. The Ol and O2 positions of the monoclinic phase are almost related by

$$\text{O1}: (x, y, z) \leftarrow \text{O2}: (\tfrac{1}{4}+y, -\tfrac{1}{4}-x, -\tfrac{1}{4}-z)$$

which corresponds to the $\bar{4}$ operation acting at $(0, \frac{1}{4}, \frac{1}{8})$ that is lost on going from $I4_1/a$ to $I2/a$.

7.11. PROTEINS

Discussion of protein structures lies outside the scope of this book, although certain aspects of space group symmetry are of particular relevance. First of all, macromolecular protein molecules are known to be chiral, and as a consequence of evolution, proteins in living organisms only occur in one configuration that immediately limits the way they can crystallize to 65 chiral space groups. It appears that the most common space groups for proteins are $P2_12_12_1(D_2^4)$ and $P2_1(C_2^2)$, whereas in small-molecule organic crystals, space group $P2_1/c(C_{24}^5)$ is preferred. The reasons for this are not entirely clear, although the evidence points to the influence of the way molecules can pack together with the maximum density.

An attempt at explaining this[7] has been made in terms of the numbers of degrees of freedom needed to assemble the first few molecules before the

[7] S. W. Wukovitz & T. O. Yeates. *Nat. Struct. Biol.* **2**, 1062 (1995).

internal structure of the crystal is completely defined. In particular it is suggested that a key factor is the minimum number of unique contacts C required to make the set of symmetry-related molecules into a connected framework.

In Fig. 7.19 the idea of contact points is illustrated by two examples in plane groups $p2$ and $p4$. In $p2$, we can see at least three distinct contact types marked by open, filled and crossed circles. A smaller number is not sufficient in this plane group when the molecule lacks internal symmetry. In plane group $p4$, two distinct contact points are found. The number of degrees of freedom D is given by

$$D = S + L - C$$

where S is the number of degrees of freedom for orienting and positioning of a single molecule in the unit cell and L is the number of independent parameters needed to describe the unit cell. In $P2_12_12_1$, there are 3 degrees of translational freedom plus 3 degrees of rotational freedom, making $S = 6$, whereas in $P2_1$ the position of the molecule along the 2-fold-direction is arbitrary, making $S = 5$. For these two space groups, $L = 3$ and 4, respectively. Table 7.1 shows some of the results of a survey of known space groups for protein crystals.

The conclusion of this study is that proteins crystallize primarily in space groups where it is easiest to achieve connectivity. On the other hand, small organic molecules form crystals in which packing is the most important feature.

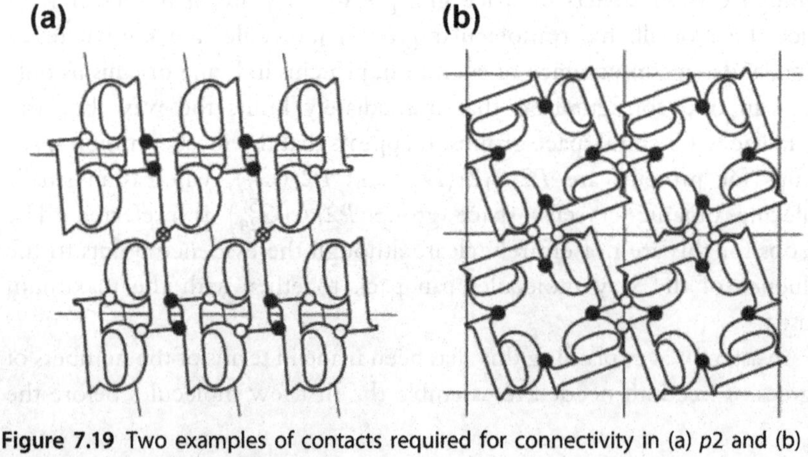

Figure 7.19 Two examples of contacts required for connectivity in (a) $p2$ and (b) $p4$ (from Wukovitz and Yeates).

Table 7.1 Percentages of observed space groups for protein crystals

Space group	*S*	*L*	*C*	*D*	%
$P2_12_12_1$	6	3	2	7	36.1
$P2_1$	5	4	3	6	11.1
$C2$	5	4	3	6	6.1
$P4_32_12$	6	2	2	6	5.7
$I4$, $P6_1$, $R3$	5	2	2	5	—
$P4_22_12$	6	2	3	5	—
$P4_2$	5	2	3	4	—

7.12. CRYSTALLOGRAPHIC INFORMATION FILE

In order for crystal structure information to be made easily exchangeable and specified in a suitable form for input into computer software, the International Union of Crystallography recommends that all crystal structure data should be submitted in the form of a **Crystallographic Information File,**[8] usually called simply a **CIF**. This is a simple text file containing heading tags followed by the data information. These days, whenever a crystallographer sends the crystal data for publication in one of the IUCr journals, submission usually includes a CIF. This then can be checked for consistency and veracity by a suite of computer programs before final acceptance. Anyone wanting to obtain the structural data can then download a CIF and use it for whatever purposes they want, for example, to enter into a crystal structure plotting program. The following is an example of a CIF, in this case for an electron density study of the structure of $CaCO_3$.

```
data_text
_audit_creation_method 'IUCr checking CIFGEN version 1.4'
_journal_date_recd_electronic 1992-11-17
_journal_date_accepted 1993-03-10
_journal_name_full 'Acta Crystallographica, Section B'
_journal_year 1993
_journal_volume 49
_journal_issue 4
_journal_page_first 636
_journal_page_last 641
_journal_paper_category FA
_journal_coeditor_code AS0621
```

[8] S. R. Hall, F. H. Allen & I. D. Brown. *Acta Cryst.* **A47**, 655 (1991).

```
loop_
    _publ_author_name
    'Maslen, E.N.'
    'Streltsov, V.A.'
    'Streltsova, N.R.'
data_as0621a
_chemical_formula_moiety ?
_chemical_formula_sum 'C Ca O3'
_chemical_formula_weight 100.09
_symmetry_cell_setting hexagonal
_symmetry_space_group_name_H-M 'R -3 c {hexagonal axes}'
_symmetry_space_group_name_Hall '-R 3 2"c'
_cell_length_a 4.991(2)
_cell_length_b 4.991(2)
_cell_length_c 17.062(2)
_cell_angle_alpha 90.0
_cell_angle_beta 90.0
_cell_angle_gamma 120.0
_cell_volume 368.1(3)
_cell_formula_units_Z 6
_exptl_crystal_density_diffrn 2.709
_exptl_crystal_density_meas ?
_exptl_crystal_F_000 300
_exptl_absorpt_coefficient_mu 2.134
_diffrn_radiation_type Mo
_diffrn_radiation_wavelength 0.71073
loop_
    _atom_site_label
    _atom_site_fract_x
    _atom_site_fract_y
    _atom_site_fract_z
    _atom_site_U_iso_or_equiv
    Ca .0 .0 .0 ?
    C .0 .0 .25 ?
    O .25729(19) .0 .25 ?
loop_
    _geom_bond_atom_site_label_1
    _geom_bond_atom_site_label_2
    _geom_bond_distance
    C O 1.2841(11)
    Ca O 2.3590(10)
    O O1b 2.2242(16)
    O O2c 3.1887(13)
    O O3d 3.2604(17)
    O O4e 3.4102(8)
```

7.13. FERROIC PHASE TRANSITIONS

We have seen that symmetry is an especially useful concept when dealing with phase transitions, where one phase is derived through simple atomic displacements or substitutions from another. Elegant examples of this are found in the broad field of ferroic phase transitions. By the term **ferroic**,[9] we refer to materials which are capable of some sort of switching behaviour between two (or more) stable states. This is a common phenomenon and so a study of ferroicity is important, not only for academic interest, but also in technology. We shall use the terminology developed by K. Aizu and others to describe the concept of ferroics. Suppose that the 'state' of a crystal can be switched from one to another by the application of an applied field; field here is taken in a general sense to mean either magnetic, electric or mechanical (stress) field. By **state**, we mean a description of a property of the crystal, such as electrical or magnetic polarity.

For example, it is well known that a **ferromagnetic** material is one in which there are magnetic domains (states or regions with a magnetization in a particular direction) that can be switched to a different orientation by an applied magnetic field. These domains remain locked in, even when the field is removed and only change when a sufficiently large magnetic field is applied in an opposing sense. Ferromagnets therefore exhibit magnetic hysteresis effects.

Similarly, there are electrical analogues to the ferromagnets, called **ferroelectrics** (the 'ferro' part does not mean the material must contain iron, but is adopted simply by analogy with ferromagnets). In these materials, the structure is electrically polar (the centers of mass for the positive and negative charges do not coincide and so domain states can be formed with the polarity in different directions). An applied electric field can then switch these domains from one state, the so-called **orientation state,** to another, just as in ferromagnets. Similarly, dielectric hysteresis is a feature of ferroelectrics. In both ferromagnets and ferroelectrics, the switching from one domain orientation to another involves work being performed on the material, and so the free energy must change from one state to the other.

Therefore it is not surprising to discover that in addition to magnetic and electrical work, mechanical work can also have a switching effect in certain materials. Such materials, called **ferroelastics**, have their elastic strains switched by an applied stress. For example, in Fig. 7.20, the structure of

[9] An excellent review of ferroicity is given by V. K. Wadhawan, *Phase Transitions,* **3**, 1 (1982) and in 'Introduction to Ferroic Materials' (Gordon & Breach, 2000) by the same author.

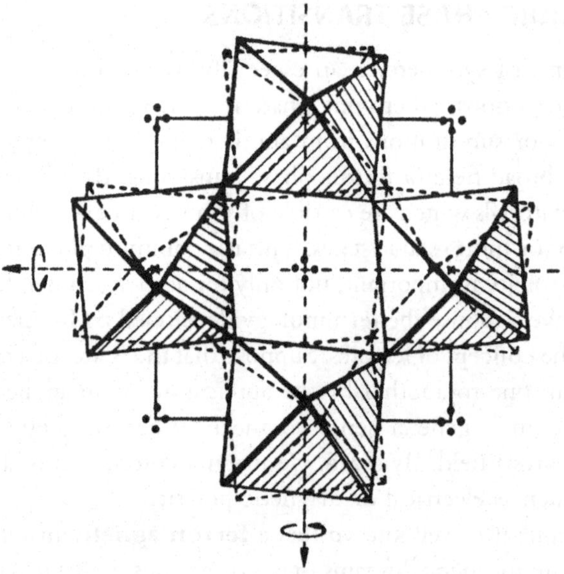

Figure 7.20 Projection of $SmAlO_3$ structure. Two projections are superimposed to show how small atomic displacements cause the unit-cell axes to appear to rotate through 90°.

the perovskite $SmAlO_3$ is shown. The structure is orthorhombic so that the **a** and **b**-axes shown in the figure are slightly different from each other. Superimposed on the figure is the same structure rotated through 90°, and it can be seen that by making small displacements of the atoms in the first structure the atomic arrangement of the superimposed structure can be formed. Thus, by small displacements of the atoms, the unit cell appears to rotate through 90°. This occurs typically under the influence of an applied stress and requires a small strain difference between the structure with these two different unit cells.

In an expression for the energy of a system, in addition to pure magnetic, electric and mechanical terms, one can also conceive of squared and cross-terms. Thus, writing the free energy difference ΔF on switching

$$\Delta F = -\Delta P_i E_j - \Delta M_i H_j - \Delta \varepsilon_{ij} \sigma_{kl} - \Delta \chi_{ij} E_i E_j - \Delta X_{ij} H_i H_j - \Delta s_{ijkl} \sigma_{ij} \sigma_{kl} - \Delta d_{ijk} E_i \sigma_{jk} - \Delta \beta_{ij} E_i H_j - \Delta \alpha_{ijk} H_j \sigma_{jk} \quad [7.1]$$

The first three work terms are the ferroelectric, ferromagnetic and ferroelastic terms, respectively, where P, M and ε are the electrical polarization, magnetization and elastic strain, and E, H and σ are the applied fields, electric, magnetic and stress, respectively. The next three quantities with field-squared terms are called **ferrobielectric**, **ferrobimagnetic** and **ferrobielastic**, respectively. These require the presence of two applied fields in order to achieve switching. Finally, the last three terms are called **ferroelastoelectric**, **ferromagnetoelectric** and **ferromagnetoelastic**. Note, in particular, that the ferroelastoelectric term defines a **piezoelectric**, where d_{ijk} are the piezoelectric coefficients. Similarly a ferrobielectric is simply a **dielectric** material, and so on. It is clear that a classification like this allows one to describe many important physical properties. In recent times the term **multiferroic**[10] has been used to describe these mixed properties, especially to materials where the magnetic structure is influenced by an electric field, and vice versa. The most studied such multiferroic at present is $BiFeO_3$.

Now the important feature of all ferroics is that they can be seen as small distortions of some higher-symmetry phase, called the **prototype phase**. Because of this, there must be some connection between the symmetries of the prototype and ferroic phases. To describe this, Aizu[11] coined the term **ferroic species**. Ferroic crystals belong to the same species when they are the same with respect to

1. Their own point symmetry
2. The point symmetry of their prototype
3. The correspondence between the elements of the ferroic point symmetry and the elements of the prototype point symmetry

The species are written in a particular form, for example:

$$4/mmmF2/m(p), 4/mmmF2/m(s), etc.$$

The capital F stands for ferroic, and the prototype point group is on the left with the ferroic point group to the right. In $4/mmmF2/m(p)$, p means 'principal' to indicate that a monoclinic $2/m$ phase is formed from a higher-symmetry $4/mmm$ phase with the 2-fold axis parallel to the 4-fold axis. In $4/mmmF2/m$(s), s stands for 'side' indicating that the ferroic 2-fold axis is perpendicular to the prototype 4-fold axis from which it is derived. Clearly the relationship between prototype and ferroic point groups is group–subgroup, and most ferroics exhibit phase transitions to their prototype phases. The number of orientation states (domains) in the ferroic phase is then given by the ratio of the orders of the prototype and ferroic point groups.

Aizu classified all of the ferroic species and for each he determined, by symmetry, which properties in principle can be exhibited. Another term

[10] H. Schmid. *Ferroelectrics.* **162**, 317 (1994).
[11] K. Aizu. *Phys. Rev.* **B2**, 754 (1970).

that is used is **full ferroic**. Consider the tetragonal-cubic phase transition in $BaTiO_3$. The ferroic species for this is $m\bar{3}mF4mm$. The space group change is $Pm\bar{3}m(O_h^1)$ to $P4mm(C_{4v}^1)$, which from the *ITA* we determine to be translationengleiche. The prototype and ferroic point groups are of order 48 and 8, respectively. Because the ratio is 6, there will be 6 possible orientation states in the ferroic phase.

These are shown schematically in Fig. 7.21; there are six possible orientations of the polarization vector with three possible unit-cell orientations. The $4mm(C_{4v})$ phase is therefore both ferroelectric and ferroelastic. These are partially coupled, since reversal of the polarization does not affect the cell deformation, but if the polarization is made to rotate through 90°, the unit cell must deform to follow. On the other hand, an appropriately directed mechanical stress affects 90° domains but not 180° domains. Because there are six ferroelectric states, the $4mm$ phase here is fully ferroelectric, but because there are only three ferroelastic states it is partially ferroelastic. By the way, note a simple rule: ferroelastic species are always those in which the crystal family (see 5.6.6) changes, since this is the only way in which a lattice distortion can be instantaneously produced. Recall that the symmetries of the crystal systems are given by the symmetries of the lattice.

By examining the symmetry properties of all possible ferroic species, Aizu determined that there are:

772 Ferroics - 212 nonmagnetic
88 Full ferroelectrics
94 Full ferroelastics
5 Pure ferrobielastics
15 Pure ferroelastoelectrics
10 Both ferrobielastic and ferroelastoelectric
35 Full ferromagnetoelastics
40 Full ferromagnetoelectrics

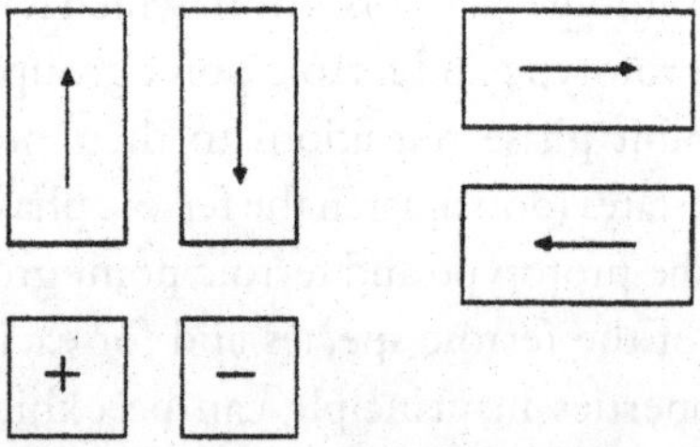

Figure 7.21 Orientation states for the species $m\bar{3}mF4mm$. The arrows and ± signs indicate polarity. Three views of a tetragonal unit cell are shown with the polarity directed along ±**c**.

7.13.1. Phase Transition in La_2CuO_4

The discovery of superconductivity in La_2CuO_4 in 1986 created a huge interest in oxide materials with high-T_c's, and led to the Nobel Prize for K.A. Müller and J.G. Bednorz. Attention was focused on its SPT with the thought that it might be related to the superconductivity mechanism. We shall consider it here because the transition and consequential structural modification illustrate well the principles of group–subgroup relationships and the ideas of ferroic phase transitions. The tetragonal high-temperature structure (Fig. 7.22) has space group $I4/mmm(D_{4h}^{17})$. When a suitable amount of Sr, Ba or Ca replaces the La, then the material is tetragonal at room temperature. However, the pure La compound is orthorhombic at room temperature. When the room-temperature structure of the pure compound was determined in 1973, it was thought to have space group $Fmmm(D_{2h}^{23})$ although the authors did state that the true structure was probably of lower symmetry, hinting

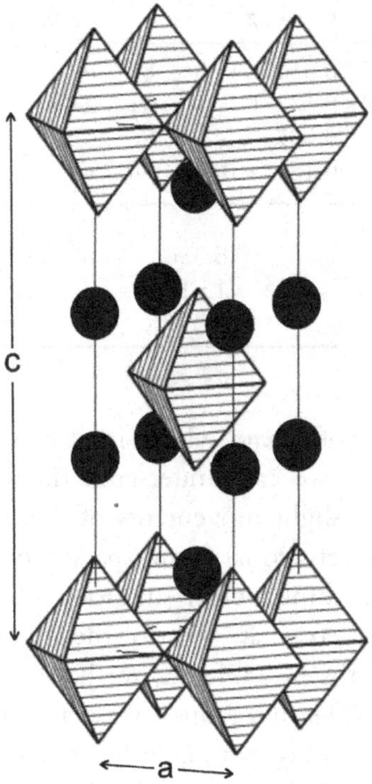

Figure 7.22 High-temperature structure of La_2CuO_4. Black: Lanthanum. Copper atoms are located at centers of oxygen octahedra.

that it might be $Pbnb(D_{2h}^{10})$. It was eventually established that the correct space group is $Cmce(D_{2h}^{18})$, with the structural parameters listed in Table 7.2.

Table 7.2 Structural models for La_2CuO_4

		$F4/mmm \equiv I4/mmm(D_{4h}^{17})$				$Fmmm(D_{2h}^{23})$			
		Site symmetry	***x***	***y***	***z***		**Site symmetry**	***x***	***y***	***z***
La	8*e*	4*mm*	0	0	0.361	8*i*	*mm*2	0	0	0.362
Cu	4*a*	4/*mmm*	0	0	0	4*a*	*mmm*	0	0	0
O_p	8*c*	*mmm*	¼	¼	0	8*e*	. . 2/*m*	¼	¼	0
O_z	8*e*	4*mm*	0	0	0.183	8*i*	*mm*2	0	0	0.182
a(Å)					5.360					5.409
b(Å)					5.360					5.363
c(Å)					13.17					13.17

		$Cmce(D_{2h}^{18})$				$Abma(D_{2h}^{18})$			
		Site symmetry	***x***	***y***	***z***		**Site symmetry**	***x***	***y***	***z***
La	8*f*	*m* . .	0	0.361	0.006	8*f*	. *m*.	0.006	0	0.361
Cu	4*a*	2/*m* . .	0	0	0	4*a*	. 2/*m* .	0	0	0
O_p	8*e*	. 2 .	¼	0.007	¼	8*e*	. . 2	¼	¼	0.007
O_z	8*f*	*m* . .	0	0.184	−0.034	8*f*	. *m* .	−0.034	0	0.184
a(Å)					5.356					5.399
b(Å)					13.17					5.356
c(Å)					5.399					13.17

Since *Cmce* is a subgroup of *Fmmm*, which, in turn, is a subgroup of *I*4/*mmm* (the same as *F*4/*mmm*), we can understand the sequence of structural changes and follow the slight movements of the atoms that cause these changes. Start with the tetragonal form (space group *I*4/*mmm*). *Fmmm* is a type I maximal non-isomorphic t-subgroup of *I*4/*mmm*, but formed with its orthorhombic axes at 45° to the tetragonal **a** and **b**-axes (Fig. 6.5e). Consequently, in Table 7.2 we have expressed the structural coordinates of the *I*4/*mmm* tetragonal phase in terms of the equivalent space group *F*4/*mmm*. This will help us to see how the structural modifications are related. The measured structural parameters in *Fmmm* are given in Table 7.2, and as can be seen, they are hardly different from those in the *F*4/*mmm* structure. From this and the observed lattice parameters, we immediately infer that the *Fmmm* structure is simply related to the *F*4/*mmm* phase by a positive distortion along **b**. As the fractional coordinates are essentially unchanged, this results in a distortion of the oxygen octahedron in this direction. Also note that the site symmetries for the atoms in the *Fmmm* phase are subgroups of those in the *F*4/*mmm* phase.

From the *ITA*, we find under space group *Fmmm*, the type IIa maximal non-isomorphic k-subgroup $Abma(D_{2h}^{18})$, which is equivalent under interchange of axes to $Cmce(D_{2h}^{18})$. The coordinates for the *Cmce* structure are given in Table 7.2 and, for easy comparison with the *F*4/*mmm* and *Fmmm* parameters, we have also transformed to *Abma*. The essential difference now is that small displacements along **a** appear for La and O_z, and along **c** for O_p. O_p is an oxygen making up the basal plane perpendicular to **c** of an octahedron, and O_z is the oxygen at the apex of the octahedron, and so we see that *the effect of these displacements is to cause the octahedron to tilt through a small angle* about the **b**-axis (Fig. 7.23). This is consistent with the oriented site symmetry information obtained from the *ITA* listed in Table 7.2. Thus O_z is at a site with symmetry given by the symbol . *m* ., (Section 6.3.8), and as the '*m*' is in the second place in this symbol, the oxygen can move in a plane perpendicular to **b**. Similarly, the oriented site symmetry for O_p is given by . . 2; the '2' in the third place

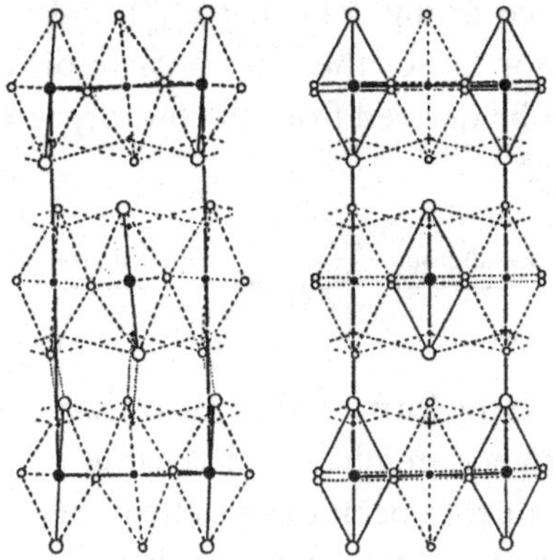

Figure 7.23 Two projections of the La_2CuO_4 structure on (010) and (100) showing how the oxygen octahedra are tilted. The tilts are similar to those found in perovskites.

means that the atom is on a 2-fold axis along **c** and is free to move along this direction.

Note that the other space group proposed for the orthorhombic phase, *Pbnb*, is not a subgroup of *Fmmm*, and so this tells us that this is unlikely to be correct for this structure.

From the above, we can see that the information available in the *ITA* is most useful in elucidating possible crystal structures produced by reversible phase transitions. In terms of ferroic species (Section 7.13), the transition would be classified as $4/mmmFmmm$. The transition is one in which the crystal family has changed and so it is ferroelastic. The subgroup relationships are:

$$F4/mmm \xrightarrow{\text{t-subgroup}} Fmmm \xrightarrow{\text{k-subgroup}} Abma$$

and this suggests that the *Fmmm* phase may be formed as an intermediate phase. The $F4/mmm$ (or $I4/mmm$) phase is the prototype phase and *Fmmm* the pure ferroelastic phase. The $F4/mmm$ to *Fmmm* transition would result in the formation of two ferroelastic domains, since the ratio of the orders of the two groups is $16{:}8 = 2$, and the *Fmmm* to *Abma* transition, because of the loss of one half of the translational elements, would result in two anti-phase domains. Both types of domains would be expected therefore in the *Abma* phase.

In $(La_{2-x}Sr_x)CuO_4$ the SPT occurs at ≈ 450 K for $x = 0$, but this temperature decreases with increasing x values. In $(La_{1.9}Ba_{0.1})CuO_4$, the tetragonal to orthorhombic phase transition occurs at ≈ 270 K; however, at ≈ 52 K, there is another transition to a structure that appears to belong to the tetragonal space group $P4_2/ncm(D_{4h}^{16})$ and so cannot be formed through a simple distortion of the intermediate orthorhombic structure. This space group can be derived from the prototype $I4/mmm$ space group as follows:

$$I4/mmm \xrightarrow{\text{k-subgroup}} P4_2/mmc \xrightarrow{\text{k-subgroup}} C4_2/amc$$

We see that, just as with the formation of the orthorhombic phase, the space group of the low-temperature phase is derived in two stages: first of all by loss of rotational and mirror operations to a hypothetical intermediate space group $P4_2/mmc$ and then by a loss of every other translational element along **a** and **b** to give a doubled cell with space group $C4_2/amc$. Now convert this *C*-centered tetragonal to primitive to obtain the reported space group

$P4_2/ncm$ with $\mathbf{a}' = \mathbf{b}' \approx \sqrt{2}\mathbf{a} \approx 5.2$ Å. It is interesting to note that group–subgroup relationships for phase transitions often infer a two-stage process, even if the intermediate space group is only hypothetical. As yet, in $(La_{2-x}Sr_x)CuO_4$, this low-temperature orthorhombic to tetragonal phase transition has not been observed but it is observed in other materials in this structural family.

Further discussion of phase transitions in terms of tilts in these, and related, structures is given in Appendix 9.

7.13.2. Ferroic Domains

A consequence of ferroic switching and the relationship between prototype and ferroic symmetries is that the ferroic phase is typified by the existence of twin domains in the crystal, which have different orientations with respect to each other. Simple rules have been established linking the number of possible domains to the symmetry of the ferroic species. In particular, for space group reduction G_0 to G, where G determines the symmetry of the transition or the so-called order parameter (recall that the order parameter is a quantity that is a measure of the distortion in the ferroic phase and that vanishes at the transition point), it is the lost symmetry operations that determine the domain structure. Thus, if mirror planes are lost at the phase transition from the prototype to the ferroic phase, these planes are likely to become the domain walls separating the ferroic orientation states.

Figure 7.24 shows schematically three possible ways in which a prototype symmetry can be lowered at a phase transition. The prototype is assumed in this example to have the plane symmetry pm. If it makes a 'klassengleiche' transition, by doubling the **a**-axis (Fig. 7.24a), symbolized here by alternating black and white objects, the domain walls that can be formed are known as **anti-phase boundaries**. On the other hand, a 'translationengleiche' transition, to say symmetry $p1$ (Fig. 7.24b), creates a **ferroic domain boundary** with the atomic arrangements in the two domains related by symmetry operations. Finally, it is possible to make a transition that changes both the point group and unit-cell translations (Fig. 7.24c), so that the domains that form are ferroic but contain also anti-phase domains. Let us consider a few examples to illustrate these ideas.

In the high to low-quartz phase transition, the prototype space group is $P6_422(C_6^5)$ and the low-symmetry phase is $P3_121(D_3^4)$. The ferroic species is therefore 622F32 and the ratio of the orders of the point groups is

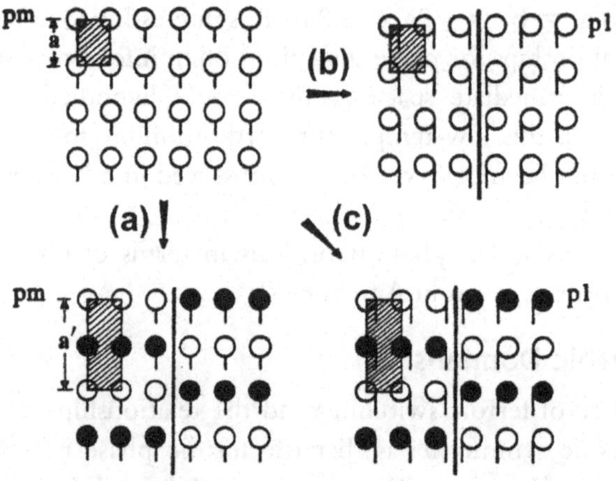

Figure 7.24 Schematic illustration of the three possible ways of lowering a prototype symmetry. Starting from a prototype symmetry of *pm*, doubling of the **a**-axis (a) retains the symmetry *pm*, but with a unit cell of twice the volume to form anti-phase domains. Changing from *pm* to *p*1 symmetry without changing the translational symmetry (b) creates ferroic domains.[12]

12:6 = 2. $P3_121$ is a maximal non-isomorphic t-subgroup of $P6_422$ and so two ferroic twin domains are expected to be formed in the lower phase. These are the famous **Dauphiné twins** of quartz.

In Cu_3Au, there is a disorder–order transition from the high-temperature phase with space group $Fm\bar{3}m(O_h^5)$ to $Pm\bar{3}m(O_h^1)$ as shown in Fig. 7.9b. This is a 'klassengleiche' transition and so only anti-phase domains appear.

In β-$Gd_2(MoO_4)_3$, there is a tetragonal $P\bar{4}2_1m(D_{2d}^3)$ high-temperature phase and the low-temperature structure has space group $Pba2(C_{2v}^8)$. The intermediate space group $Cmm2(C_{2v}^{11})$ is a maximal non-isomorphic t-subgroup of $P\bar{4}2_1m$ and also a minimal k-supergroup of $Pba2$, and so both ferroic and anti-phase domains are expected in β-$Gd_2(MoO_4)_3$.

In $BaTiO_3$, the prototype space group is $Pm\bar{3}m$ and the low-temperature phases have space groups $P4mm$, $C2mm$ and $R3m$. Each of these phases has $Pm\bar{3}m$ as its prototype. The ratios of the orders of the groups for the ferroic species are:

$$m\bar{3}m\text{F}4mm \quad 48:8=6$$

$$m\bar{3}m\text{F}2mm \quad 48:4=12$$

$$m\bar{3}m\text{F}3m \quad 48:6=8$$

12 For further discussion see H. Wondratschek & W. Jeitschko. *Acta Cryst.* **A32**, 664 (1976).

Thus, in the tetragonal, orthorhombic and rhombohedral phrases, there are 6, 12 and 8 ferroelectric/ferroelastic possible domains, respectively. The group–subgroup relationships are all 'translationengleiche', and hence there are no anti-phase domains.

Finally, in the tetragonal–orthorhombic transition in the high-T_c superconductor $YBa_2Cu_3O_{7-\delta}$ the change is from $P4/mmm(D_{4h}^1)$ to $Pmmm(D_{2h}^1)$. The ratio of the orders of the point groups is 16:8 = 2 and *Pmmm* is a maximal non-isomorphic t-subgroup of P4/*mmm*. The ferroic species is 4/*mmmFmmm*, and because the crystal system changes without any polarization effects, it is pure ferroelastic. Thus two ferroelastic domains are found in the orthorhombic phase. These are separated by domain boundaries along [110].

7.14. SURFACE STRUCTURE PLANE AND LAYER GROUPS

We have seen how three-dimensional structures can be classified in terms of the 230 space groups. These three-dimensional space groups were discussed in Chapter 5, and the 17 **two-dimensional space groups** or **plane groups** were briefly discussed in Section 5.7 and listed in Table 5.4. The reason that the latter were discussed so briefly is that they are simple to deal with once the three-dimensional space groups are understood. We now consider the use of two-dimensional space groups in more detail.

For completeness, first consider planar structures with periodicity in two dimensions and a finite or semi-infinite extent in the third dimension. The surface of a macroscopic crystal illustrates the semi-infinite case. A free-standing crystalline film might be an example of such a structure with finite extent. Then, there is the possibility of having a limited set of symmetry operations in the third dimension. For example, a mirror reflection across the film or a 180° rotation about an axis in the film might be symmetry operations. However, a screw axis perpendicular to the film could not possibly be a symmetry operation. For such films the 80[13] layer groups have been identified.

There is a large branch of solid state science, called **surface science**, where the properties of clean, 'perfect' surfaces of crystals are studied, such as the surfaces of the planar structures described earlier. By **surface**, we mean the atomic layers next to the vacuum interface which differ from

[13] See E. A. Wood. *J. Appl. Phys.* **35**, 1306 (1964).

the bulk crystal. The process of obtaining atomically clean, perfect surfaces is not easy and requires UHV conditions so that the surface remains clean long enough to do an experiment. The structures of these surfaces have been extensively studied by **low-energy electron diffraction** (**LEED**) techniques and also by ion scattering and surface extended X-ray absorption fine structure (SEXAFS). Low-energy electrons typically scatter coherently from the surface to a depth of 5–10 Å; the diffracted beam intensities as a function of energy and angle of incidence are used to deduce the structure. Another diffraction technique for finding static and dynamic properties of surfaces uses low-energy helium beams. The interaction of a He atom having the appropriate energy for diffraction (1 Å ~ 0.02 eV) is so strong that the He does not penetrate the surface. The closest distance of approach is just a few Ångstroms. For LEED and He surface measurements, symmetry in the third dimension plays no role, and the 17 plane groups describe the symmetry of the diffraction patterns.

7.14.1. Special Projections

In Section 7.14.3, we consider the structures of adsorbates on clean surfaces and the **reconstructions** of clean surfaces: that is, a change of the structure (and plane group) from that expected if the bulk solid were simply terminated by truncation. However, first let us consider the information that is directly obtainable from the *ITA*, namely, the 'symmetry of special projections' given for each space group.

Consider any structure that has either the $Fm\bar{3}m(O_h^5)$ or $Fd\bar{3}m(O_h^7)$ space groups. Cu with the fcc structure in Fig. 7.1a and Si with the diamond structure in Fig. 7.2f serve as useful models. The *ITA* gives the plane groups for projections in a few special directions. For example, when projected along a [110] direction (in the cubic system, [110] is perpendicular to the (110) plane), any structure with either of these space groups will have the symmetry of the *c2mm* plane group with $\mathbf{b}' = \mathbf{c}$ and $\mathbf{a}' = ½(-\mathbf{a} + \mathbf{b})$. Figure 7.25a shows part of the page from the *ITA* with the relevant information for these two space groups. Projections for the fcc structure and diamond structures are shown in Fig. 7.25b and c, respectively, where for both structures we also show views of the three-dimensional structures that clarify how the projections along [110] are obtained.

For the projection of the fcc structure (Fig. 7.25b), the centered cell is obvious for the *c2mm* plane group; the plane group cell has dimensions $\mathbf{b}'$ and $\mathbf{a}'$, or $\mathbf{a}$ and $\sqrt{2}\mathbf{a}/2$, where $\mathbf{a}$ is the edge length of the conventional

Symmetry of special projections $Fm\bar{3}m(O_h^5)$

Along [001] $p4mm$
$a' = \frac{1}{2}a$ $b' = \frac{1}{2}b$
Origin at 0,0,z

Along [111] $p6mm$
$a' = \frac{1}{6}(2a-b-c)$ $b' = \frac{1}{6}(-a+2b-c)$
Origin at x,x,x

Along [110] $c2mm$
$a' = \frac{1}{2}(-a+b)$ $b' = c$
Origin at x,x,0

Symmetry of special projections $Fd\bar{3}m(O_h^7)$

Along [001] $p4mm$
$a' = \frac{1}{4}(a-b)$ $b' = \frac{1}{4}(a+b)$
Origin at 0,0,z

Along [111] $p6mm$
$a' = \frac{1}{6}(2a-b-c)$ $b' = \frac{1}{6}(-a+2b-c)$
Origin at x,x,x

Along [110] $c2mm$
$a' = \frac{1}{2}(-a+b)$ $b' = c$
Origin at $x,x,\frac{1}{8}$

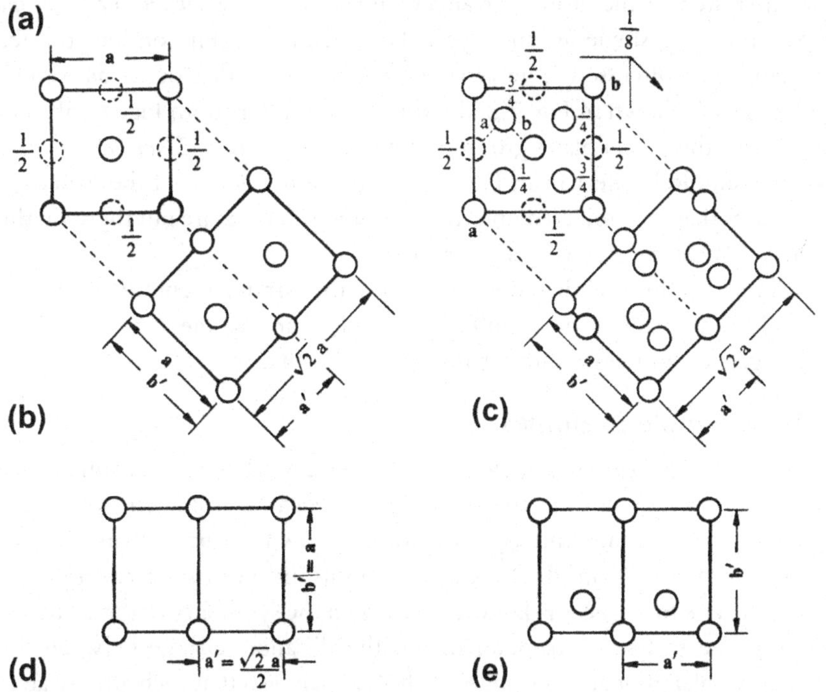

Figure 7.25 (a) For $Fm\bar{3}m$ and $Fd\bar{3}m$, we show parts of the page from the *ITA* indicating the plane groups for special projections. (b) The projection onto the (110) face of the fcc structure. The (001) view of the three-dimensional structure is also shown to help in understanding the projection along [110]. The atoms at height $z = ½$ are shown dashed. (c) Similar to (b), but for the diamond structure. (d) The atoms on the (110) face of an fcc structure. (e) Similar to (d), but for the diamond structure.

three-dimensional unit cell. The plane group unit cell is indicated by the dashed line. For the projection of the diamond structure, the same plane group and unit cell is obtained (dashed line), but the result is more complicated. This arises because the basis in the three-dimensional structure is more involved, being composed of atoms at (0, 0, 0) and (¼, ¼, ¼). As a result, for the two-dimensional projection, the basis has atoms at (0, 0) and (½, ¼) with respect to the new unit cell whose dimensions are $\mathbf{a}' \times \mathbf{b}'$ (Fig. 7.25c).

7.14.2. Very Simple Surfaces

The projection of the three-dimensional structure along the directions studied by surface scientists is not typically what is of interest. Instead, they are interested in the atomic arrangements that lie just *on the surface of the crystal* (i.e. in a single atomic layer). For example, again consider the fcc structure just discussed. The arrangement of atoms that occur *on a single (110) plane* is shown in Fig. 7.25d. Note how it differs from Fig. 7.25b, and that it has the *p2mm* plane group symmetry, rather than *c2mm*. The reason for this should be clear from Fig. 7.25b; in the projection of the structure, the atoms that are projected onto the centered positions are not those on the actual (110) surface that we have chosen.

Similarly, for the diamond structure, the arrangement of atoms on a single (110) plane is shown in Fig. 7.25e. This arrangement also has *p2mm* plane group symmetry with a basis of Si at (0, 0) and (½, ¼).

7.14.3. Simple Examples

The atoms on or very near a clean surface of a crystal need not maintain the atomic positions found in the bulk. This is because atoms on the surface (taken as the *xy*-plane) interact with atoms below the surface plane but not above. Deviations from the bulk atomic positions give rise to two types of changes. The first type is **relaxation** in which the symmetry of the atoms on or near the surface is maintained, but the distances change between the successive planes of atoms parallel to the *xy*-plane (so that each atomic layer has the same unit cell but the surface layers translate with respect to the bulk). For most surfaces of most metals, the first interplanar distances are compressed compared with those deeper in the bulk. In the second type, there can be a change in the symmetry of the surface atoms, known as a **reconstruction** (so that the unit cell of the surface atoms is changed).

Si(001): The Si(001) surface provides an example of an extensively studied reconstructed surface. The unreconstructed Si(001) surface is shown in Fig. 7.26a, where the open circles are the Si atoms in the topmost layer on which we are focusing; the small dashed rectangular unit cell connects the atoms on the surface. Counting just the atoms on the surface, the symmetry is that of the plane group *p4mm*. (Also shown dashed, at 45° to the surface unit cell, is a unit cell of the conventional fcc lattice; this is included for

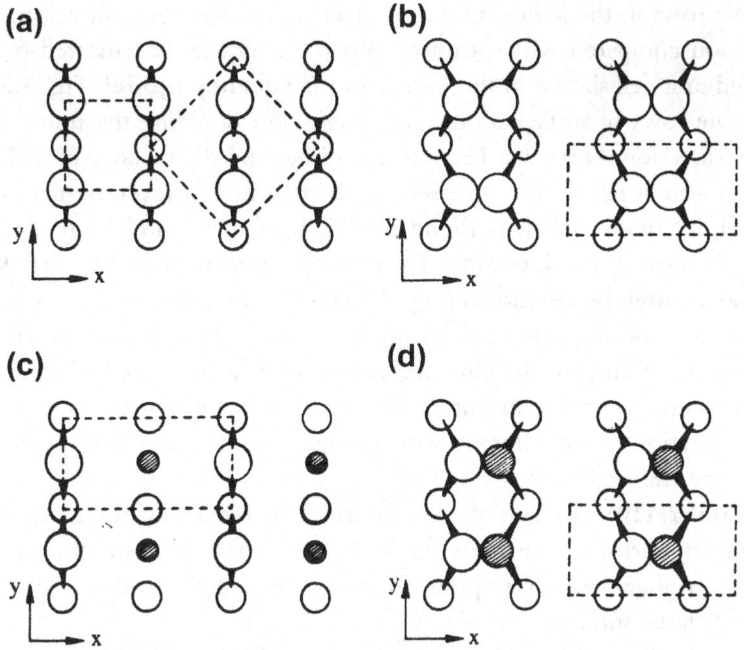

Figure 7.26 Si(001) surface both unreconstructed and reconstructed. (a) An unreconstructed surface where the atoms **a**/4 below (**a** is the conventional unit cell dimension) are shown as smaller circles, (b) the dimer model for the (2 × 1) reconstruction of this surface, (c) the vacancy model for the same (2 × 1) reconstruction. The second and fourth columns of Si surface atoms are missing. Thus the dashed, smaller Si atoms in the second and fourth columns (in the *y*-direction) are **a**/2 below the surface and (d) the buckled dimer model for the same (2 × 1) reconstruction.

orientation purposes.) Such a surface is called the Si(001)-(1 × 1) unreconstructed surface; (1 × 1) means that the unit cell is one by one times the size of the unit cell for the atoms on the surface when they maintain their bulk positions. A more careful look at this surface shows that if the atoms on the surface as well as the bonds to the atoms below are considered, the plane group is actually *p2mm*. Since these bonds lead to a different symmetry from that of the surface atoms, we might expect a reconstruction. Study of this surface shows that it is actually made up of domains with (2 × 1) and (1 × 2) unit cells. That is, the experiments indicate that the unit cell has the same dimension in one direction but is twice as long in the other direction.

From LEED patterns, it is easy to determine the symmetry of the surface structure, but an LEED intensity analysis is required to obtain a model of the atomic arrangement that gives rise to this symmetry. In the **dimer model** (Fig. 7.26b), adjacent rows of atoms pair together to form double rows in the x-direction. This pairing doubles the unit cell in that direction compared with an unreconstructed surface, as indicated by the dashed unit cell shown in the figure. In the **vacancy model** (Fig. 7.26c), alternate rows of surface atoms are absent, also doubling the unit cell in the x-direction. Note that for both the dimer and the vacancy model, the plane group is *p2mm*, as can be seen in the figures. However, by analysis of the LEED intensities, it is found that both of these models are unsatisfactory. Instead, the intensities are in better agreement with a **buckled dimer model**. In this model (Fig. 7.26d), the unit cell is again doubled in the x-direction, but the plane group is *pm* because the two atoms on the surface are at different heights; the plane group would be *p*1 if these two atoms were to move in the y-direction with respect to each other. Consideration of the different plane group symmetries could play a role in deducing the best model.

S on Ir(110): We have already discussed the reconstruction of a surface with no impurities. Let us consider the Ir(110) surface reconstruction as well as the arrangement of sulphur (S) atoms that are adsorbed onto this reconstructed surface.

For the fcc structure, a projection along [110] yields the *c2mm* plane group (Fig. 7.25b), while the plane group for the topmost layer of atoms is *p2mm*. Figure 7.27a again shows the (110) plane of an fcc structure with the unit cell outlined; this has dimensions $\mathbf{a} \times \mathbf{a}\sqrt{2}$.

When the (110) surfaces of the free metals Ir, Pt and Au are studied by LEED intensity analysis, a 1 × 2 reconstructed **missing-row model** for the surface is found. This reconstruction is shown in Fig. 7.27b. Parallel to the [100] direction, a row of atoms from the topmost layer is removed. The unit cell for this layer is the same size in the x-direction as for the unreconstructed surface. However, in the y-direction, the unit cell is twice as large, and hence the label Ir(110)–(1 × 2). The experimental evidence for the missing-row model for the Ir(110)-(1 × 2) reconstruction is good and it can be directly observed with the scanning-tunnelling microscope. Thus this model is now generally accepted. The plane group is *p2mm*.

There are interesting structural effects when sulphur (S) is adsorbed onto the Ir(110)-(1 × 2) surface. The diffraction spots from this surface indicate that the (1 × 2) pattern from the Ir(110) substrate remains (implying that the missing rows of Ir atoms remain) but, in addition, a (2 × 2) pattern develops. This means that the unit cell is now twice as large in each direction as the unreconstructed cell.

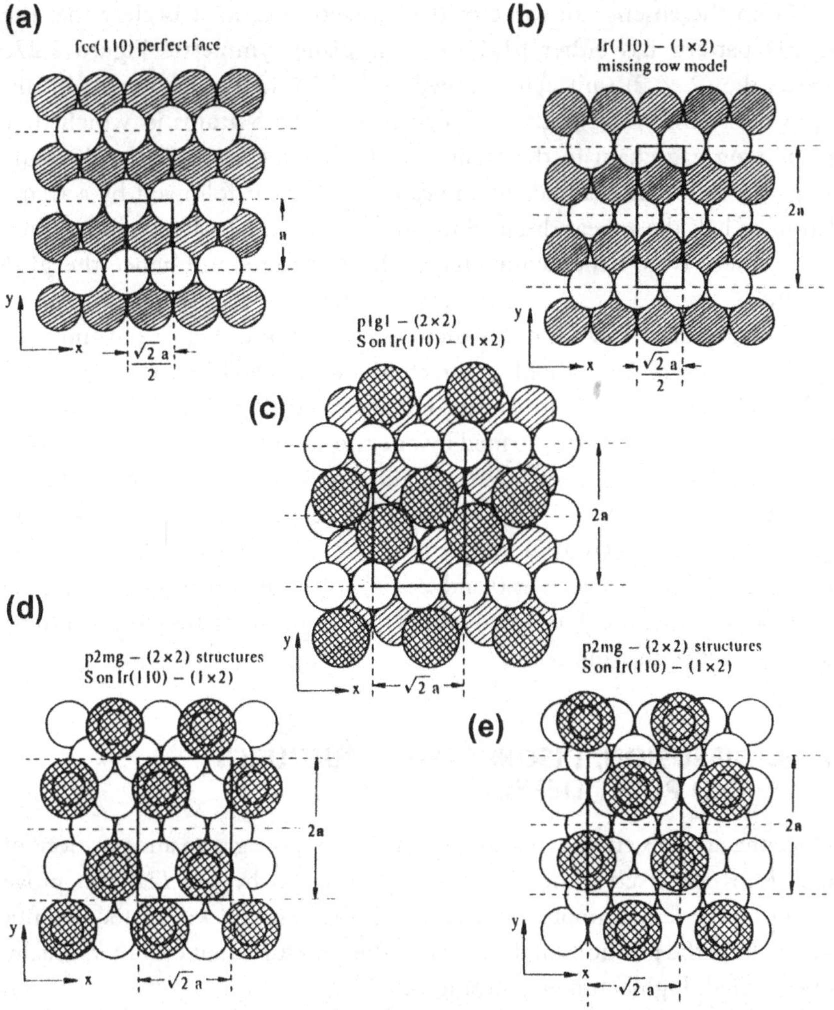

Figure 7.27 The Ir(110) surface. Ir has the fcc structure, where **a** is the conventional fcc bulk unit cell dimension, and the diagrams are drawn using the touching radius for the atoms. (a) A 'perfect' fcc (110) surface before any reconstruction. The unit cell is outlined and can be compared with Fig. 7.27b. (b) The missing-row Ir(l 10) reconstructed surface. The (1 × 2) unit cell is shown. (c) A possible *p*1*g*1-(2 × 2) structure for sulphur on Ir(110). The S atoms are marked by the large cross-hatched circles. (d) and (e) Two possible *p2mg*-(2 × 2) structures for sulphur on Ir(110). The inner circle on the S atoms shows the covalent radius of 1.0 Å.

From the absence of some of the diffraction spots it is clear that the LEED pattern has either *p*1*g*1 or *p*2*mg* plane symmetry. Figure 7.27c shows the (2 × 2) unit cell outlined by a thick line. Also shown in this figure is a possible *p*1*g*1-(2 × 2) structure of the S atoms in which they form a zig-zag chain in the trough of the missing row. The glide plane at $y = ½$ is drawn (dashed) to indicate a reflection followed by a translation of half the repeat distance in the *x*-direction. Apart from the glide plane, there is no other symmetry in this structure, and hence the *p*1*g*1 symbol.

Two possible *p*2*mg*-(2 × 2) structures are shown in Fig. 7.27d and e. In the first structure, S is located on threefold sites formed by two top-layer and a second-layer Ir atom. In the second structure (Fig. 7.27e) S atoms are located on three-fold sites formed by one top-layer and two second-layer Ir atoms. Note how both of these structures possess twofold axes at (½, ½) and (¼, ½), respectively, and a mirror plane parallel to the *y*-axis and intersecting the *x*-axis at ¼ and ¾ (Fig. 7.27d); in Fig. 7.27e the mirror plane intersects the *x*-axes at 0 and ½. Finally, both structures have a glide plane intersecting the *y*-axis at ½ with a translation along the *x*-direction by half a repeat in that direction.

7.15. DIFFUSION, DISORDERED STRUCTURES AND POINT DEFECTS

Once the crystal structure is known, so that the space group and positions of the atoms are fixed, other information provided by the *ITA* can prove useful. In the few examples discussed here, the concern will be focused not so much on the positions of the atoms in the structures, but on other, usually unoccupied, high-symmetry positions.

7.15.1. Diffusion

When considering diffusion in a crystal, one usually has in mind a crystal with a very small number (roughly 1 in 10^6) of the atoms 'out of place'. For either impurity or self-diffusion, the atom may start from a **substitutional site** or an **interstitial site.** The former is normally occupied by a host atom. The latter is a site that is not normally occupied.

When the lowest-energy interstitial sites for diffusing impurity or host atoms are determined, it is found that they are most often high-symmetry sites (i.e. those whose site-symmetry group contains a relatively large number of operations). Likewise, when diffusion pathways are calculated, it is often found that the diffusing atom travels along a high-symmetry path.

For example, consider diffusion in a silicon Si crystal when an atom (an impurity or Si atom) is ejected by another atom from a substitutional site and the ejected atom diffuses as an interstitial. The ejection is called a **kick out process**. Ultimately, the kicked out atom may eject another Si atom, allowing the process to repeat. For some diffusion processes, the lowest-energy diffusion pathway is along a <111> direction (Fig. 7.28). Si atoms occupy the 8*a* sites in the $Fd\overline{3}m(O_h^7)$ space group. In this figure, the 8*a* positions at (0, 0, 0), (¼, ¼, ¼) are indicated by solid circles; a solid circle is also shown at the end body diagonal, at (1, 1, 1), with distance $\sqrt{3}\mathbf{a}$ from (0, 0, 0). The site symmetry of the occupied 8*a* sites belongs to the tetrahedral point group $\overline{4}3m(T_d)$.

Figure 7.29 shows the calculated potential energy for a positive boron ion (B^+) diffusing along the [111] direction in a Si crystal. The positions marked 'T' correspond to the 8*b* sites, located at (½, ½, ½) and (¾, ¾, ¾), which also have the tetrahedral site symmetry $\overline{4}3m(T_d)$. When the B^+ ion enters at a site normally occupied by a Si atom, the Si is 'kicked out' into a T site. BC stands for bond center, the 16*c* position at (⅛, ⅛, ⅛) with site symmetry $\overline{3}m(D_{3d})$, which lies directly between the two tetrahedrally bonded Si atoms. This site is a center of inversion, $\overline{1}(i)$. Similarly the 16*d* site, marked 'H', at (⅝, ⅝, ⅝) has site symmetry $\overline{3}m(D_{3d})$, and therefore also coincides with a center of inversion. Note that along the entire [111] direction the site symmetry is at least $3m(C_{3v})$ because a general point along this line is at the Wyckoff 32*e* site (see Fig. 7.6). Thus the diffusion path is

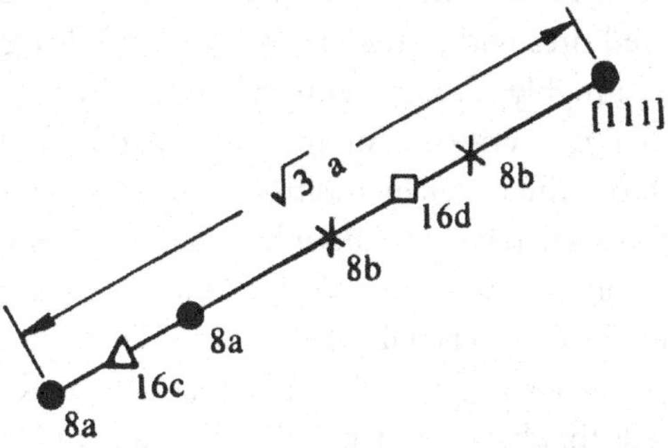

Figure 7.28 The special positions along the body diagonal of a unit cell in the space group. The positions are labelled using the Wyckoff notation and the coordinates of these positions can be found in the *ITA* or in Fig. 7.6. Silicon has the diamond structure and so all 8*a* positions are occupied by Si.

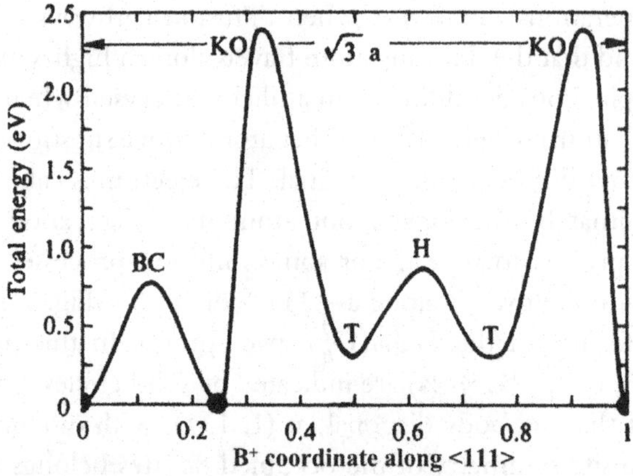

Figure 7.29 Theoretical calculations of the potential energy versus distance of a positively charged boron impurity along the [111] direction in Si crystal (C. S. Nichols, private communication).

along a line of fairly high symmetry and the energy extrema usually occur at points of higher symmetry.

Symmetry is important in the theoretical calculations because, as the diffusing atom traverses this path, the neighbouring Si atoms must be allowed to relax. However, this relaxation (or movement of the surrounding atoms) is radial to the diffusion path, so it does not alter the symmetry traversed by the diffusing atom. These are called **symmetry-preserving relaxations**, a concept found in many areas of solid state science.

7.15.2. Disordered Structures

In our discussion of symmetry in crystals, we have tacitly assumed that all the atoms lie in ordered sites within the crystal structure (**long-range order**). This is, of course, a highly idealized concept, since there is always a certain amount of disorder in a crystal. For instance, in reality all atoms exhibit thermal motion about their average positions, increasingly so as temperature increases. Therefore, when we talk about long-range order or **translational periodicity** we really are referring to the time-averaged positions of the atoms and molecules in the crystal structure.

As examples of other types of disorder that can be found, we discuss two cases of crystals that have a normal periodic structure (time-averaged!) for some of the atoms, but a disordered arrangement for others. Both of these crystals are **fast-ion conductors**, or **solid electrolytes**; that is, they have ionic conductivity of about the same order of magnitude as a liquid electrolyte at the same temperature (the acid solution in a Pb-acid car battery is a liquid electrolyte). We shall see how a study of the structures and their space groups makes these physical properties easy to understand.

From room temperature up to 147 °C, AgI has the wurtzite structure and is an ordinary (poor) ionic conductor. However, at this temperature, there is a phase transition to a new high-temperature α-AgI phase and the Ag^+ ionic conductivity increases by ~10^4. From X-ray diffraction it is found that the structure can be classified according to the $Im\overline{3}m(O_h^9)$ space group (see the appropriate page of the *ITA*). The two I atoms in the conventional unit cell are on the Wyckoff 2*a* sites at (0, 0, 0) and (½, ½, ½). But then where are the two Ag atoms in the unit cell? In this space group, there are no other possible unique positions for the other two atoms. Once the 2*a* site is occupied, the site with the fewest number of positions is 6*b*, and so six atoms are required to fill these positions completely.

For this space group, Fig. 7.30 shows some of the allowed positions labelled with the Wyckoff notation. The large open circles represent the I atoms at the 2*a* positions. In the original (1935) X-ray structure refinement of α-AgI, it was reported that the two Ag ions were uniformly and randomly distributed over the 6*b*, 12*d* and 24*h* sites, which means that the two Ag atoms per unit cell can randomly occupy any of the 42 positions. Thus there appears to be a large randomness in the possible positions of any given Ag atom, and this can occur because of a very low potential energy barrier for Ag atom diffusion (i.e. the fast-ion conductor character of this material).

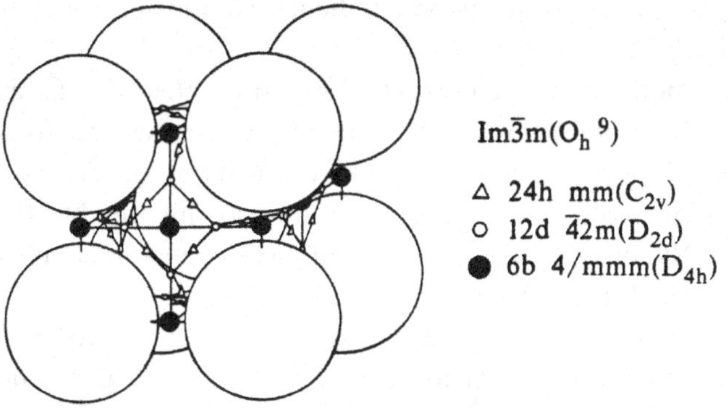

Figure 7.30 Some special positions of the space group $Im\overline{3}m(O_h^9)$. The large open spheres represent I atoms on the 2*a* positions with point symmetry $m\overline{3}m(O_h)$.

Let us examine the site symmetry and coordination of the Ag atoms at these different possible positions. The 6*b* position has site symmetry $4/mmm(D_{4h})$ with six first-neighbour I atoms, two at a distance of $\mathbf{a}/2$ and four at $\sqrt{2}\mathbf{a}/2$. Thus a Ag atom at this site is surrounded by six I atoms in the form of a compressed octahedron. A 12*d* site has symmetry $\bar{4}2m(D_{2d})$ and in this structure this site is surrounded by four I atoms at a distance of $\sqrt{5}\mathbf{a}/4$. Thus an atom at the 12*d* site has four nearest-neighbour atoms in the form of a compressed tetrahedron.

More recent diffraction work has shown that the two Ag atoms are not randomly distributed over all of the 42 positions; instead, to a better approximation, they are randomly distributed only over the 12*d* sites. Thus there are still many more possible positions in the unit cell than there are Ag atoms, and hence it is not surprising that there is a very low activation energy for diffusion. With hindsight, this position might have been expected to be favoured over the 6*b* position because, in this structure, an atom at a 12*d* site is at the center of a distorted tetrahedron. The tetrahedral coordination is just like the coordination in the room-temperature wurtzite phase, and so in the α-AgI phase, the Ag atoms find positions that keep the local coordination similar.

CuI at room temperature has the zinc-blende structure (Section 7.6) and is a poor ionic conductor. At 407 °C, it undergoes a phase transition to the so-called α′-CuI structure which is a fast-ion conductor, the conduction being due to mobile Cu ions. This structure is described by the $Fm\bar{3}m(O_h^5)$ space group (Fig. 7.3), where the I atoms are on the 4*a* sites; that is, all of the face-centered sites and the four Ca atoms per unit cell randomly occupy the 8*c* sites of this space group. Thus this structure closely resembles the CaF_2 structure (Fig. 7.2c) with the I atoms at the Ca positions and the Cu atoms randomly occupying one half of the F positions. Note that in the α′-CuI structure, the Cu atoms maintain tetrahedral coordination as in the room-temperature phase.

Thus, in both the α-AgI and α′-CuI structures, the ratio of available sites to positive ions is at least two, and in these structures this leads to a relatively low potential energy barrier for positive-ion diffusion (fast-ion conductivity). Other well-known fast-ion conductors are the Na β-aluminas, where the Na ions can rapidly diffuse in a two-dimensional plane, which also has many more available sites than Na atoms.

A ratio of available sites-to-ions that is greater than unity appears to be a necessary condition for fast-ion conductivity. However, it is not a sufficient condition. There are many disordered structures with vacant positions in which the ionic conductivity is poor (low). In such materials, the disordered atoms may be in deep potential wells making diffusion to other sites difficult. For example, Al_2ZnS_4 has the wurtzite structure, and the three metal atoms are randomly arranged on four possible positions. However, this material is a poor ionic conductor.

7.15.3. Point Defects

If the position of a point defect in a crystal is known, then the Wyckoff symbol is a useful way to label its site. This was already used in our discussion of diffusion and fast-ion conductors. Furthermore, since the defect only senses the point symmetry due to its surroundings (the site symmetry), then from the Wyckoff symbol, we immediately know the site symmetry.

To perform **electron spin resonance** (ESR) experiments, impurity atoms are deliberately added to a crystal. For example, Fe^{3+} or Mn^{2+} ions in a crystal usually have unpaired 3d electrons and ESR can be detected by standard techniques and the spectra fitted to a spin Hamiltonian, the form of which is determined by the (site) symmetry seen by the impurity atom.

For example, if paramagnetic ions, such as Fe^{3+} and Mn^{2+}, are added to the spinel structure (Section 7.5), then their positions can be found because the spin Hamiltonian for a defect on an $8a$ site must be cubic, whereas it must have axial symmetry if the ion is on a $16d$ site.

Another example is PZT, which is an acronym for $Pb(Zr_xTi_{1-x})O_3$, a material with the perovskite structure. PZT in its room-temperature phase is the most commonly used piezoelectric ceramic material. Recall that the general form of the prototype phase (Section 7.13) is cubic with

$$\begin{aligned} &\text{A}: \quad \text{at } 1a \text{ sites} \quad (0,0,0) \\ &\text{B}: \quad \text{at } 1b \text{ sites} \quad (\tfrac{1}{2},\tfrac{1}{2},\tfrac{1}{2}) \\ &\text{O}: \quad \text{at } 3d \text{ sites} \quad (\tfrac{1}{2},0,0); \quad (0,\tfrac{1}{2},0); \quad (0,0,\tfrac{1}{2}) \end{aligned}$$

of the $Pm\bar{3}m(O_h^1)$ space group. Both the $1a$ and $1b$ sites have the full point group symmetry $m\bar{3}m(O_h)$, so that either could sensibly be chosen to be at the origin (see Figs 7.5 and 7.14). By writing the chemical formula for PZT as we have, it is (correctly) implied that both Zr and Ti atoms occupy the $1b$ sites.

PLZT is an acronym for $(Pb,La)(Zr,Ti)O_3$, a material closely related to PZT. Since Pb^{2+} is partially replaced by La^{3+}, vacancies (□) are introduced.

Are the vacancies on the 1*a* site, the 1*b* site, or distributed over both? If the vacancies are all on the 1*a* site, the formula will be

$$(Pb_{1-3y/2}La_y\ W_{y/2})\ (Zr_xTi_{1-x})O_3$$

If the vacancies are all in the 1*b* site, then the formula will be

$$(Pb_{1-y}La_y)\ ([Zr_xTi_{1-x}]_{1-y/4}\ W_{y/4})O_3$$

Note that the Pb/La ratio is different in the two cases and this can be measured to determine which situation is more appropriate. For most preparative conditions of PLZT, the result is closer to the former situation with most of the vacancies on the 1a site. However, there are some vacancies on the 1*b* site. Thus, for some structures, various mixed systems result in unexpected site occupancies and the Wyckoff symbols are useful in designating the sites.

7.16. EUCLIDEAN NORMALIZERS

It is not commonly realized that for most crystal structures a variety of different sets of atomic coordinates can be chosen that are symmetry-equivalent. Thus a list of coordinates for a certain material in one book may appear at first sight to look different from that given elsewhere. To the novice, this can be extremely confusing. Moreover, given different materials one might like to know if their structures are related to one another. The *ITA*, fortunately, provides tabulated information that helps to decide if two descriptions really are equivalent. To do this, the concept of Euclidean normalizers can be applied.

Consider a group $\mathcal{G}$ and one of its supergroups **S**. Both can be uniquely related to another intermediate group $\mathcal{N}_\varepsilon(\mathcal{G})$ that is called the **normalizer of $\mathcal{G}$ with respect to S**. The normalizer group $\mathcal{N}_\varepsilon(\mathcal{G})$ is defined by the set of all elements of **S** that map $\mathcal{G}$ onto itself by conjugation $\mathbf{S}^{-1}\mathcal{G}\mathbf{S}$. The normalizer may coincide with $\mathcal{G}$ or with **S** or it may be a proper intermediate subgroup. The normalizer of a space group $\mathcal{G}$ with respect to the group of all Euclidean mappings (motions, isometries) is called the **Euclidean normalizer** of $\mathcal{G}$.

In general two unique groups $\mathcal{K}(\mathcal{G})$ and $\mathcal{L}(\mathcal{G})$ exist intermediate between $\mathcal{G}$ and $\mathcal{N}_\varepsilon(\mathcal{G})$, such that

$$\mathcal{G} \leq \mathcal{K}(\mathcal{G}) \leq \mathcal{L}(\mathcal{G}) \leq \mathcal{N}_\varepsilon(\mathcal{G})$$

$\mathcal{K}(\mathcal{G})$ is a class-equivalent supergroup of $\mathcal{G}$ that is at the same time a translation-equivalent subgroup of $\mathcal{N}_{\mathcal{E}}(\mathcal{G})$. The group $\mathcal{L}(\mathcal{G})$ differs from $\mathcal{K}(\mathcal{G})$ only if $\mathcal{G}$ is non-centrosymmetric, while $\mathcal{N}_{\mathcal{E}}(\mathcal{G})$ is centrosymmetric. It then belongs to the same Laue class as $\mathcal{G}$. If $\mathcal{N}_{\mathcal{E}}(\mathcal{G})$ is non-centrosymmetric, then $\mathcal{L}(\mathcal{G})$ does not exist. In general, therefore, the three relationships between these four possible groups can be described by the ratios of their numbers of symmetry operations, known as the **index**.

Euclidean normalizers were first introduced into crystallography by Hirshfeld[14] who called them 'Cheshire' groups, because, like the famous Cheshire cat in *Alice in Wonderland* who vanished leaving just its grin, such groups can be found by removing all the atoms from a crystal structure and leaving behind only its symmetry elements. The resulting symmetry always includes all the symmetry elements of the space group $\mathcal{G}$, and so the Cheshire group is necessarily a supergroup of $\mathcal{G}$.

As an example, consider the symmetry elements in the space group $I4_122(D_4^{10})$ (Fig. 7.31). Note that the 4_1-axis at position marked A is the mirror image of the 4_3-axis at position B. Therefore, we can mark in mirror planes between them (marked in gray), and then this forces the 4_1 and 4_3 to be replaced by 4_2 axes (as in the space group diagram for $P4_2/nnm$). In addition, we then see that the 2[001] axes pass through the intersection of two mirror planes and so this gives a center of symmetry at this position.

Since now A B C and D are equivalent, they are 4-fold related, and this 4-fold-axis also passes through the 2[001] axis and its center of symmetry, thus creating a $\bar{4}$-axis. Continuing in this way, we see that the symmetry elements of $I4_122$ can be described by an arrangement corresponding to $P4_2/nnm$, its normalizer group. In the *ITA*, the normalizer for $I4_122$ is given as follows:

Space Group $\mathcal{G}$	**Normalizer $\mathcal{N}_{\mathcal{E}}(\mathcal{G})$**		**Additional generators of $\mathcal{N}_{\mathcal{E}}(\mathcal{G})$**			
Symbol	**Symbol**	**Basis vectors**	**Translations**	**Inversion through center at**	**Further generators**	**Index of $\mathcal{G}$ in $\mathcal{N}_{\mathcal{E}}(\mathcal{G})$**
$I4_122$	$P4_2/nnm$	½(**a** − **b**), ½(**a** + **b**), ½**c**	0, 0, ½	¼, 0, ⅛		2 ·2·1

14 *F.L. Hirshfeld, Acta Cryst.*, **A24**, 301 (1968).

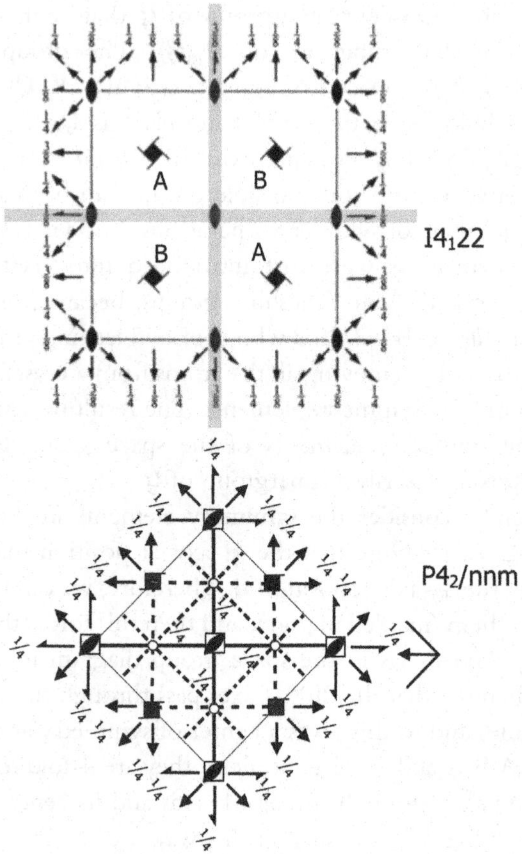

Figure 7.31 The relationship between $I4_122(D_4^{10})$ and $P4_2/nnm(D_{4h}^{12})$ (rotated through 45°).

The basis vectors give the axes of the normalizer group $\mathcal{N}_{\mathcal{E}}(\mathcal{G})$ with respect to the space group $\mathcal{G}$. Following this we see the additional generators that are needed to for $\mathcal{N}_{\mathcal{E}}(\mathcal{G})$, in this case a translation through 0, 0, ½ to form $\mathcal{K}(\mathcal{G})$ and inversion through a point at (¼, 0, ⅛) to form $\mathcal{L}(\mathcal{G})$. Finally the index of $\mathcal{G}$ in $\mathcal{N}_{\mathcal{E}}(\mathcal{G})$ is given as $2 \cdot 2 \cdot 1$.

Another example is that of $Pmmm(D_{2h}^1)$, shown in the following table. In this case, the normalizer group depends on the metric relationships between the axial lengths (note that no inversion through a center is involved here and so we have omitted this column for the sake of not overcrowding the table). We see that when the axes are unequal in length the normalizer has the same symmetry as the space group *Pmmm*, which can be described in $8 \times 1 \times 1 = 8$ different settings. However, if two axes are forced to be equal, a tetragonal normalizer group *P*4/*mmm* is obtained. The space group *Pmmm* can then be described by 8 × 1 × 2 = 16 different settings. Finally if all axes are constrained to be equal in length, the normalizer is the cubic group $Pm\bar{3}m$, and the space group *Pmmm* can be described in 8 × 1 × 6 = 48 different settings.

Space Group $\mathcal{G}$	Cell metric	Normalizer $\mathcal{N}_\varepsilon(\mathcal{G})$		Additional generators of $\mathcal{N}_\varepsilon(\mathcal{G})$		
Symbol		Symbol	Basis vectors	Translations	Further generators	Index of $\mathcal{G}$ in $\mathcal{N}_\varepsilon(\mathcal{G})$
Pmmm	a ≠ b ≠ c	*Pmmm*	½**a**, ½**b**, ½**c**	½,0,0; 0,½,0; 0,0,½		8·1·1
	a = b ≠ c	*P*4/*mmm*	½**a**, ½**b**, ½**c**	½,0,0; 0,½,0; 0,0,½	*y*,*x*,*z*	8·1·2
	a = b = c	$Pm\bar{3}m$	½**a**, ½**b**, ½**c**	½,0,0; 0,½,0; 0,0,½	*z*,*x*,*y*; *y*,*x*,*z*	8·1·6

As an example of the practical use of Euclidean normalizers, consider the tetragonal crystal structure of $BaTiO_3$. This crystallizes in space group $P4mm(C_{4v}^1)$ whose Euclidean normalizer is described by

Space Group $\mathcal{G}$	Normalizer $\mathcal{N}_\varepsilon(\mathcal{G})$		Additional generators of $\mathcal{N}_\varepsilon(\mathcal{G})$			
Symbol	Symbol	Basis vectors	Translations	Inversion through center at	Further generators	Index of $\mathcal{G}$ in $\mathcal{N}_\varepsilon(\mathcal{G})$
P4*mm*	$P^1 4/mmm$	½(**a** − **b**), ½(**a** + **b**), ε**c**	½, ½, 0; 0, 0, *t*	0, 0, 0		$(2^\infty)\cdot 2\cdot 1$

The normalizer group is given as $P^1 4/mmm$, where the superscript 1 indicates that one direction (the **c**-axis here) is not constrained. This is also seen in the basis vectors by ε**c**. In order to generate the normalizer, there are two translations, ½, ½, 0 and 0, 0, *t*, where *t* can be any distance. Furthermore inversion through the origin is specified. The index of $\mathcal{G}$ in $\mathcal{N}_\varepsilon(\mathcal{G})$ is $(2^\infty)\cdot 2\cdot 1$.

In Fig. 7.32, we show the way in which $\mathcal{G} = P4mm$ and $\mathcal{N}_\varepsilon(\mathcal{G}) = P4/mmm$ are related. Consider now the coordinate description in *P*4*mm* for the Ba, Ti, O1 and O2 atoms as in column (a) of Table 7.3. If we now apply the translation 0, 0, ½ and inversion through the origin in each case, we get the four alternative descriptions in the table corresponding to the index

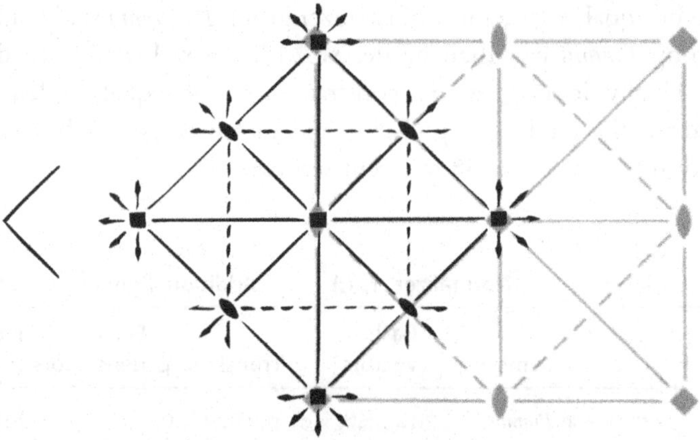

Figure 7.32 Relationship between $P4mm(D_{2h}^1)$ and $P4/mmm(D_{4h}^1)$.

Table 7.3 Four equivalent descriptions of the structure of $BaTiO_3$.

	(a)		(b)		(c)		(d)	
Ba	4*a*	0,0,0	4*a*	0,0,0	4*a*	½, ½, 0	4*a*	½, ½, 0
Ti	4*b*	½, ½, 0.51	4*b*	½, ½, 0.49	4*b*	0, 0, 0.51	4*b*	0, 0, 0.49
O1	4*b*	½, ½, −0.02	4*b*	½, ½, 0.02	4*b*	0, 0, −0.02	4*b*	0, 0, 0.02
O2	2*c*	½, 0, 0.485	2*c*	½, 0, 0.515	2*c*	0, ½, 0.485	2*c*	0, ½, 0.515

$2 \times 2 \times 1$ of $\mathcal{G}$ in $\mathcal{N}_{\mathcal{E}}(\mathcal{G})$. Figure 7.33 shows pictures of the structure corresponding to each description in Table 7.3.

In addition, the translation 0, 0, *t* means that to any of the four descriptions in the table, one can arbitrarily add any value along the **c**-axis, so that in reality one can envisage an infinite number of equivalent descriptions of this structure. Nonetheless, it can be seen that given, say, two sets of coordinates, for example (a) and (c) in Table 7.3, the use ofEuclidean normalizers makes it immediately apparent that both sets of coordinates give the same crystal structure. Therefore Euclidean normalizers are very useful in determining if alternative coordinate sets are really describing the same crystal structure, or if two structures are similar. This is particularly useful in the study of SPTs where one needs to compare the structures of two different phases that have different space group descriptions and possibly specified on differently oriented axes. The reader is especially encouraged to try out the program NORMALIZER at the Bilbao Crystallographic Server.[15]

15 http://www.cryst.ehu.es/.

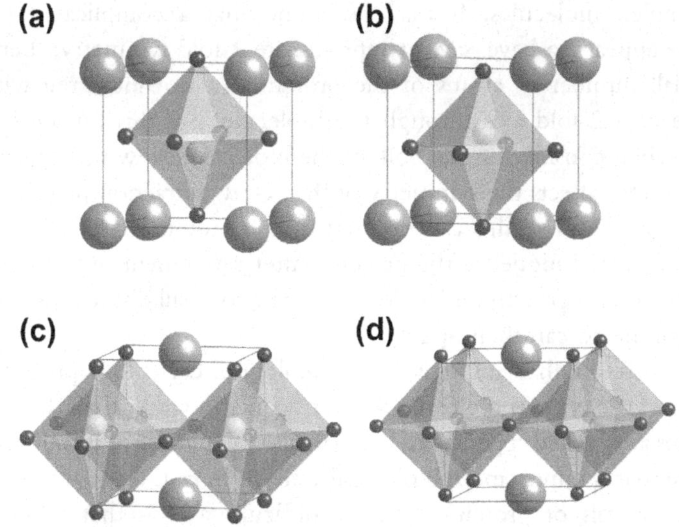

Figure 7.33 Four equivalent views of the tetragonal structure of $BaTiO_3$ (the atomic displacements have been exaggerated for clarity).

7.17. NON-CRYSTALLOGRAPHIC SYMMETRY

So far in this book, we have discussed crystals with translational symmetry. We might call these crystals **normal crystals**. In this and the next two sections, we discuss certain aspects of relaxing the stringent symmetry conditions found in normal crystals. In this section, the primary concern is relaxation of the rotation conditions; in Section 7.20 the translational symmetry is relaxed; in Section 7.22 both symmetries are relaxed.

For normal crystals, long-range translational symmetry is only consistent with 1, 2, 3, 4 and 6-fold rotational symmetry (Section 2.4). However, molecular crystals (those composed of tightly bonded atoms forming molecules which in turn are weakly bonded to each other), virus crystals and other types of crystals may appear to have extra symmetry operations that are not contained within the point or space group of normal crystals. Such **non-crystallographic symmetry** may not be exact, nor indeed can it be from a theoretical point-of-view. However, if the long-range, symmetry-compatible forces within the crystal are too weak to disrupt noticeably the non-crystallographic symmetry, then to all intents and purposes this symmetry is *effectively* exact.

Non-crystallographic symmetry is very important for protein crystallographers, who are interested in determining the crystal structures of large and complex molecules. It may be found that a complicated protein molecule appears to have, say, an approximate 2-fold symmetry; there may be a small number of atoms of the protein that do not agree with the presence of a 2-fold axis through the molecule, but this can be ignored initially when considering the bulk of the molecule. Now this apparent 2-fold symmetry affects the symmetry of the X-ray diffraction pattern of this protein crystal and so this can be used by the protein crystallographer in developing a trial model of the protein. After refinement of the structure, other parts of the protein molecule, giving rise to small distortions from the 2-fold symmetry, can then be included.

It is worth emphasising that for the biological crystallographer it is the structure of the protein molecule that is of concern rather than the way the molecules are packed together in the crystal structure, since it is the protein molecule that is important in biological activity. In fact, the main reason for growing crystals of proteins (often a difficult task) is that when these extremely large molecules are regularly arranged with long-range order, the resulting diffraction pattern consists of sharp spots whose intensities can be used in the solution of the crystal structure and, hence, of the molecule. If the molecules were randomly arranged, as in say a liquid solution, the diffraction pattern would be essentially featureless and it would be impossible with conventional methods to determine much about the structure of the molecule. Thus crystallization provides the necessary long-range translational symmetry which is in turn critical for determining the structures of the molecules themselves.

7.18. STRUCTURES WITH Z′ > 1

In recent years interest has developed, particularly within the organic molecular crystallographic community, in those structures with unusually high numbers, Z', of formula units divided by the number of general equivalent positions.[16] Normally, molecules pack in the densest form possible, and in most cases $Z' \leq 1$. For instance, consider the crystal structure of the molecule of naphthalene which crystallizes in space group $P2_1/a(C_{2h}^5)$. The molecule is centrosymmetric, and the molecular center coincides with the center of inversion of the space group. The result is that the structure is uniquely determined by the positions of half of the atoms in the molecule. These atoms are in the 4*e* general positions, and this then leads

[16] See http://www.dur.ac.uk/zprime/.

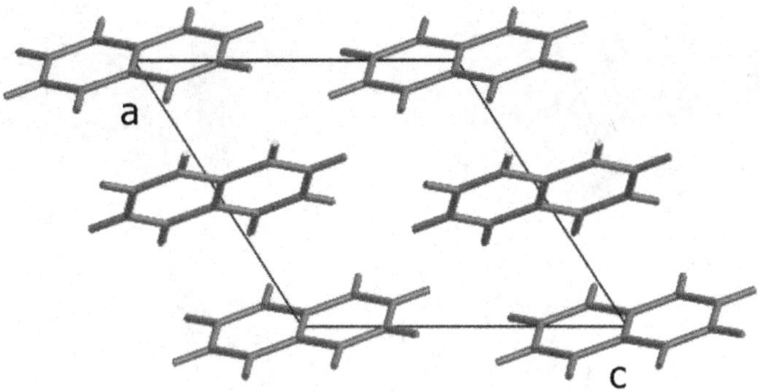

Figure 7.34 The crystal structure of naphthalene projected on (010).

to two complete molecules (4 × ½) in the unit cell. Figure 7.34 shows a projection of the structure where the two molecules can be seen (remember to displace the origin of the unit cell in order to count the unit cell contents). Thus in this case $Z' = 2$ molecules/4 = 0.5.

On the other hand, consider the case of benzidine ($C_{12}H_{12}N_2$).[17] This material crystallizes in four different polymorphic forms (a polymorph is usually considered to be where the same chemical entity crystallizes in different structural arrangements). Figure 7.35 shows projections of the four forms, I, II, III and IV, and Table 7.4 lists the unit cell information. In the table, Z is equal to the number of molecules in the unit cell. You can check this for yourself by counting them in the figure.

Then, for Forms I and IV there are 18 molecules with 4 general equivalent positions giving $Z' = 18/4 = 4.5$. In Form II, there are 6 molecules in 2 general positions making $Z' = 6/2 = 3$. Finally in Form III, there are 6 molecules in 4 general positions giving $Z' = 6/4 = 1.5$. Values as high as 4.5 are extremely rare, and yet we see that two forms appear in which a large number of molecular entities co-crystallize together without a symmetry operation connecting them.

It seems that high values of Z' are found with several chiral molecules such as nucleosides, nucleotides, steroids and other natural products. The reasons for this are not clear, although it has been suggested that this phenomenon arises from frustration between two or more competing interactions.

Although this topic is normally of concern to the organic small-molecule and macromolecular community, there are examples of this

[17] M. Rafilovich & J. Bernstein. *J. Amer. Chem. Soc.* **128**, 12185 (2006).

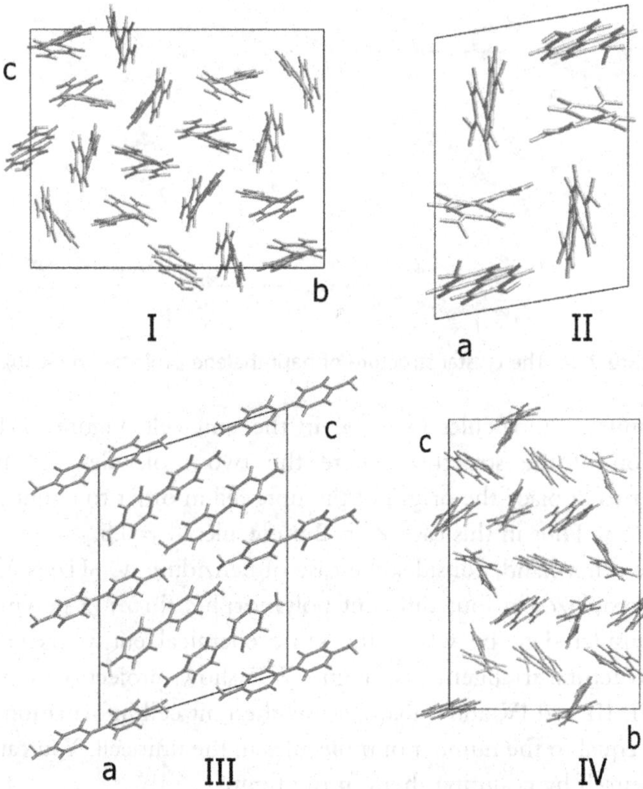

Figure 7.35 Four polymorphic forms of benzidine.

Table 7.4 Unit cell data for the four polymorphic forms of benzidine.

Space group	1Form I $P2_1/n(C_{2h}^5)$	Form II $P\bar{1}(C_i^1)$	Form III $P2_1/c(C_{2h}^5)$	Form IV $P2_1/n(C_{2h}^5)$
a (Å)	11.318	9.744	14.735	11.631
b (Å)	22.552	11.444	5.533	16.189
c (Å)	18.076	13.979	19.200	24.173
α (°)		108.847		
β (°)	99.966	97.681	106.827	99.153
γ (°)		95.018		
V (Å^3)	4544.0	1481.2	1498.4	4493.7
Z	18	6	6	18
Z′	4.5	3	1.5	4.5

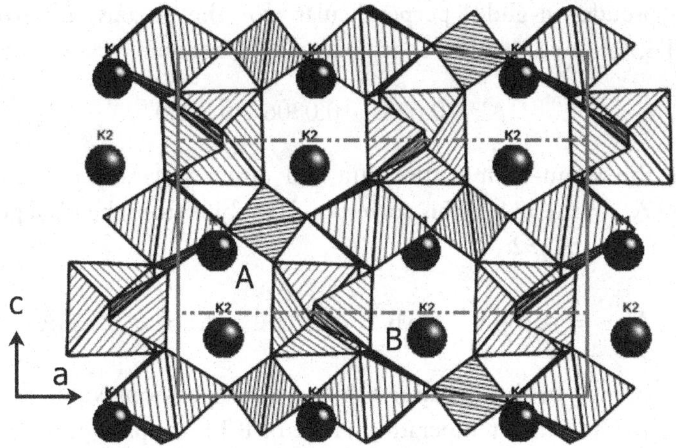

Figure 7.36 Projection of the KTP structure on (010).

phenomenon in the inorganic field too. A particularly interesting case is that of $KTiOPO_4$ (or KTP for short), a material of considerable importance for its non-linear optical properties. This crystallizes in the polar space group $Pna2_1(C_{2v}^9)$ with PO_4 tetrahedra and TiO_6 octahedra with all atoms on general equivalent positions. The K atoms (K1 and K2) are independent of each other and occupy spaces between these polyhedra. In this space group there are 8 formula units and 4*a* general positions, so that $Z' = 2$. Figure 7.36 shows the structure projected on (010), where the origin of the unit cell has been displaced by (¼,¼,¼). With this choice of origin it is found that, for example, K1 (marked A) and K2 (marked B) are not far from being related by an *n*-glide perpendicular to the **c**-axis. This observation suggested[18] that this structure could transform at some higher temperature to one in supergroup $Pnan(D_{2h}^6)$, where the crystal would then become non-polar and where $Z' = 1$. In this space group, the K atoms become symmetry-related, occupying 8*e* positions. This was subsequently verified by experiment.

At room temperature, in space group P*na*2$_1$, and with the origin displaced by (¼,¼,¼), the fractional coordinates of K1 and K2 are

A:	0.1281	0.5306	0.4380
B:	0.6447	0.0501	0.1820

18 P. A. Thomas, A. M. Glazer & B. E. Watts. *Acta Cryst.* **B46**, 333 (1990).

The pseudo *n*-glide perpendicular to the **c**-axis is given by $\{m[001]|½,½,½\}$, so that the coordinates for A become

A': 0.6281 0.0306 0.0620

Since the room-temperature unit cell dimensions are $a = 12.819$ Å, $b = 6.399$ Å, $c = 10.584$ Å, this corresponds to shifts from the ideal positions in *Pnan* of (B − A')/2:

$$\Delta x\text{a} = 0.106\text{Å}, \quad \Delta y\text{b} = 0.062\text{Å}, \quad \Delta z\mathbf{c} = 0.635\text{Å}$$

Therefore we see that the largest shift in the K positions is along the polar **c**-axis. Thus the room-temperature phase of KTP in space group $Pna2_1$ is said to be **pseudosymmetrically** related to the high-temperature **prototypic** phase in *Pnan*.

7.19. ICOSAHEDRAL SYMMETRY

Let us consider an important non-crystallographic symmetry, one possessed by nearly all small viruses, and by some complex metal alloys as well. Icosahedral groups are characterized by six 5-fold axes subtending 63.43°. Each 5-fold axis is surrounded by five 3-fold axes and five 2-fold axes. There are two icosahedral point groups, which are given the symbols 235(I) and $m\bar{3}\bar{5}(I_h)$ (the naming convention used is the same as in the cubic system, Section 4.2.7). Stereograms for these two point groups are given in Fig. 7.37 to illustrate an icosahedron viewed along its 3, 5 and 2-fold axes in turn. From these diagrams it can be seen that there are operations belonging to the cubic point groups 23(T) and $m\bar{3}\bar{5}(I_h)$, respectively. In fact, these two cubic point groups are maximal subgroups of the respective icosahedral groups. This can be seen by comparing Fig. 7.37a and e, where in Fig. 7.37e the letter 'c' has been marked on the four faces that contain the four 3-fold axes of a cube. Correspondingly, in the stereogram in Fig. 7.37a, the triangles denoting these particular 3-fold axes have been drawn larger than the other 'non-crystallographic' 3-fold axes. Likewise, the three 2-fold axes corresponding to 23(T) cubic point symmetry are drawn larger than the others.

7.19.1. Viruses

The crystal structures of viruses provide important examples of non-crystallographic symmetry. Many small viruses have icosahedral symmetry,

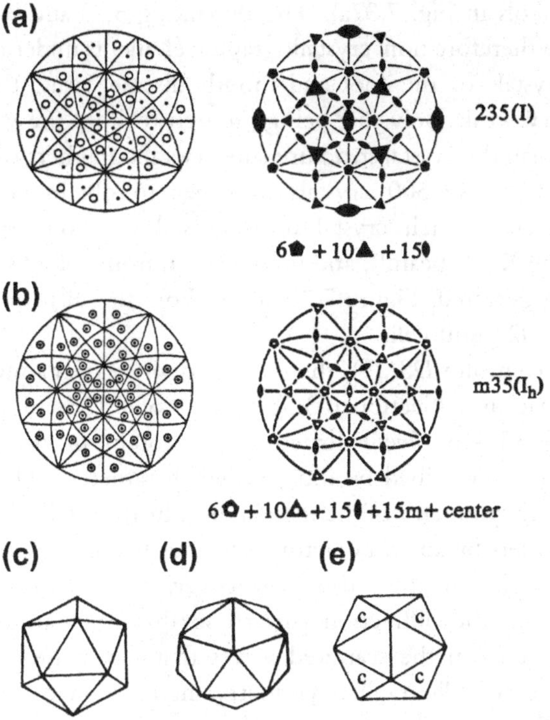

Figure 7.37 Stereograms for the icosahedral point groups (a) 235(I) and (b) $m\overline{3}\,\overline{5}$($I_h$). For 235, we mark those symmetry elements pertaining to the subgroup 23(I) slightly larger than the others. (c)–(e) Drawings of an icosahedron viewed along the 3, 5 and 2-fold axes.

such as the foot-and-mouth disease virus. This virus occurs in the form of a regular icosahedron, with protein molecules arranged on the surface according to the symmetry of the icosahedron, but with a single RNA (ribonucleic acid) molecule lying in the interior. Although we know that 5-fold rotational symmetry is incompatible with long-range translational symmetry, nevertheless it is possible to crystallize such a virus. Because a virus is a very tightly bound entity and the forces between the viruses are weak, it is not surprising to discover that the 5-fold icosahedral symmetry is to all intents and purposes maintained within the crystal structure. The foot-and-mouth disease virus crystallizes in the cubic space group $I23(T^3)$, with two virus units per unit cell, and with a cell edge $\mathbf{a} = 345$ Å. In the icosahedron, four of the 3-fold and three of the 2-fold axes are oriented with respect to the 3-fold and 2-fold axes of the $I23$ unit cell (see the enlarged symbols in Fig. 7.37a). The remaining 5, 2 and 3-fold axes of symmetry are therefore non-crystallographic elements under the *I*23 space group. In crystals of the foot-and-mouth disease virus, the intervirus forces are very weak, only just being enough to hold the virus particles together to form the crystal structure (the crystallographers who solved its structure had to use 500 crystals in order to obtain enough X-ray diffraction data, as each crystal decomposed in about 10 min when exposed to the X-ray beam!), and no deviation from 5-fold symmetry has actually been detected. Figure 7.38 shows how the virus icosahedra are packed in the *I*23 unit cell.

In the space group $I23(T^3)$, there are two viruses per unit cell, and so the asymmetric unit (Section 6.3.5) consists of 1/12th of a virus. This means that in order to solve its structure, the proteins making up 1/12th of the virus will be described by independent parameters. The extra non-crystallographic symmetry of the icosahedron further reduces the number of free parameters by an extra factor of 5, thus making it easier to locate the protein units in the virus. Moreover, the non-crystallographic symmetry affects the diffraction pattern so that one can assume phase relationships between the scattered waves that otherwise would not be apparent. Non-crystallographic symmetry therefore helps in establishing the phases of X-ray reflections, which means that the structure can be solved more easily.

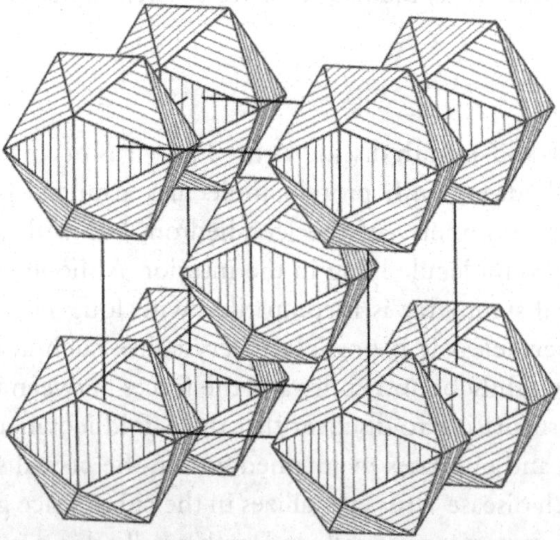

Figure 7.38 The packing of icosahedral foot-and-mouth disease virus in a crystal with space group $I23(T^3)$. The middle virus is in the body-centered position.

Although at present the main purpose of crystallization of a virus is to enable its structure to be determined, it may be that a study of the entire crystal structure will one day be of interest in understanding the biological activity of the virus. The active receptors of viruses like the foot-and-mouth disease virus are located on the surface and so a study of the weak hydrogen bonds that link them together to form a crystal might be relevant to the understanding of how the virus attaches itself to proteins in an organism.

There are many other viruses with similar icosahedral structures. For instance, the tomato bushy stunt virus and the polyoma virus also crystallize in $I23(T^3)$. The satellite tobacco necrosis virus is found in space group $C2(C_2^3)$. Interestingly in this case none of the 2-fold axes of the icosahedron are related to the 2-fold axis of this space group; yet the virus itself maintains its icosahedral symmetry.

Icosahedral-like symmetry is also found in the atomic coordination arrangements in metallic alloys and in many boron-containing compounds. Typical examples include WAl_{12}, Fe_3C, $Cr_{23}C_6$, $MgCu_2$, $MgZn_2$, α-Mn and B_6H_6. A good illustration is given by $Mg_{32}(Al,Zn)_{49}$. In this structure, the space group is $Im\overline{3}(T_h^5)$ with a $= 14.16$ Å. All of the 98 zinc and aluminium atoms in the unit cell have icosahedral coordination. Note that the point group of this space group is $m\overline{3}(T_h)$, a subgroup of the icosahedral point group $m\overline{3}\,\overline{5}(I_h)$.

7.20. INCOMMENSURATE MODULATIONS

Translational symmetry has played a central role in the development of solid state science. Therefore deviations from translational symmetry are of distinct interest. Incommensurate modulations, found in some materials, are examples of such deviations. Some incommensurate modulations are found to be associated with a phase transition, so that above a temperature T_i we have a normal crystal. However, below T_i an *additional periodic modulation* of the basic structure is found whose periodicity is *not commensurate* with that of the basic structure.

Let us consider this in more detail. First, the prototype structure ($T > T_i$) has periodicity along **a**, **b** and **c**-axes, as usual in a normal crystal. Then, below T_i there occurs a change so that along typically one of these directions, the **c**-direction, say, some local property such as electric polarization, magnetization, charge density, mass density or chemical composition is modulated with an additional incommensurate periodicity d, such that d/c is not a rational number:

$$\frac{d}{c} \neq \frac{m}{n}, \quad \text{where } m, n = 1, 2, 3, \ldots \qquad [7.2]$$

Normal crystals give rise to a diffraction pattern with fundamental reflections due to the prototype structure. Extra reflections appearing near the fundamental reflections are called **satellite reflections.** A commensurate superstructure, usually called simply a superstructure, can be detected by the appearance of 'extra' spots in diffraction (reciprocal) space on both sides of the fundamental reflections at commensurate positions, such as ½ or ⅓ (Appendix 8). These extra reflections appear in one or more reciprocal lattice directions (Appendix 8). When the structure has an incommensurate modulation, these satellite reflections occur at positions given by irrational fractions of the reciprocal lattice repeat.

In summary, normal crystals exhibit long-range translational order (Eqn [2.1]), and this implies a periodic spacing for the lattice. When a crystal is incommensurate, there is a superimposed additional translational order, making the crystal **quasiperiodic**, characterized by two or more periodicities that are incommensurate with each other (Eqn [7.2]).

We have introduced incommensurate crystals through association with a prototype structure for $T > T_i$. Indeed, many of the incommensurate crystals that have been studied do fit into this category. In fact, at lower temperature, they usually undergo another phase transition at T_c, the so-called **lock-in transition** to a structure that has a superstructure with respect to the high-temperature prototype structure. However, there are some incommensurate crystals that remain stable even up to their melting or decomposition temperatures. This occurs for many materials that have an incommensurate chemical modulation. Also, there are some crystals that have a T_i but remain incommensurate down to 0 K.

Everything said so far applies to insulators as well as to metals. But there is also a special incommensurate modulation that is only found in metals, where the incommensurate modulation is primarily due to the conduction electrons. Such an effect, called a **charge density wave** (**CDW**), is discussed later.

7.20.1. Superstructures and Incommensurate Structures

Certain materials are found to crystallize with superstructures that repeat over huge distances. For instance the so-called **polytypes** of SiC show a large variety of repeats, some of which occur over hundreds of Ångstroms.

Over 250 varieties (or so-called **polymorphs**) of SiC are known! It is evident from this that such crystals must contain symmetry operations that are not contained with those specified within the normal space groups. Similar considerations apply for many magnetic structures.

Superstructures are closely related to some of the incommensurate modulations, especially those with a low-temperature lock-in transition. As an example, consider the material $NaNbO_3$, which at high temperatures has the cubic perovskite structure (Fig. 7.5) with $a \approx 4$ Å. As the temperature is lowered it passes through a series of phase transitions. Two of the phases, known as P and R, are illustrated schematically in Fig. 7.39a and b, respectively. In phase P, along one of the cube directions, there is a 4-fold repeat with respect to the cubic unit cell (i.e. a repeat of ≈16 Å), while in phase R the repeat is six times the cubic length (i.e. a repeat of ≈24 Å). This is symbolized simply by Nb atoms with a transverse displacement from the centers of the pseudocubic perovskite unit cells. In the actual structure, the Na atoms are also displaced with the same periodicity and the oxygen octahedra are tilted. Note that the main characteristic of the phase P structure is that the Nb displacements are given by the sequence Δ, Δ, $-\Delta$, $-\Delta$ (Fig. 7.39a), while in phase R they are Δ, Δ, 0, $-\Delta$, $-\Delta$, 0 (Fig. 7.39b). These displacements result in crystal structures with a regular repeat of four and six times the prototype pseudocubic 4-Å cell.

In phase P, the diffraction pattern shows weak extra reflections along one of the reciprocal lattice directions whose indices are given by ¼ and ¾

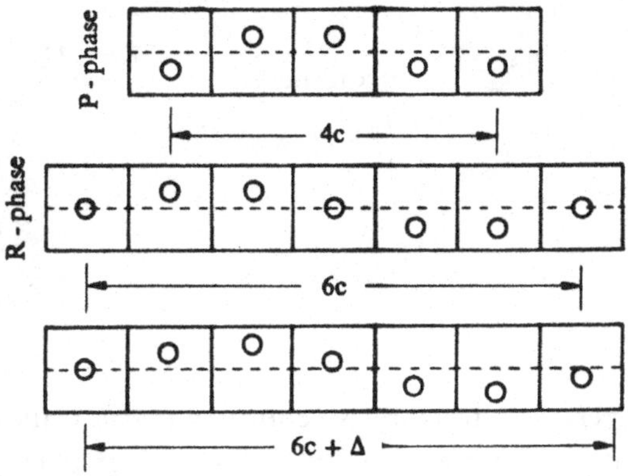

Figure 7.39 A schematic view of possible displacements in $NaNbO_3$. Each square represents a prototype unit cell with ≈4 Å repeat. (a) Nb displacements which result in a superstructure along the **c**-axis with a repeat of four times the cubic repeat. (b) Similar to (a) but here the repeat is six times. (c) Similar to (b) but here the repeat is slightly more than six times which results in an *incommensurate* modulation of the structure.

within each reciprocal unit cell. These fractions are rational, and hence the diffraction pattern is that of a *true superstructure.* Similarly, the diffraction pattern for phase R shows extra reflections at reciprocal lattice positions given by 1/6, 2/6, 4/6 and 5/6 in each reciprocal unit cell, again symptomatic of a true superstructure with a sixfold repeat.

Thus the phases P and R of $NaNbO_3$, as described earlier, are superstructures of the high-temperature cubic perovskite structure. It is worth considering the relationship between such superstructures and a possible incommensurate modulation of a crystal. Consider phase R, but suppose further that the Nb displacements were not quite as regular as shown in Fig. 7.39b. In other words the pattern of displacements would not exactly repeat over six unit cells but would have the periodicity as shown in Fig. 7.39c. This structure can be described by the product of the original unperturbed prototype structure (i.e. with no displacements) and a sinusoidal wave, whose wavelength is not commensurate with the cubic repeat of ≈4 Å. As shown in Appendix 8, any additional sine wave modulation in real space leads to two satellite peaks in reciprocal space; these two satellite peaks occur equally spaced around the diffraction spots due to the prototype structure. For this case, the spacings of the satellites are not at rational fractions of the reciprocal lattice repeat.

An example of an incommensurate modulation is illustrated by K_2SeO_4, a material that has been much studied and is now well understood. Its structural properties are summarized as follows:

$P2_1/c(C_{2h}^5)$	$\overset{T_c\approx 93\text{ K}}{\Leftrightarrow}$	Incommensurate	$\overset{T_i\approx 127\text{ K}}{\Leftrightarrow}$	$Pnma(D_{2h}^{16})$
$a'\approx a$				$a = 7.7$ Å
$b'\approx b$				$b = 6.0$ Å
$c'\approx 3c$				$c = 10.5$ Å
$\beta\approx 90°$				

The room-temperature structure is centrosymmetric orthorhombic with four formula units per unit cell ($Z=4$). The Bragg reflections in a diffraction experiment are given in terms of reciprocal lattice vectors (Appendix 8) by

$$\mathbf{Q} = h\mathbf{a}^* + k\mathbf{b}^* + l\mathbf{c}^* \quad [7.3]$$

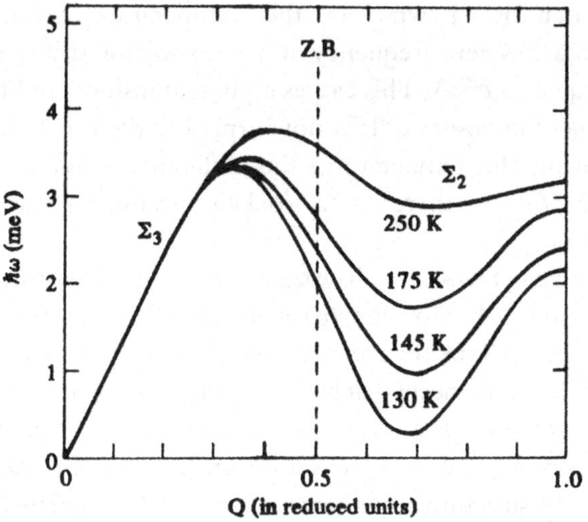

Figure 7.40 The phonon dispersion curve of the soft branch of K_2SeO_4 in the extended zone scheme along the **c*** direction. The branch is shown in the high-temperature phase at various temperatures as the incommensurate temperature is approached from above.[19]

Below the lock-in transition at T_c, a superstructure of the room-temperature structure is found with

$$\mathbf{Q} = h\mathbf{a}^* + k\mathbf{b}^* + l\mathbf{c}^* + n\mathbf{c}^* \qquad [7.4]$$

where $n = 1/3$ or $2/3$. Hence, in real space the low-temperature structure has a superstructure of the room-temperature prototype structure with a three-fold repeat along **c**.

For temperatures $T_c < T < T_i$, satellite reflections in reciprocal space are found close to but not equalling the 1/3, 2/3 positions. These positions can be described with Eqn [7.4] by a component $n = 1/3 - \delta$, where δ is a small quantity, varying from ≈ 0.02 near T_i to ≈ 0.005 near T_c.

Let us consider the driving force for this type of phase transition. Condensation of infrared-active soft phonons at the Brillouin zone center leads to ferroelectric phases (Section 7.10.2). Condensation at the Brillouin zone boundary gives a superstructure (unit cell doubled). Condensation of a phonon at a position intermediate in the Brillouin zone given by a rational fraction, such as 1/3, also leads to a superstructure of appropriate order (three in this case). In Fig. 7.40 a soft acoustic

19 For more details, see J. D. Axe, M. Iizumi & G. Shirane. *Phys. Rev.* B. **22**, 3408 (1980).

phonon branch is shown. As the temperature is lowered, the phonon approaches zero frequency at a wave-vector that is close to but not exactly equal to $\mathbf{c}^*/3$. This causes a phase transition and the resulting structure is incommensurate. It is not surprising then to observe that on further lowering the temperature, the difference from $\mathbf{c}^*/3$ becomes smaller until finally there is a lock-in transition to the 3-fold superstructure.

It is interesting to follow this phase transition using the information given in the *ITA* (Section 6.4). For the space group of the prototype structure $Pnma(D_{2h}^{16})$ there is a Type IIc maximal isomorphic subgroup of lowest index: $[3]Pnma$ ($\mathbf{c}' = 3\mathbf{c}$). This means that by eliminating two-thirds of the translation operations along the **c** repeat, a space group with the same symmetry operations is obtained but with $\mathbf{c}' = 3\mathbf{c}$. However, because of high-order interatomic forces, the structure can distort further. It is observed that space group $P112_1/a(C_{2h}^5)$ is listed as a Type I maximal non-isomorphic subgroup of *Pnma*, and this is the space group of the phase below the lock-in temperature. Therefore, in order to understand the space group change on going from the high-temperature prototype structure to the low-temperature superstructure, first imagine a hypothetical phase with the same space group as that of the high-temperature phase, but with two-thirds of the translation operations removed along the **c** repeat. It seems then that the incommensurate phase can be thought of as the means by which the crystal responds in order to achieve this reduction in translational symmetry.

There are many closely related materials in the A_2BX_4 family with characteristics similar to those described here. Some of these are Rb_2ZnCl_4, $(NH_4)_2BeCl_4$, $[N(CH_3)_4]_2MnCl_4$, $[N(CH_3)_4]_2ZnCl_4$ and $[P(CH_3)_4]_2CuCl_4$.

In regard to the use of space groups, it is possible to describe incommensurate structures in terms of a higher dimensional symmetry than 3. Thus, in the case of a structure with a 1-dimensional modulation, as indicated in Eqn [7.4] (the most common type), a four-dimensional **superspace group** is used, which when projected back onto three dimensions gives rise to nodes that are incommensurate with respect to the underlying three-dimensional reciprocal lattice.[20] As a full description of superspace groups would take too much space in this book, we refer the reader to the

[20] A good introductory account can be found in the paper by P. M. de Wolff, T. Janssen & A. Janner. *Acta Cryst.* **A37**, 625 (1981), and more recently by T. Janssen, Acta Cryst., A68, 667 (2012).

International Tables, Volume C, for a complete explanation and listing of all superspace groups.

7.21. CHARGE DENSITY WAVE

This term has been used somewhat loosely for metals that have an incommensurate modulation (or a commensurate superstructure). It should be used for metals that have a modulation *primarily driven by the conduction electrons.*

In general, in a metal the conduction electron density is modulated in real space where the modulation is related to the lattice periodicity. However, for some metals, below a temperature T_i, it is found that there is an additional periodicity (CDW) which is unrelated to the lattice periodicity. Instead, the additional periodicity is determined by the dimensions of the Fermi surface of the conduction electrons in reciprocal space; in particular it is given by $2\mathbf{k}_F$, where $\mathbf{k}_F$ is the Fermi wave-vector. Since the value of $\mathbf{k}_F$ has little to do with the repeat distance of the direct space structure, the occurrence of a CDW often leads to an incommensurate modulation. For one-dimensional metals, Rudolf Ernst Peierls showed that at low temperatures the electron conduction band should spontaneously distort and create a periodic modulation with wave-vector $2\mathbf{k}_F$. This creates an energy gap at the Fermi energy, transforming the metal to a narrow-gap insulator. The Peierls distortion mechanism may be applicable to two- or three-dimensional metals if they have large areas of their Fermi surfaces that are near-parallel to each other in **k**-space.

The tetragonal, partially oxidized compound $K_2Pt(CN)_4Br_{0.3}\bullet(3.2)H_2O$, usually called KCP, has an electrical conductivity $\approx 10^5$ times larger along the **c**-axis than perpendicular. This anisotropy is consistent with the existence of Pt chains along the **c**-axis and so this is one of the few metals that might be considered to be one-dimensional. At room temperature, the spacing of the Pt atoms along the **c**-axis is uniform. Peierls distortion occurs when the energy gained from the distortion of the conduction electron system is large enough to overcome that lost to the elastic interactions between the atoms (which tend to keep the atoms equally spaced). At room temperature, the latter is apparently larger than the former. However, below $T_i \approx 200$ K the resistance increases sharply (because of the opening of a small band gap in the electron states) and the Pt spacing becomes non-regular. Thus this material appears to show a charge density wave along the **c**-axis. Naturally, the conduction electrons interact with the ion cores in the metal, and this interaction must be small in order for the extra CDW modulation to form. Consequently, CDWs should be expected to form more easily in metals with small elastic moduli.

The key-signature of the sine-wave modulation of the electron density of a CDW is the appearance of two sharp features in reciprocal space convoluted with the Bragg peaks due to the normal structure (Appendix 8). This result is like that observed in an incommensurate modulation (or superstructure) in an insulating crystal. As the periodicity of the CDW is determined by $2\mathbf{k}_F$, the separation of the extra peaks in reciprocal space is $2\mathbf{k}_F$. For KCP, this corresponds to an incommensurate modulation of about 6.6 times the room-temperature, normal **c**-axis repeat.

CDWs appear to occur in a number of two-dimensional layer compounds such as TaS_2, which have rather special features in the electronic band structure. However, TaS_2 remains metallic at low temperatures (unlike KCP) probably because the energy gap is only produced for electrons moving along the direction corresponding to the distortion, but in two dimensions the electrons have the freedom to move in other directions (avoiding the energy gap). The modulation in TaS_2 is incommensurate between $\approx$350 K and 200 K, but below $\approx$ 200 K there is a lock-in transition to a commensurate superstructure.

7.22. QUASICRYSTALS

Till now we have considered normal crystal structures that can be generated by a periodic repetition of a unit cell, each with an identical shape. Such an assumption automatically leads to long-range translational periodicity and orientational order, which results in sharp, periodic spots in a (X-ray, neutron or electron) diffraction pattern.

However, what if instead of a single type of unit cell, we allowed more than one type of cell to be stacked together to fill all space? As far back as 1619, Kepler showed how to fill a two-dimensional space with different 5-fold symmetric tiles, and in 1981 Alan Lindsay Mackay showed how to obtain a 5-fold-symmetric diffraction pattern using a non-periodic pattern of points. Figure 7.41 shows an example of tiling in two dimensions, where two different cells are used; this figure shows the so-called **Penrose tiling**. We see that with appropriate rules for stacking, fat and thin rhombs can be placed together to fill two-dimensional space. The thin rhomb has internal angles of θ (= 36°) and 4θ, while the fat rhomb has internal angles of 2θ and 3θ, which allows for the filling of space, because $10\theta = 360°$. Such an

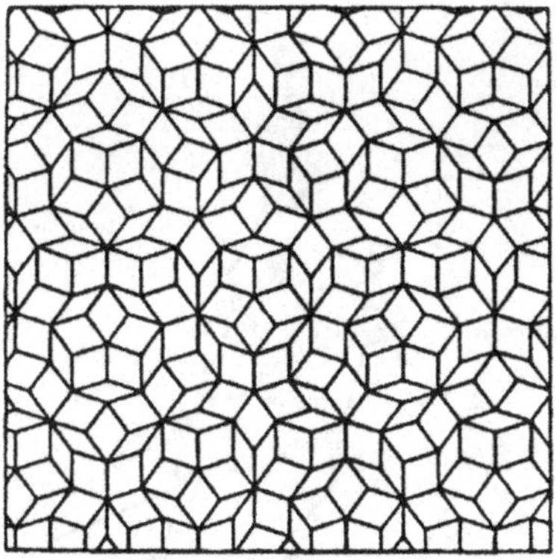

Figure 7.41 An example of Penrose tiling consisting of fat and thin rhombs. Note the local 5-fold symmetries.

arrangement has no long-range periodicity, although it does show local orientational symmetry of a 5-fold kind, and so one may ask if such an arrangement could be found in nature.

In 1984, electron diffraction patterns of a rapidly quenched metallic Al_4Mn alloy were observed by Dan Shechtman (Nobel prize winner in Chemistry for 2011) to show sharp spots arranged with a 10-fold symmetry, a phenomenon that has since been seen in many other complex alloys. This immediately raised the possibility that these materials in some way violated the rule that crystal lattices (with long-range translational symmetry) could not show 5-fold symmetry (10-fold symmetry is observed because diffraction patterns are always centrosymmetric, thus turning 5-fold into apparent 10-fold axes). Such materials were termed **quasiperiodic crystals** or **quasicrystals**, and they clearly could not be explained by conventional crystal symmetry ideas. Originally this discovery was met with some scepticism, Shechtman's original paper being rejected, and with even the Nobel Laureate Linus Pauling claiming that the effect could be explained as a form of multiple twinning. However, for the alloy $Al_{63}Cu_{24}Fe_{13}$ it is possible to grow faceted, single quasicrystals up to 1 mm^3 in size. The positional order measured by high-resolution X-ray diffraction of this material shows that the correlation length is greater than 7000 Å. The perfection of these quasicrystals appears to establish that quasiperiodic crystals exist as a phase of matter. As a consequence, the International Union of Crystallography in 1992 revised the definition of a crystal to mean 'any solid having an essentially discrete diffraction diagram'.

One of the important questions to settle is the nature of the actual structure (the positions of all of the atoms) in a quasicrystal, and not just the

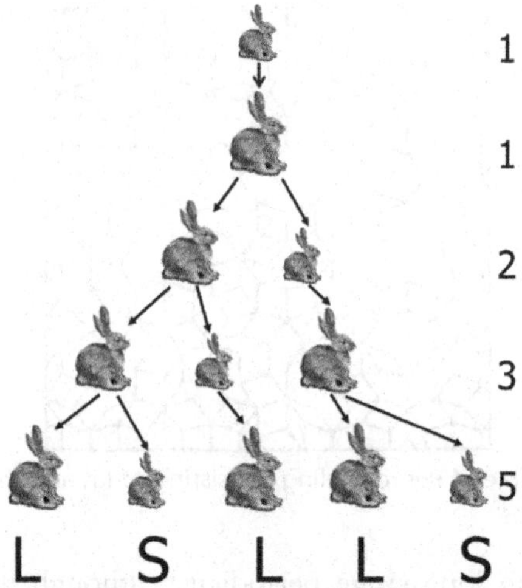

Figure 7.42 The generation of the Fibonacci series 1, 1, 2, 3, 5, ... in terms of long (L) and short (S) spacings, according to the rules L = >LS, and S = >L. This is illustrated here by female rabbits. A baby rabbit grows up to become an adult. This remains an adult but also produces a small offspring. The offspring grows to become a large rabbit, and the entire process is repeated.

symmetry. It is important also to understand how such an arrangement, with no long-range periodicity can generate sharp diffraction spots. One theory which leads to an understanding of the diffraction patterns is to imagine a lattice of points in a six-dimensional space. If a cut is then made through such a lattice and projected onto two or three-dimensional space, obviously the projection will consist of points. If the cut and projection is made in the right way it turns out that an array of points can be obtained with 5-fold symmetry. In addition, the positional ordering of the points along particular directions is found to obey the sequence known as a **Fibonacci Series.**[21] Such a sequence is also observed in a series of spots in the diffraction patterns. A Fibonacci series can be obtained in the following way. Consider points separated by either long (L) or short (S) distances. To generate the Fibonacci series, start with a long period L and replace it by LS. Then replace the L again by LS and the S by L, and continue the process indefinitely. Figure 7.42 shows part of this process down to where the series generated is L S L L S ....

21 Named after the mathematician Leonardo Pisano Bigollo (c1170–c1250), also called Leonardo of Pisa or simply Fibonacci.

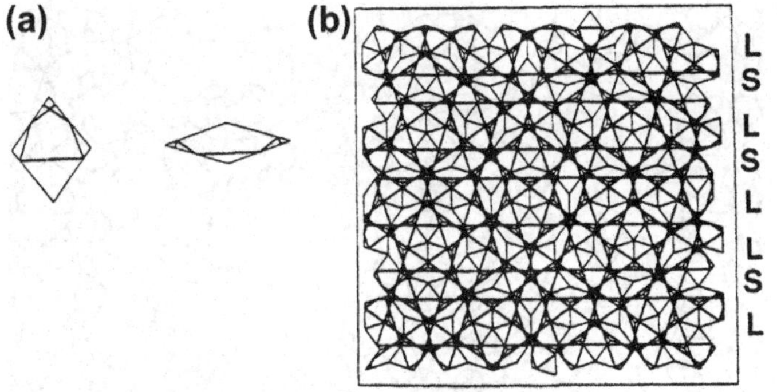

Figure 7.43 (a) Special lines marked on the fat and thin rhombs of a Penrose tiling. (b) The effect of these special lines is to generate a Fibonacci sequence of spacings in five directions at 72° to one another.

Now return to the Penrose tiling of Fig. 7.41. If special lines are drawn on the fat and thin rhombs as indicated in Fig. 7.43a, the pattern shown in Fig. 7.43b is obtained. We now observe that sets of parallel planes have been derived (marked by lines) that are arranged at 72° to one another, that is related by a 5-fold rotation! In addition, by considering the separations of these planes, we find that they are either long or short and that the sequence in any direction is the Fibonacci sequence! In this case it is interesting to note that the ratio $L/S = \tau$, is the 'golden mean' $= (1 + \sqrt{5})/2 = 1.61803$. We therefore have some new kind of 'lattice' (a **quasilattice**) consisting of long and short spacings, known also as an **Ammann quasilattice**. Such a lattice is said to be **quasiperiodic**. It is possible to conjecture that by adding atoms (a basis) to such a quasilattice, a process called **decoration**, one could produce a model for the actual quasicrystal.

Figure 7.44 shows an example of a quasicrystal structure determined for the alloy $Al_{72}Ni_{20}Co_6$, in which atoms at $z = 0.25$ and 0.75 are indicated by open and filled circles, respectively. The thin black lines show Penrose tiling, while the thick white lines indicate hexagon, boat and star (HBS) tiling. The thick black circle shows atom positions within a decagon of 4.68 Å.

Remember, in normal crystals the interplay between the orientational and positional long-range order limits the possible rotational axes to 1, 2, 3, 4 and 6-fold. In a quasicrystal any rotation symmetry is allowed. However, the choice of rotational symmetry defines the quasiperiodicity. For example, in a 5-fold Penrose tiling, the quasiperiodicity is give by the Fibonacci sequence. A tiling with different symmetry would require a different quasi-periodic sequence, uniquely defined by the rotational symmetry.

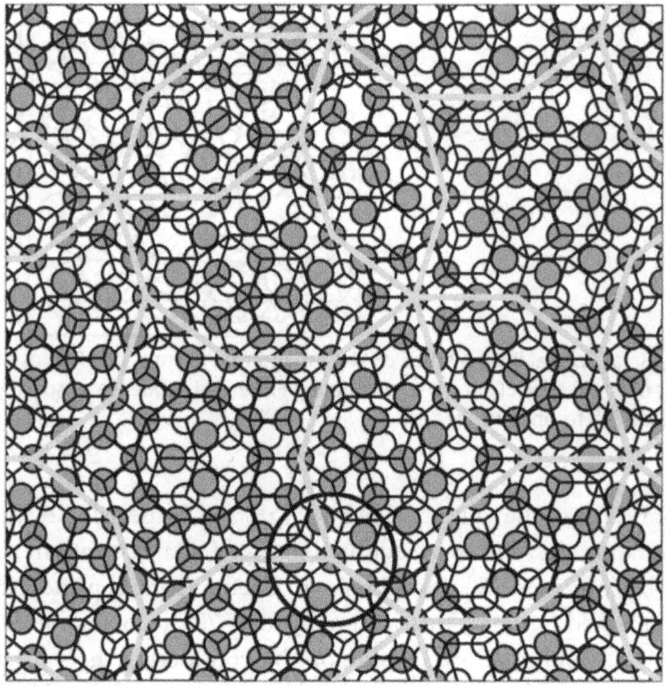

Figure 7.44 Projection of the atom positions (41 × 41 Å) for $Al_{72}Ni_{20}Co_6$ along the *c*-axis.[22]

Returning to the question of the sharp spots in the diffraction pattern, it is not surprising to find that the quasilattice in real space, consisting of Fibonacci sequences, should in reciprocal space also produce a quasiperiodic distribution of Bragg peaks. The 5-fold symmetry will be preserved in converting from real to reciprocal space, and vice-versa. In fact, it can be experimentally shown that if a mask is made consisting of holes at the intersections of the lines making up the Penrose tiles (Fig. 7.45a), and light is diffracted by the holes, a sharp diffraction pattern of the right type is indeed obtained (Fig. 7.45b).

This system can also be shown rigorously by mathematics to yield infinitely sharp diffraction spots of the type shown in Fig. 7.45b.

Optical simulation can also be used to give further clues as to the origin of the sharp scattering from quasicrystals. Let us consider, for the moment, a totally different type of structure, one with partial order. Figure 7.46a shows a mask consisting of holes arranged in the form of triangles of different sizes, but all pointing in the same direction. Such a structure is

22 H. Takakura, A. Yamamoto & A. P. Tsai. *Acta Cryst.* **A57**, 576–585 (2001).

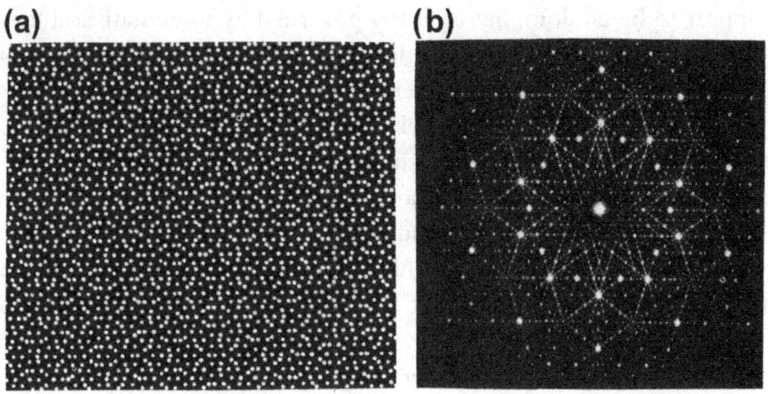

Figure 7.45 Optical simulation of the diffraction pattern from a structure arranged according to the Penrose tiling model (from T. R. Welberry).

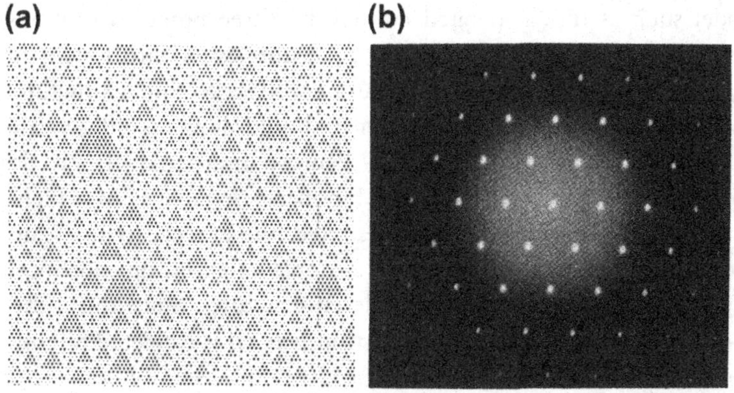

Figure 7.46 Optical simulation of the diffraction pattern from a partially disordered structure in which neighbouring sites are governed purely by three-point correlations. (a) The model and (b) the diffraction pattern (from T. R. Welberry).

clearly not fully periodic, but it does possess local orientational symmetry. Now, as explained in Appendix 8, a partially ordered structure gives rise to a diffraction pattern consisting of the superposition of two terms:

$$I = I_{Bragg} + I_{diffuse} \qquad [7.5]$$

where, in general, I_{Bragg} consists of sharp spots reflecting the *average* crystal structure, and $I_{diffuse}$ is diffuse background relating to the disorder in the structure. Although the structure in Fig. 7.46a is disordered, on average it consists of points arranged in a triangular fashion with a fixed separation, unlike the quasiperiodic structures discussed earlier. The disordering does not appear to be random, but is clearly governed by some statistical 'rules', to give the local orientational ordering. From Appendix 8, we would therefore expect a diffraction pattern consisting of sharp spots from the average structure plus a diffuse background. Because of the obvious local ordering, we might expect that the diffuse background should not be featureless, but should have some patterning, and yet, as seen in Fig. 7.46b, it appears to be smooth and continuous, rather as one would expect for a random gas. The reason for this surprising observation is that this particular structural arrangement was generated by computer using a probability distribution function which consisted purely of three-body interactions. Owing to the fact that the intensity of diffraction is given by the product of the structure factor and its complex conjugate (Appendix 8), a diffraction pattern can only provide information about one-point (site occupation) or two-point (pair correlation) properties of a crystal. Thus, in a model such as this, generated entirely by three-point correlations, the diffuse diffraction pattern appears to be featureless, even though there is a high degree of short-range order. As a corollary to this, a structure governed *entirely by perfect two-point correlations* should show no diffuse scattering at all, but purely sharp spots. The Penrose tiling model was derived according to the strict rules of the Fibonacci sequence, in which each separation determines the next (once started, the exact sequence is derived without error), and so it is a perfect example of exact, but quasiperiodic, two-point correlation (i.e. it is an ordered structure). Since this is the only type of correlation present, the diffraction pattern is expected to consist of sharp spots alone.

Problems

1. An important structure in which many simple metals, such as Be, Mg, Ti, Zr, Os, Zn and Cd crystallize is the so-called hcp structure, with space group $P6_3/mmc(D_{6h}^4)$.. The atoms are on $2c$ sites at $\pm(⅓, ⅔, ¼)$. Each atom is surrounded by 12 neighbours, 6 in the same **ab**-plane, 3 above and 3 below.

(a) Show that the 12 neighbours are equidistant if $c/a = (8/3)^{½} = 1.633$. This value is the ideal c/a ratio.

(b) In this structure, the three atoms above the plane are directly over the three atoms below the plane, looking down the **c**-axis. Consider now an fcc lattice with one atom per lattice point. Look down the [111] direction of the cubic unit cell. Then, surrounding any atom, which we may take at the origin of the unit cell, there are 12 equidistant neighbours (6 in the plane of the origin, 3 above and 3 below). Show that the 3 above are not directly over the 3 below. In fact they are rotated by 60°. This means that in a hcp structure the stacking sequence of atoms can be described by ABABABABAB... along [001], whereas in the **cubic close-packed** (**fcc**) case it is ABCABCABCABC... along [111].

(c) ZnS, in the wurtzite structure, belongs in space group $P6_3mc(C_{6v}^4)$ with the atoms at $2b$ sites: (⅓, ⅔, u), where $u = 0$ and $\approx$⅜ for Zn and S, respectively. Show that, for u = ⅜ and the ideal c/a ratio, regular tetrahedra are formed by Zn and its four S neighbours.

2. At high temperatures (above $\approx$385 °C) CuAu is disordered with the fcc structure, that is, Cu and Au atoms randomly occupy the (0, 0, 0), (½, ½, 0) (0, ½, ½) and (½, 0, ½) sites. Below $\approx$385 °C, the atoms order such that the Cu atoms occupy the first two positions and the Au atoms occupy the second two positions. What is the space group in each case?

3. Consider a $(GaAs)_1/(AlAs)_1$ superlattice, and draw a diagram of the structure. Outline the unit cell of the structure and show that the space group is $P\bar{4}m2(D_{2d}^5)$. Show why the space group is not $P\bar{4}2m(D_{2d}^1)$. (Hint: The order of the symbols in point and space groups is explained in Appendix 6.)

4. Consider the cubic ABO_3 perovskite structure. If the crystal is compressed along the **c**-axis, what is the resulting space group? If in the uncompressed structure the B atom at (½, ½, ½) moves along the **c**-axis to (½, ½, ½ + Δ), what is the space group? Which of these two structures can be ferroelectric; how do the space group symbols show this?

5. Consider the cubic ABO_3 perovskite structure where the B atom is replaced by equal amounts of Sc^{3+} and Ta^{5+}, for example in $Pb_2(Sc,Ta)O_6$. Assume that the Sc and Ta orders so that the cell is doubled in all three directions. The space group is then $Fm\bar{3}m(O_h^5)$. Determine the atom positions.

6. Consider the following ferroic species:

$$m\bar{3}mF222 \quad 4F1$$

$$4mmF2mm \quad \bar{6}F3$$

$$4/mmmF4/m \quad 6/mFm$$

Find the number of domains in the ferroic phase in each case and determine if the ferroic phase is ferroelectric or ferroelastic.

7. It has been observed that the space groups of all natural proteins and viruses contain neither a center of symmetry nor mirror planes. Can you explain why this should be? (Hint: consider both the fact that proteins consist of helical arrangements of molecular sub-units with a particular chirality and that organic life has arisen through a process of evolution).
8. Crystals of alum $KAl(SO_4)_3$ occur in space group $P321(D_3^2)$ with the K, Al and S atoms located at

   K: 1*a* 0, 0, 0
   Al: 1*b* 0, 0, ½
   S: 2*d* ⅓, ⅔, 0.222

   By consulting the table of Euclidean normalizers in the *ITA* or by using the Bilbao Crystallographic Server, determine the normalizer group and write out all the other possible coordinate descriptions for these atoms.

CHAPTER 8

Antisymmetry

Contents

It was the best of times, it was the worst of times,
it was the age of wisdom, it was the age of foolishness,
it was the epoch of belief, it was the epoch of incredulity,
it was the season of Light, it was the season of Darkness,
it was the spring of hope, it was the winter of despair,
we had everything before us, we had nothing before us,
we were all going direct to Heaven, we were all going direct the other way.

Charles Dickens (1812–1870) 'A Tale of Two Cities'

BICOLOR SYMMETRY

In this chapter, we shall look at the concept of two-color symmetry and how it can be described. There is a long and confusing history associated with this subject, and what follows is intended to provide a simple introduction. The groups that use this type of symmetry are often called magnetic groups, mainly because this is where they are most likely to be applied. However, two-color groups can have wider applications.

Let us consider systems where a particular property can take one of the two possible values. For instance, in a ferroelectric crystal, particular atoms

Space Groups for Solid State Scientists
http://dx.doi.org/10.1016/B978-0-12-394400-9.00008-3

may be displaced in a certain direction to form an electrically polar structure whose polarity can be switched by an electric field. Thus such a system has two states available. If the atomic displacements are parallel then the material can in principle be **ferroelectric**, whereas, on the other hand, if for each atomic displacement there is an equivalent displacement in the opposite direction to cancel the overall polarization, this is known as an **antiferroelectric**. Similarly, in a **ferromagnet**, the magnetic moments are aligned parallel to one another and can be switched to the opposite sign by an applied magnetic field. Again, if for every magnetic moment there is an opposite moment in the crystal so that there is no net magnetic moment, we call this an **antiferromagnet**.

In order to describe the symmetric properties of such binary properties, a new type of symmetry operator, the **antisymmetry operation** (we shall denote this by the symbol $1'$), was introduced by Heinrich Heesch in 1929 and by H.J. Woods in 1935. This operation has the effect of reversing the sign of the property in question (polarization in a ferroelectric or antiferroelectric system and magnetization in a ferromagnetic or antiferromagnetic system). There are many examples where the concept of antisymmetry can be seen to act in the world around us, for instance, the on-off positions of a light switch, the 1 0 binary code used in computers and even the famous Taoist symbol of Yin-Yang (Fig. 8.1).

At the time it was not realized that the operation of antisymmetry had practical application. It was Lev Vasilyevich Shubnikov (1901–1937) in Russia, many years later, who developed this concept through the introduction of so-called **black and white symmetry**. Thus we may choose, say, black for a positive moment and white for a negative moment. A full set

Figure 8.1 The Taoist Yin-Yang symbol, corresponding to the point group $2'$.

of point group and space group diagrams of these so-called Shubnikov groups was published in 1966 by Vladimir Aleksandrovich Koptsik (1924–2005), but because this book was written in Russian, it has not been widely read (Fig. 8.2).

Lev Landau and Evgeny Lifshitz interpreted the change in color in terms of **time-reversal** symmetry and applied it to magnetic systems. The reason for this is that magnetic moments arise from electron spin, and we can think of the spin motion in time as being clockwise or anti-clockwise. It should be obvious that when we consider the effect of time, there are two possible symmetries that need to be considered. Suppose for instance that we make

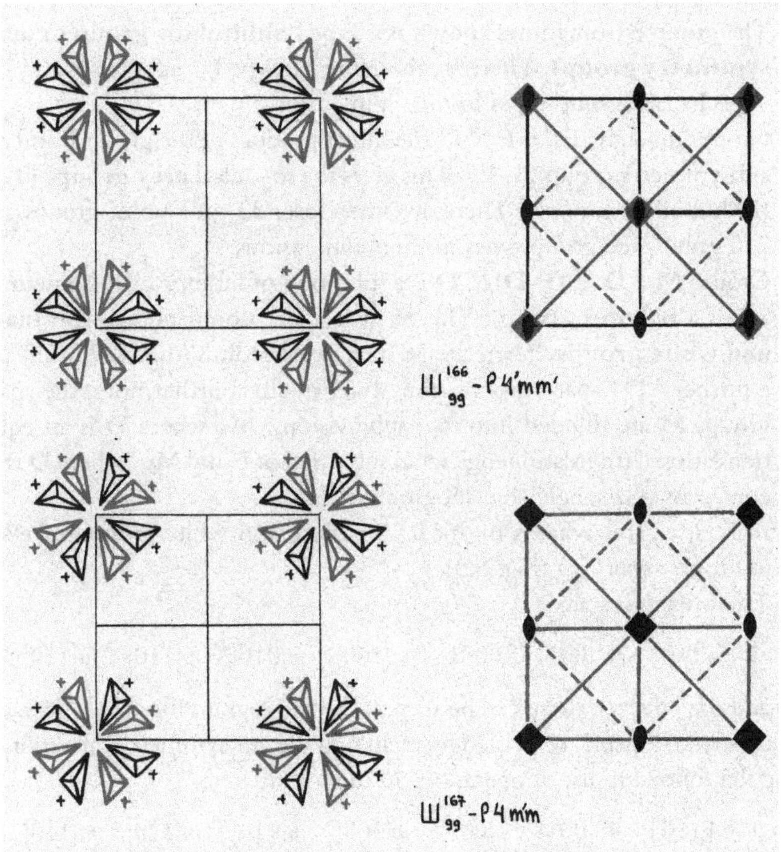

Figure 8.2 Example of a page from Shubnikov Groups (Шубниковские Группы) by V.A. Koptsik, edited by N.V. Belov and published by Moscow State University in 1966. Note the small error in the bottom diagram labelled *P4m′m* instead of *P4m′m′*.

a film of a pendulum swinging back and forth: assuming that this is an idealized pendulum that does not experience any frictional forces slowing it down, we would not be able to determine if the film were run backwards or forwards. Such a system therefore is **time symmetric**. On the other hand, a film of a horse race clearly looks different when run forwards rather than backwards, and so this would be **time antisymmetric**.

8.1. BLACK AND WHITE ANTISYMMETRY GROUPS

Let F be a group (point group or space group) that does not contain the operator 1′. As we have already seen, in three dimensions, this leads to 32 point groups and 230 space group types. We can construct an antisymmetric super family that consists of the following:

1. The group **F** (sometimes known as a Type I **Shubnikov group or antisymmetry group).** There are therefore 32 Type I point groups and 230 Type I space group types in three dimensions.
2. Groups given by $\mathbf{F1'} \equiv \mathbf{F} \times \mathbf{1'}$, the direct product of the group **F** and the antisymmetry group $\{1, 1'\}$. This gives the so-called **grey groups** (Type II Shubnikov groups). There are, therefore, 32 grey point groups and 230 grey space group types in three dimensions.
3. Groups $\mathbf{M} \equiv \mathbf{D} + (\mathbf{F}\text{-}\mathbf{D})\mathbf{1'}$. **D** is a subgroup of index 2 of **F** (sometimes called a **halving group**). The **M** groups are isomorphous with **black and white groups.** There are 58 such point groups (thus 122 in all) and a further 1191 space groups (thus 1651 in all). Furthermore, the space groups **M** are divided into two subdivisions: $\mathbf{M_T}$ where **D** is an equi-translational (translationengleiche) subgroup of **F** and $\mathbf{M_R}$, where **D** is an equi-class (klassengleiche) subgroup of **F**.

To make it clearer what is meant by the black and white groups, take the point group $4mm(C_{4v})$ (Fig. 8.3).

The operations are

$$1 \quad 4[001] \quad 4^3[001] \quad 2[001] \quad m[100] \quad m[010] \quad m[110] \quad m[\bar{1}10]$$

and the effect of these can be seen in the stereogram for $4mm$. Now, the group marked $4mm1'$ is formed by including the antisymmetry operation to give the following list of operations in the group

$$1 \quad 4[001] \quad 4^3[001] \quad 2[001] \quad m[100] \quad m[010] \quad m[110] \quad m[\bar{1}10]$$

$$1' \quad 4'[001] \quad 4'^3[001] \quad 2'[001] \quad \mathrm{m}'[100] \quad \mathrm{m}'[100] \quad \mathrm{m}'[100] \quad \mathrm{m}'[\bar{1}10]$$

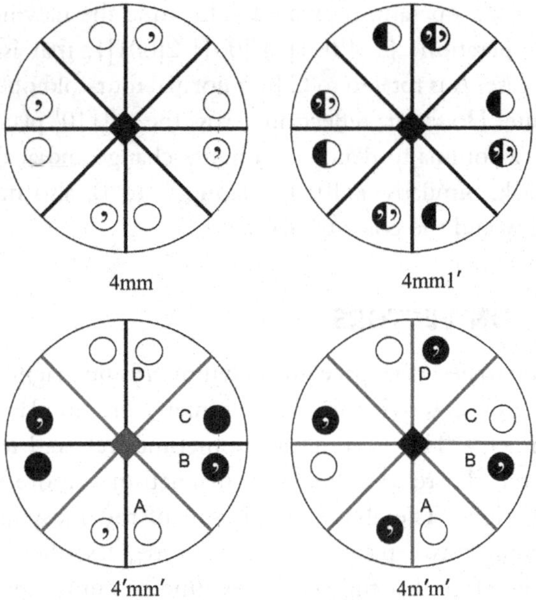

Figure 8.3 Black and white symmetry in point group *4mm*. Note that in some texts, instead of primes, underlines are used: $\underline{4}m\underline{m}$ and $4\underline{mm}$.

where the primed operations denote the combination of a normal operation with 1′. Thus, for example

$$m' \equiv 1'm$$

The $4mm1'$ point group contains twice as many operations as in $4mm$ and is an example of a grey group. In the figure, each white object has accompanying it a black object, as indicated by the split circles.

The point group $4'mm'$ has the following operations:

$$1 \quad 4'[001] \quad 4^3[001] \quad 2[001] \quad m[100] \quad m[010] \quad m'[110] \quad m'[\bar{1}10]$$

and is an example of a black and white group. Starting with the object at A, the 4′ operation (marked in red) rotates A anti-clockwise through 90° to C and changes the color to black. Similarly, the object at A is reflected by the $m'[\bar{1}10]$ mirror (marked in red) to B, thus, adding a comma and at the same time has its color changed to black. Notice that in this group half the operations are primed and half not, so that the halving group **D** is given by $\{1\ 2[001]\ m[100]\ m[010]\}$ and is a group of type $mm2$.

The point group $4m'm'$ has the operations

$$1 \quad 4[001] \quad 4^3[001] \quad 2[001] \quad m'[100] \quad m'[010] \quad m'[110] \quad m'[\bar{1}10]$$

again with half the operations primed. This time the halving group **D** is given by the operations {1 4[001] 4^3[001] 2[001]}, that is point group type 4. In this case, A is rotated to C by a normal four-fold operation and so it remains white. However, reflection across the $m'[\bar{1}10]$ plane (marked in red) to B adds a comma to denote a chirality change and at the same time changes to black. Similarly, $m'[010]$ relating C to B, also marked in red, changes chirality and the color to black.

8.2. EFFECT ON VECTORS

Before going on to describe space group symmetry and magnetism, we need to ask the question as to what role antisymmetry groups play in describing magnetic symmetry. These groups are sometimes referred to as magnetic groups: however, the relationship with magnetism is more subtle than it appears at first sight. The above description of the antisymmetry groups relies on switching between two states, in this case, described by two colors. It is obvious therefore that this type of symmetry can be applied to properties that can be described in terms of functions that take one of two values, for example up and down or plus and minus. However, magnetic moments are created by electron spin and this is a property that is described by the use of **axial vectors**. We can formulate this by considering a classical space group operation $\{R|\mathbf{t}\}$ and its time-reversed operation $\{R|\mathbf{t}\}'$ acting on an axial vector function $\mathbf{S}(\mathbf{r})$ representing a magnetic moment thus:

$$\begin{aligned} \{R|\mathbf{t}\}\mathbf{S}(\mathbf{r}) &= \delta_{\mathrm{R}} R\mathbf{S}(\{R|\mathbf{t}\}^{-1}\mathbf{r}) \\ \{R|\mathbf{t}\}'\mathbf{S}(\mathbf{r}) &= -\delta_{\mathrm{R}} R\mathbf{S}(\{R|\mathbf{t}\}^{-1}\mathbf{r}) \end{aligned} \qquad [8.1]$$

where $\delta_{\mathrm{R}} = 1$ if R is a proper rotation and $\delta_{\mathrm{R}} = -1$ if R is an improper operation. The operations $\{R|\mathbf{t}\}$ and $\{R|\mathbf{t}\}'$ determine the magnetic (Shubnikov) groups.

In Fig. 8.4, the effect of applying proper and improper operations (Eqn [8.1]) on axial spin vectors is illustrated. Starting at the top left, it is seen that applying a 2-fold operation has no effect on the sense of spin rotation but only moves the vector to a symmetry-related position through 180°: the sense of spin rotation is unaltered for any proper rotation. On the other hand, as shown at the right, application of a $2'$ operation not only rotates the vector to its symmetry-related position but also changes the sense of the spin. The axial vector then points in the opposite direction. However, when dealing with improper operations more care is needed. Thus, in the middle

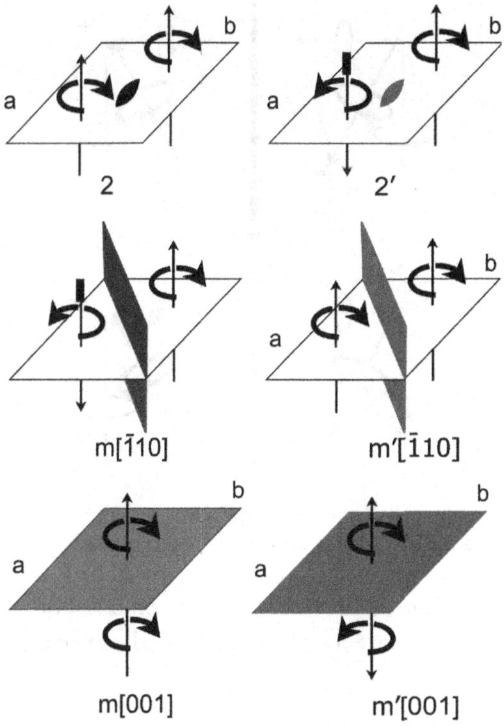

Figure 8.4 Effect of proper and improper operations acting on magnetic moment vectors.

at the left we show the effect of a reflection $m[\bar{1}10]$ on an axial vector that is parallel to the mirror plane: this clearly changes the sense of spin rotation, thus converting the up axial vector direction to down. Similarly, the effect of the antisymmetry operator $m'[\bar{1}10]$ is to first apply the reflection, thus changing the spin rotation, followed by applying the antisymmetry operation, thus reconverting the axial vector back to its original state (the order in which these two operations is carried out is irrelevant). Finally, at the bottom we see the effect of a reflection perpendicular to the axial vector: for $m[001]$ the sense of rotation is unchanged and so the direction of the axial vector remains the same. On the other hand, $m'[001]$ causes the spin rotation to be changed, thus reversing the axial vector direction.

In Fig. 8.5, we show how the reflection operations affect an axial vector pointing in a general direction. The plus and minus signs indicate whether the axial vectors point out of the plane towards the reader or away, respectively. We can see that by taking components (dashed) the effect of the m operation on the horizontal components is to leave them unchanged, but

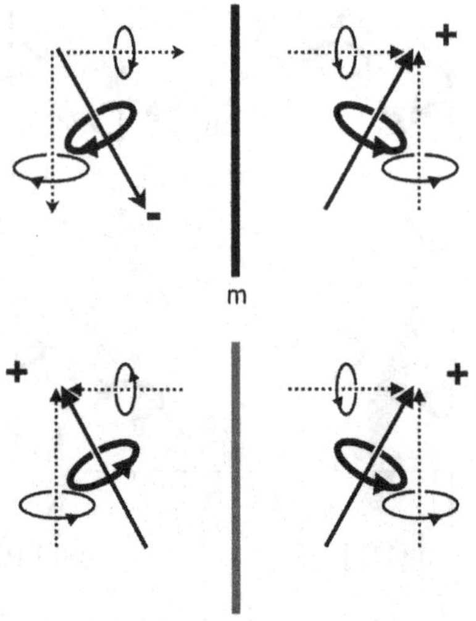

Figure 8.5 Reflection acting on an inclined axial vector.

to reverse the direction of the vertical component. For the m' operation the horizontal component is reversed but the vertical component is unchanged.

8.3. MAGNETIC POINT GROUPS

Let us consider the two black and white groups $4'mm'$ and $4m'm'$. This time instead of using black and white we shall draw the equivalent symmetry for a set of axial vectors (Fig. 8.6).

In these diagrams, the representative points are colored red to indicate that they lie above the plane of the paper. The arrows denote the directions of axial vectors and the plus and minus signs denote whether they have components pointing above or below the plane of projection. Note that if

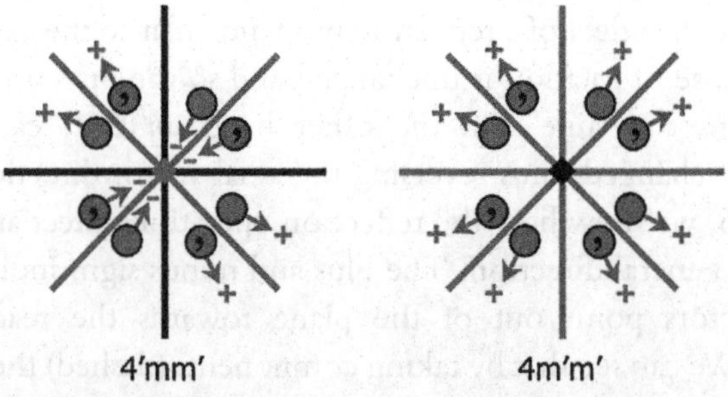

Figure 8.6 Stereograms for point groups $4'mm'$ and $4m'm'$.

for $4'mm'$ we replace the regions marked with a plus by the color black and those marked by a minus with white, we again obtain a black and white group of the same group type but this time with the symbol $4'm'm$, that is the axes appear to have been rotated through 45° about the **c**-axis. In the case of $4m'm'$, all the vectors have a component pointing above the plane of projection.

Thus, this is a point group for a so-called **weak ferromagnet** (a normal full ferromagnet would have all moments pointing in the same direction: here the components in the x–y plane form an antiferromagnetic array, but all the moments have a ferromagnetic component in the z direction).

It has been shown that there are 31 full ferromagnetic point groups. These are as follows:

	Monochrome			Black and white		
1		$\bar{1}$	$2'$	m'		$2'/m'$
2	m	$2/m$	$2'2'2$	$m'm'2$	$m'm2'$	$m'm'm$
3		$\bar{3}$	$32'$	$3m'$		$\bar{3}m'$
4	$\bar{4}$	$4/m$	$42'2'$	$4m'm'$	$\bar{4}2'm'$	$4/mm'm'$
6	$\bar{6}$	$6/m$	$62'2'$	$6m'm'$	$\bar{6}m'2'$	$6/mm'm'$

In Fig. 8.7, are shown the seven tetragonal magnetic point groups from this table. Here, the color blue indicates points below the plane of the paper. Note that in all these examples, they are antiferromagnetic in the x–y plane with a z-component in the same direction. Thus, if the moments were arranged in this way the material would be a weak ferromagnet. However, if the moments are directed only along $+z$ they would be normal full ferromagnets.

Similarly, Litvin[1] has shown that there are 31 ferroelectric groups. In this case, if we define $\mathbf{P}(\mathbf{r})$ to be a polar vector function

$$\begin{aligned} \{R|\mathbf{t}\}\mathbf{P}(\mathbf{r}) &= R\mathbf{P}(\{R|\mathbf{t}\}^{-1}\mathbf{r}) \\ \{R|\mathbf{t}\}'\mathbf{P}(\mathbf{r}) &\doteq R\mathbf{P}(\{R|\mathbf{t}\}^{-1}\mathbf{r}) \end{aligned} \qquad [8.2]$$

	Monochrome		Black and white			Grey
1	m		m'		$1'$	$m1'$
2	$mm2$	$2'$	$m'm'2$	$m'm2'$	$21'$	$mm21'$
3	$3m$		$3m'$		$31'$	$3m1'$
4	$4mm$	$4'$	$4m'm'$	$4'mm'$	$41'$	$4mm1'$
6	$6mm$	$6'$	$6m'm'$	$6'mm'$	$61'$	$6mm1'$

1 D.B. Litvin, *Acta Cryst.*, **A42**, 44 (1986).

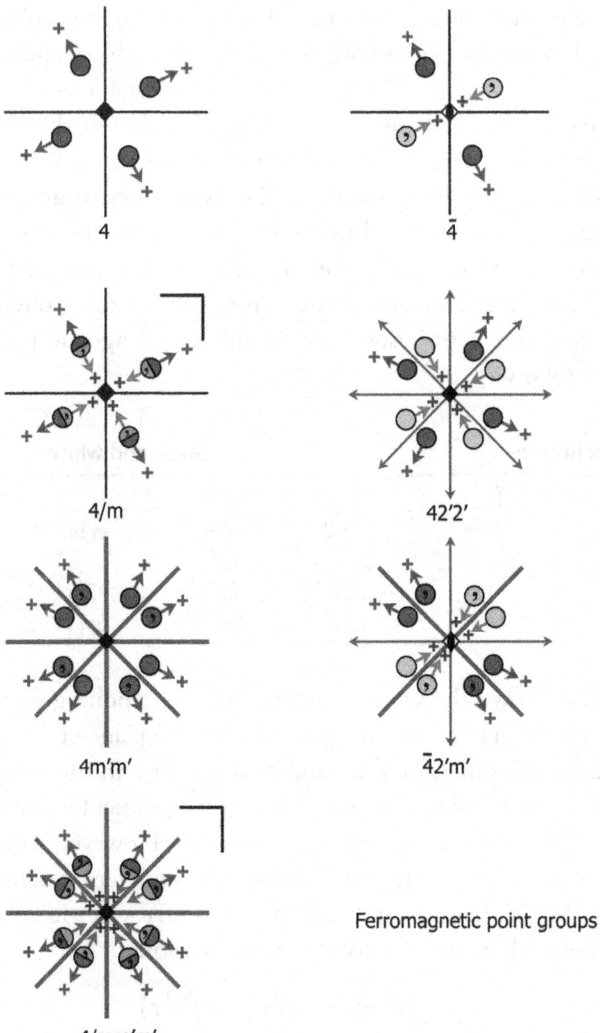

Figure 8.7 The seven tetragonal ferromagnetic groups.

Note that there are 17 point groups compatible with ferromagnetism and ferroelectricity.

8.4. TRANSLATIONAL SUBGROUPS OF MAGNETIC GROUPS

In order to define the magnetic space groups, it is first necessary to consider the number and types of unique lattices that can be defined, just as was done

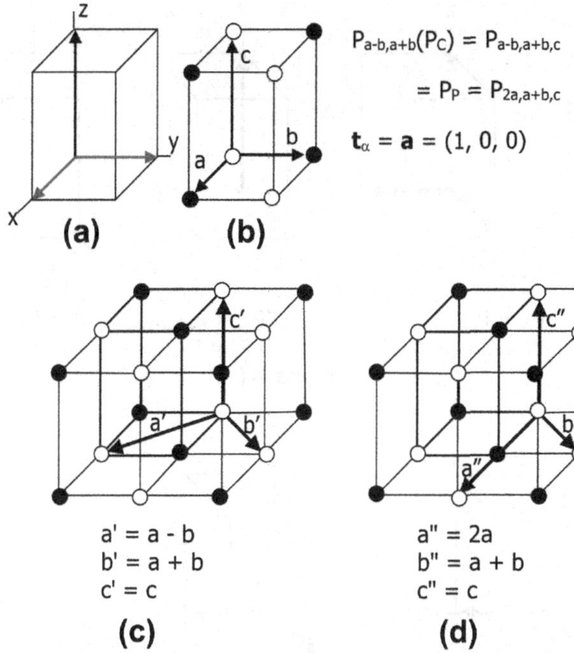

Figure 8.8 Details of the P_C magnetic lattice.

earlier for the Bravais space lattices. The difference here is that it is necessary to include the effect of time-reversal symmetry into the lattice and to establish a notation to describe it. It is worth mentioning here that, in a monumental piece of work, Litvin has derived and presented all the 1651 magnetic space groups in the style of the *ITA*, and the reader is strongly encouraged to download them.[2]

The lattices can be divided into two types. In the first type, the magnetic groups **F**, $\mathbf{F1}'$ and $\mathbf{M_T}$ are denoted by a single letter without subscripts. A second symbol gives the generators of the translation group in the subscript (these generators are also shown as black arrows in the diagrams). An example of this is the translational subgroup $I = I_{a,b,½(a+b+c)}$ in the tetragonal system (Fig. 8.9).

Translational subgroups of $\mathbf{M_R}$ groups, on the other hand, have symbols consisting of a letter with a second letter or numeral as subscripts. The translational subgroups of these groups are of the form $\mathbf{TM_R} = \mathbf{T}^D + \mathbf{t}_\alpha{}'\mathbf{T}^D$,

[2] Thanks to D. Litvin, these excellent tables are currently freely available at http://www.bk.psu.edu/faculty/litvin/Download.html.

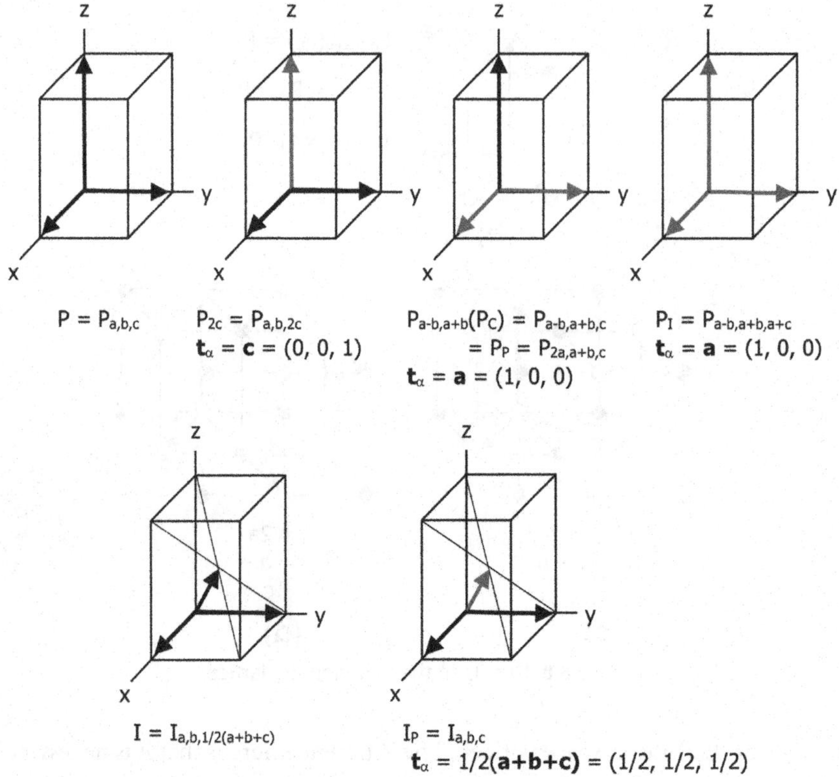

Figure 8.9 Tetragonal magnetic lattices.

where $\mathbf{T}^{\mathrm{D}}$ is the halving subgroup of unprimed (not coupled with time inversion) translations of $\mathbf{TM}_{\mathrm{R}}$. t_α' is a translation of $\mathbf{TM}_{\mathrm{R}}$ not in $\mathbf{T}^{\mathrm{D}}$, while the translation of $\mathbf{T}^{\mathrm{D}}$ is denoted by $\mathbf{t}_\alpha$.

As an example of how the lattices are presented, consider Fig. 8.8. In Fig. 8.8a, we show one of the tetragonal unit cells drawn as in the space group tables of Litvin. The red arrows indicate in this example that time-reversal symmetry is applied to two of the axes along x and y but not along z. In Fig. 8.8b, we show this same unit cell, labelled **a**, **b** and **c**, with black and white points to make this clearer. Thus, the planes perpendicular to **c** are identical, but those perpendicular to **a** and **b** alternate. In Fig. 8.8c, several such unit cells are drawn together: starting at one of the white points we find three axes $\mathbf{a}'$, $\mathbf{b}'$ and $\mathbf{c}'$ that define a primitive unit cell with a black point at the center of the C-face, giving the notation $P_{a-b,a+b}(P_C)$.

Alternatively, another primitive unit cell can be defined by $\mathbf{a}''$, $\mathbf{b}''$ and $\mathbf{c}''$ as shown in Fig. 8.8d, and this can be called P_P. Figure 8.9 shows all the magnetic lattices for the tetragonal system and Appendix 10 shows all of the lattices.

8.5. BLACK AND WHITE SPACE GROUPS

Just as we did earlier, the different lattice types can be combined with symmorphic and non-symmorphic operations to form space groups. The difference here from the normal space group types is that we now have to include the antisymmetry operator $1'$ which turns black into white and *vice versa*, and then we generate the 1651 three-dimensional (80 two-dimensional) **black and white space groups**.

To illustrate this, Fig. 8.10 shows an example of a two-dimensional black and white structure whose space group is described by the plane group symbol $p_{2b}1m'1$. At the bottom left the symmetry operators are shown, in this case with the **a**-axis horizontal and the **b**-axis vertical. The p_{2b} symbol tells us that the **b**-axis is antisymmetric (marked by a red arrow) and indeed

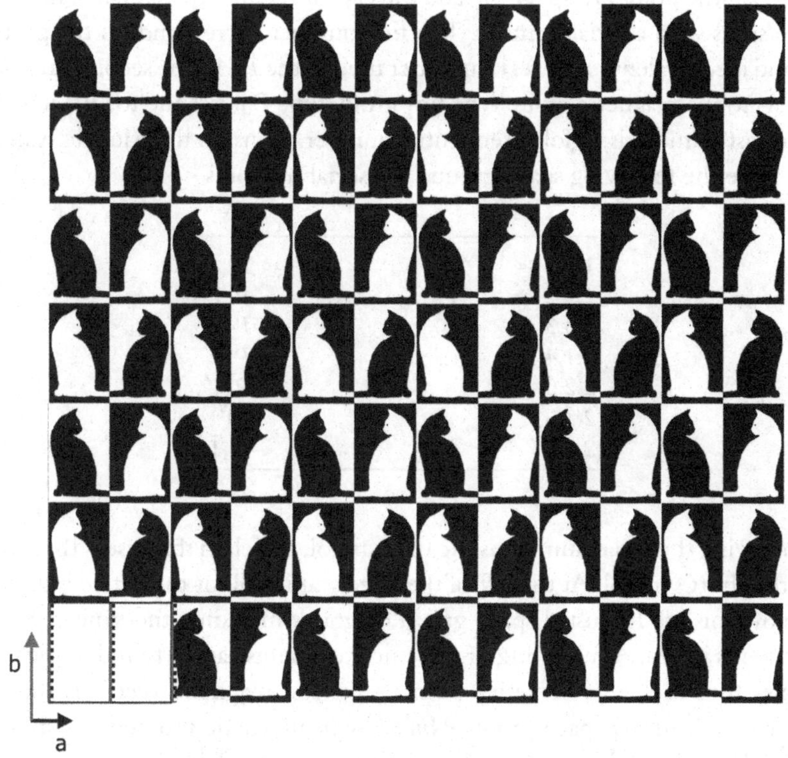

Figure 8.10 An example of black and white two-dimensional $p_{2b}1m'1$ plane group symmetry.

we can see that the cats along this direction alternate between black and white. Furthermore, the m' antireflection (vertical red lines) reflects black cats to white cats that are at the same time mirror related. Note that coinciding with the m′ elements glide lines are also produced.

8.6. MAGNETIC SPACE GROUPS

Since the main use of antisymmetry is to describe magnetic structures, the antisymmetry operator $1'$ is taken to be the time-reversal operator and the symmetry operations now act on spin vectors. In order to describe magnetic structures, therefore, it is important to have a compilation of the 1651 space groups equivalent to the 230 normal space groups. As an example, we show the space group pages for $P4m'm'$ (Fig. 8.11), where it can be seen that they do resemble the layout for the *ITA*, although there are some notable differences.

Reading at the top from left to right, we see the short space group symbol: $P4m'm'$, the point group: $4m'm'$ and the crystal system: Tetragonal. The next line starts with a serial number. The first number corresponds to the group **F** and is equivalent to the serial number used in the *ITA*. The second number refers to the magnetic group type (for group type **F** it is 1 and for **F1′** it is 2). The last number is a global sequential number. Thus, in the triclinic system we have the following space groups and serial numbers:

1.1.1	$P1$
1.2.2	$P11'$
1.3.3	$P_{2s}1$
2.1.4	$P\bar{1}$
2.2.5	$P\bar{1}1'$
2.3.6	$P\bar{1}'$
2.4.7	$P_{2s}\bar{1}$

Following the serial number is the full symbol, which in this case is the same as the short symbol. At the left of the page a unit cell for the lattice is given. Below this are the usual space group diagrams including the time-reversal elements in red. A full listing of these new elements can be found in Litvin's tables. For each representative point, a general magnetic vector is shown (a structure in the space group $P4m'm'$ with magnetic moments as shown would be classified as a potential weak ferromagnet). Below the diagrams we see the origin specification followed by the symmetry operations. Litvin also includes in the list of operations corresponding Seitz symbols, albeit in a slightly different notation from that used in this book.

On the following page are the space group generators as usual followed by the list of multiplicities, Wyckoff letters, site symmetries and coordinates. The main difference to note here is that in addition to the positional parameters (x, y, z), the directions $[uvw]$ of the magnetic moments are added. Note that if the magnetic structure is such that the magnetic atoms are on $1a$, $1b$ or $2c$ sites the moments are confined to [00w] directions and thus, the material is a potential full ferromagnet.

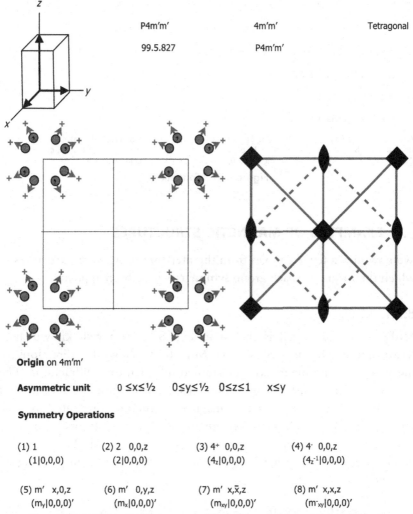

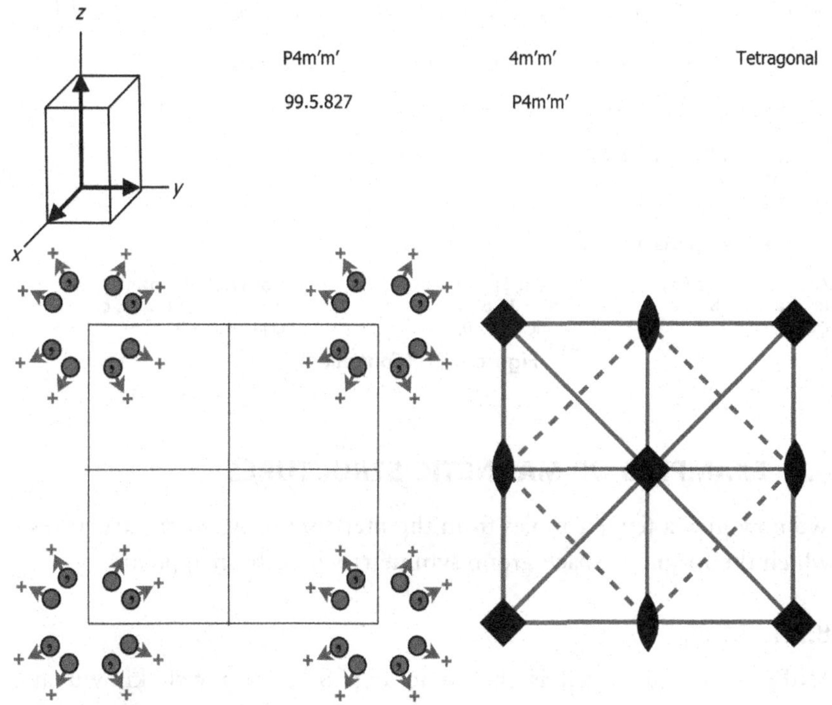

Origin on 4m′m′

Asymmetric unit $0 \leq x \leq ½$ $0 \leq y \leq ½$ $0 \leq z \leq 1$ $x \leq y$

Symmetry Operations

(1) 1	(2) 2 0,0,z	(3) 4^+ 0,0,z	(4) 4^- 0,0,z
(1|0,0,0)	(2|0,0,0)	(4_z|0,0,0)	(4_z^{-1}|0,0,0)
(5) m′ x,0,z	(6) m′ 0,y,z	(7) m′ x,$\bar{x}$,z	(8) m′ x,x,z
(m_y|0,0,0)′	(m_x|0,0,0)′	(m_{xy}|0,0,0)′	($m_{\bar{x}y}$|0,0,0)′

Figure 8.11 Magnetic space group *P4m′m′* (copied from Litvin's tables).

Generators selected (1): t(1,0,0); t(0,1,0); t(0,0,1); (2); (3); (5).

Positions

Multiplicity, Wyckoff letter, Site Symmetry			Coordinates			
8	g	1	(1) x,y,z [u,v,w]	(2) $\bar{x},\bar{y}$,z [$\bar{u},\bar{v}$,w]	(3) $\bar{y}$,x,z [$\bar{v}$,u,w]	(4) y,$\bar{x}$,z [v,$\bar{u}$,w]
			(5) x,y,z [u,$\bar{v}$,w]	(6) $\bar{x}$,y,z [u,v,w]	(7) y,$\bar{x}$,z [v,$\bar{u}$,w]	(8) y,x,z [v,u,w]
4	f	.m′.	x,1/2,z [u,0,w]	$\bar{x}$,1/2,z [$\bar{u}$,0,w]	1/2,x,z [0,u,w]	1/2,$\bar{x}$,z [0,$\bar{u}$,w]
4	e	.m′.	x,0,z [u,0,w]	$\bar{x}$,0,z [$\bar{u}$,0,w]	0,x,z [0,u,w]	0,$\bar{x}$,z [0,$\bar{u}$,w]
4	d	..m′	x,x,z [u,u,w]	$\bar{x},\bar{x}$,z [$\bar{u},\bar{u}$,w]	$\bar{x}$,x,z [$\bar{u}$,u,w]	x,$\bar{x}$,z [u,$\bar{u}$,w]
2	c	2m′m′.	1/2,0,z [0,0,w]	0,1/2,z [0,0,w]		
1	b	4m′m′	1/2,1/2,z [0,0,w]			
1	a	4m′m′	0,0,z [0,0,w]			

Symmetry of Special Projections

Along [0,0,1] p4m′m′	Along [1,0,0] p1m1	Along [1,1,0] p1m1
a* = a b* = b	**a* = b b* = c**	**a* = (-a+b)/2 b* = c**
Origin at 0,0,z	Origin at x,0,0	Origin at x,x,0

Figure 8.11 (*Continued*).

8.7. EXAMPLES OF MAGNETIC STRUCTURES

We now give a few examples from the literature of magnetic structures to which the magnetic space group symmetries have been applied.

8.7.1. MnF_2

MnF_2, whose unit cell is shown in Fig. 8.12, is a well known antiferromagnet. The arrows on the Mn atoms show the directions of the magnetic moments as determined by neutron diffraction. The structure then consists of an atomic arrangement (just the locations of the Mn and F atoms) and a magnetic structure (indicated by the arrows). Thus, the magnetic symmetry is seen to introduce an extra operation into the space group symmetry of this crystal. The crystallographic space group of MnF_2 is $P4_2/mnm(D_{2h}^{14})$. The F atoms are located at 4*f* positions:

$$x, x, 0 \quad \bar{x}, \bar{x}, 0 \quad \bar{x} + ½, x + ½, ½ \quad x + ½, \bar{x} + ½, ½$$

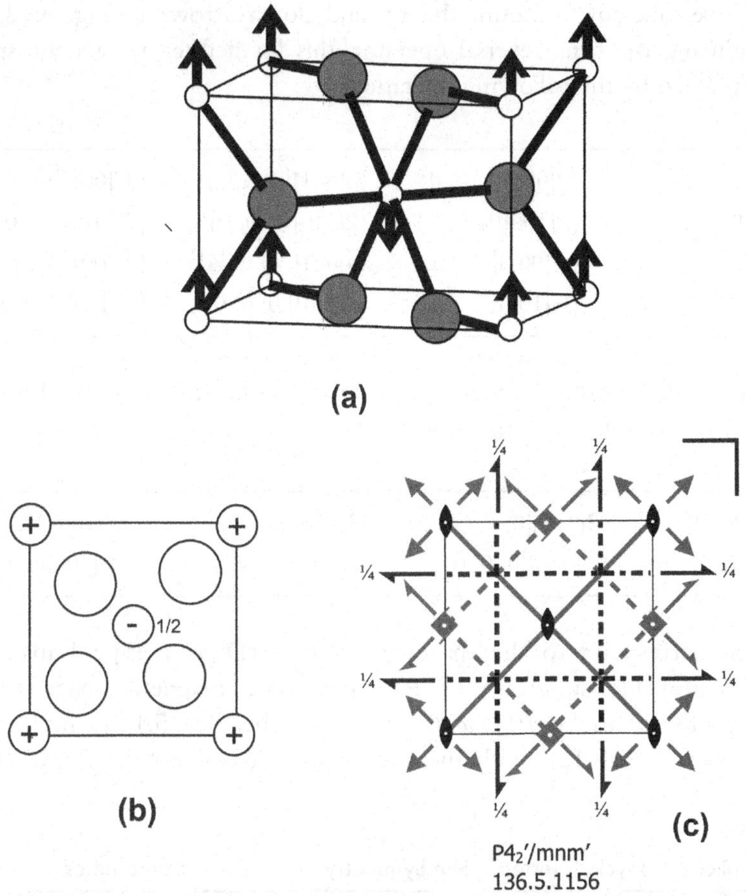

Figure 8.12 (a) Crystal structure of MnF_2 showing directions of magnetic moments. Small spheres are Mn atoms (b) structure projected on (001): + and − signs indicate directions of magnetic moments; (c) elements of magnetic space group $P4_2'/mnm'$.

The Mn atoms are located at the 2*a* sites with the coordinates (0, 0, 0) and (½, ½, ½) (note that this structure is not body-centered because the F atoms are not related by body-centering). Now, the space group operations in Seitz notation, as specified from a common origin, are

{1|0, 0, 0}	{2[001]|0, 0, 0}	{4[001]|½,½,½}	$\{4^3[001]\|½,½,½\}$
{2[010]| (½,½,½}	{2[100]|½,½,½}	{2[110]|0, 0, 0}	$\{2[\bar{1}10]\|0, 0, 0\}$
$\{\bar{1}\|0, 0, 0\}$	{*m*[001]|0, 0, 0}	$\{\bar{4}[001]\|½, ½, ½\}$	$\{\bar{4}^3[001]\|½, ½, ½\}$
{*m*[010]|½,½,½}	{*m*[100]|½,½,½}	{*m*[110]|0, 0, 0}	$\{m[\bar{1}10]\|0,0,0\}$

If we take into account the up and down arrows in Fig. 8.12, by introducing the time-reversal operator, this list defines a magnetic space group given by the following operations:

$\{1\|0, 0, 0\}$	$\{2[001]\|0, 0, 0\}$	$\{4[001]\|½,½,½\}'$	$\{4^3[001]\|½,½,½\}'$
$\{2[010]\|½,½,½\}$	$\{2[100]\|½,½,½\}$	$\{2[110]\|0, 0, 0\}'$	$\{2[\bar{1}10]\|0,0,0\}'$
$\{\bar{1}\|0,0,0\}$	$\{m[001]\|0, 0, 0\}$	$\{\bar{4}[001]\|½, ½, ½\}'$	$\{\bar{4}^3[001]\|½, ½, ½\}'$
$\{m[010]\|½,½,½\}$	$\{m[100]\|½,½,½\}$	$\{m[110]\|0, 0, 0\}'$	$\{m[\bar{1}10]\|0,0,0\}'$

Note that half of the symmetry operations have been replaced by $1'\{R|\mathbf{t}\}$. The halving group **D** then consists of the operations

$\{1\|0, 0, 0\}$	$\{2[001]\|0, 0, 0\}$	$\{2[010]\|½,½,½\}$	$\{2[100]\|½,½,½\}$
$\{\bar{1}\|0,0,0\}$	$\{m[001]\|0, 0, 0\}$	$\{m[010]\|½,½,½\}$	$\{m[100]\|½,½,½\}$

which correspond to the space group $Pnnm(D_{2h}^{12})$, a Type I maximal non-isomorphic subgroup of $P4_2/mnm$. The magnetic space group is then found to be $P4_2'/mnm'$ (serial number 136.5.1156 in Litvin's tables, see Fig. 8.12), with the Mn atoms situated on the $2a$ Wyckoff sites:

Multiplicity	**Wyckoff letter**	**Site Symmetry**	**Coordinates**	
2	*a*	$m.\,m'\,m'$	0, 0, 0 $[0,0,w]$	½, ½, ½$[0, 0, \bar{w}]$

Thus, we find the Mn atoms at (0, 0, 0) and (½, ½, ½) with magnetic moments directed along $[00w]$ and $[00\bar{w}]$, respectively, as shown in Fig. 8.12.

In order to emphasize the difference between magnetic space groups and black and white space groups, consider Fig. 8.13, where instead of magnetic vectors attached to each Mn atom, the Mn atoms have been colored black and white. In this case, we see that the Mn atom at (0, 0, 0) is related to the Mn at (½, ½, ½) by the $4_2'$ operation as before, in this case causing a change from black to white. However consider the 2-fold screw axis positioned at *x*, ¼, ¼. In space group $P4_2'/mnm'$ this operation is $\{2[100]|½, ½, ½\}$ which takes the atom at (0, 0, 0) to (½, ½, ½) but does

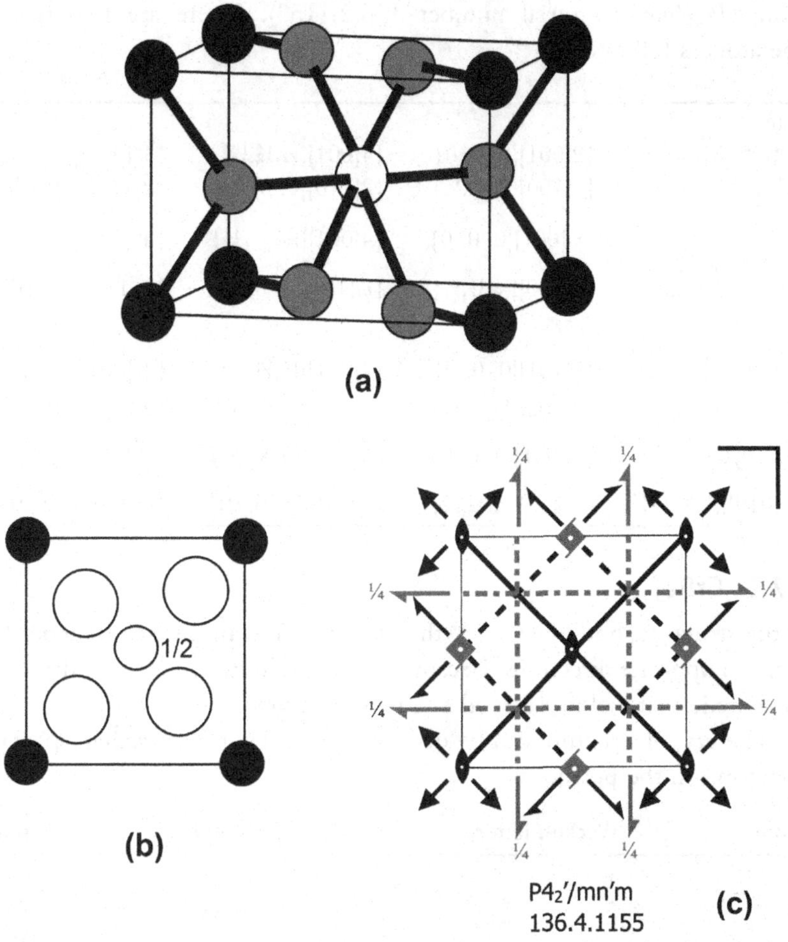

Figure 8.13 (a) MnF_2 structure with Mn atoms alternately colored black and white; (b) structure projected on (001); (c) symmetry elements of magnetic space group $P4_2'/mnm'$.

not *change the color*! In order for this to happen the screw operation must include the $1'$ operation to give $\{2[100]|½, ½, ½\}'$. The result is that the space group description for the black and white structure has changed from $P4_2'/mnm'$ to $P4_2'/mn'm$ (serial number 136.4.1155) with the Mn atoms at $2a$ sites.

What about the high-temperature paramagnetic structure? In this case, the up and down spins are disordered and occur on average at each of the two Mn sites. In this case, the space group symmetry is given by the grey group $P4_2/mnm1'$ (serial number 136.2.1153). There are two sets of operators as follows:

1 set			
$\{1\vert 0,0,0\}$	$\{2[001]\vert 0,0,0\}$	$\{4[001]\vert ½,½,½\}$	$\{4^3[001]\vert ½,½,½\}$
$\{2[010]\vert ½,½,½\}$	$\{2[100]\vert ½,½,½\}$	$\{2[110]\vert 0,0,0\}$	$\{2[\bar{1}10]\vert 0,0,0\}$
$\{\bar{1}\vert 0,0,0\}$	$\{m[001]\vert 0,0,0\}$	$\{\bar{4}[001]\vert ½,½,½\}$	$\{\bar{4}^3[001]\vert ½,½,½\}$
$\{m[010]\vert ½,½,½\}$	$(m[100]\vert ½,½,½\}$	$\{m[110]\vert 0,0,0\}$	$\{m[\bar{1}10]\vert 0,0,0\}$
$1'$ set			
$\{1\vert 0,0,0\}'$	$\{2[001]\vert 0,0,0\}'$	$\{4[001]\vert ½,½,½\}'$	$\{4^3[001]\vert ½,½,½\}'$
$\{2[010]\vert ½,½,½\}'$	$\{2[100]\vert ½,½,½\}'$	$\{2[110]\vert 0,0,0\}'$	$\{2[\bar{1}10]\vert 0,0,0\}'$
$\{\bar{1}\vert 0,0,0\}'$	$\{m[001]\vert 0,0,0\}'$	$\{\bar{4}[001]\vert ½,½,½\}'$	$\{\bar{4}^3[001]\vert ½,½,½\}'$
$\{m[010]\vert ½,½,½\}'$	$(m[100]\vert ½,½,½\}'$	$\{m[110]\vert 0,0,0\}'$	$\{m[\bar{1}10]\vert 0,0,0\}'$

8.7.2. $CrPt_3$

In this structure, both the Cr and the Pt atoms carry magnetic moments that point in opposite directions. Because they are of different magnitudes, such a magnetic material is classified as a **ferrimagnet**.

The crystal structure can be described in space group $Pm\bar{3}m(O_h^1)$ with the atoms in the positions

Atom	Wyckoff letter	Site Symmetry	Position
Cr	1*a*	$m\bar{3}m$	0, 0, 0
Pt	3*c*	$4/mm.m$	0, ½, ½
			½, 0, ½
			½, ½, 0

However, when the magnetic moments are considered (Fig. 8.14) one direction (the **c**-axis) is differentiated from the other two, so that the magnetic structure is tetragonal. The arrangement of spins corresponds to magnetic space group $P4/mm'm'$ with the following arrangement:

Atom	Wyckoff letter	Site Symmetry	Position	Spin direction
Cr	1*a*	$4/mm'm'$	0, 0, 0	$[00w_1]$
Pt	1*c*	$4/mm'm'$	½, ½, 0	$[00w_3]$
Pt	2*e*	$mm'm'.$	0, ½, ½	$[00w_2]$
			½, 0, ½	

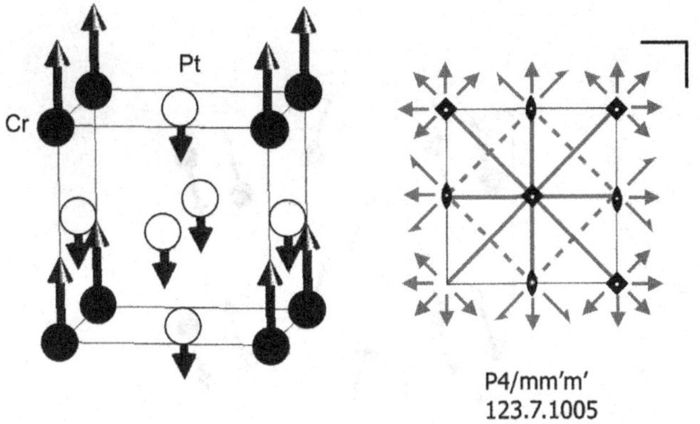

Figure 8.14 Crystal and magnetic structures of $CrPt_3$ together with space group diagram for $P4/mm'm'$.

It can be seen from this that there are three possible symmetry-independent spin directions along $\pm\mathbf{c}$. It appears from the literature that only one moment is given for the Pt atoms of $-0.27\ \mu_B$/atom with $+2.33\ \mu_B$/atom for the Cr atoms. The fact that $w_2 = w_3$ is accidental as far as symmetry is concerned: moreover, symmetry does not dictate that Pt on $1c$ sites should have the same magnitude of magnetic moment as Pt on $2e$ sites. In principle, therefore, it should be possible to measure two different moments for the Pt atoms: as far as is known this has not been done. This is a nice example of how magnetic space group symmetry can be used to reveal effects that might otherwise be missed.

8.7.3. $CsCoCl_3.2H_2O$

The structure of this crystal belongs in the orthorhombic space group $Pcca(D_{2h}^8)$ with $a = 8.89$Å, $b = 7.10$ Å and $c = 11.31$ Å at a temperature of 1.3 K. The Co atoms, which carry magnetic moments, are situated on $4c$ sites:

$0, y, \frac{1}{4}$	$\frac{1}{2}, \bar{y}, \frac{1}{4}$	$0, \bar{y}, \frac{3}{4}$	$\frac{1}{2}, y, \frac{3}{4}$

with $y \approx 0.47$. A study by neutron scattering revealed the arrangements of magnetic moments on the Co atoms as shown in Fig. 8.15, where it can be seen that this is an example of a canted antiferromagnet with moments aligned at approximately 17° to the $\mathbf{c}$-axis.

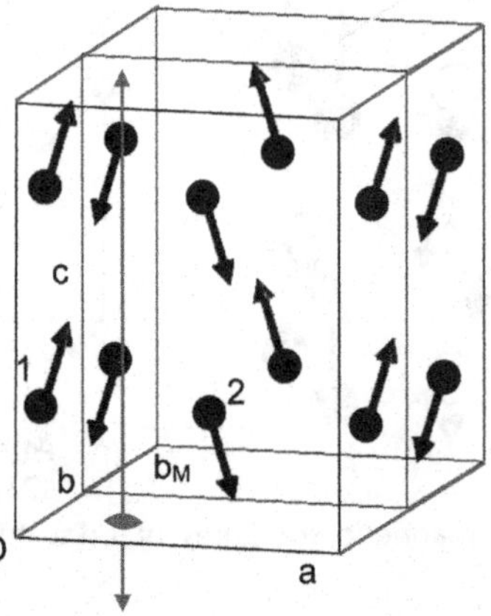

Figure 8.15 The magnetic structure of $CsCoCl_3 \cdot 2H_2O$. Only the cobalt atoms are shown. The magnetic moments are confined to the **ac** planes. The magnitude of the magnetic moment is found to be $2.5\mu_B$.[3]

The magnetic space group is $P_{2b}cca'$ (serial number 54.12.439) and the pages from Litvin's tables are shown in Fig. 8.16. In this space group, the Co atoms are now listed on 8*c* sites:

(0,0,0) + (0,1,0) ′

$0,y,\frac{1}{4}[u,0,w]$	$\frac{1}{2},\bar{y},\frac{1}{4}[\bar{u},0,w]$	$0,\bar{y},\frac{3}{4}[\bar{u},0,\bar{w}]$	$\frac{1}{2},y,\frac{3}{4}[u,0,\bar{w}]$

which when written out in full is

$0,y,\frac{1}{4}[u,0,w]$	$\frac{1}{2},\bar{y},\frac{1}{4}[\bar{u},0,w]$	$0,\bar{y},\frac{3}{4}[\bar{u},0,\bar{w}]$	$\frac{1}{2},y,\frac{3}{4}[u,0,\bar{w}]$
$0,1+y,\frac{1}{4}[\bar{u},0,\bar{w}]$	$\frac{1}{2},1-y,\frac{1}{4}[u,0,\bar{w}]$	$0,1-y,\frac{3}{4}[u,0,w]$	$\frac{1}{2},1+y,\frac{3}{4}[\bar{u},0,w]$

In $CsCoCl_3.2H_2O$, the spin moments are in the directions <*u*0*w*> with $u \cong 0.4w$ to form the canted antiferromagnetic arrangement shown in Fig. 8.15.

[3] After A.L. Bongaarts & B. van Laar, *Phys. Rev.*, **B6**, 2669 (1972).

Notice that the magnetic unit cell axes $\mathbf{a}_M$, $\mathbf{b}_M$ and $\mathbf{c}_M$ are related to the crystallographic axes **a**, **b** and **c** according to

$$\mathbf{a}_M = \mathbf{a}$$
$$\mathbf{b}_M = 2\mathbf{b}$$
$$\mathbf{c}_M = \mathbf{c}$$

and the magnetic unit cell is shown in Fig. 8.15. Consider the cobalt atom marked 1 and its magnetic moment direction at

Co1 0,0.47, ¼ [*u*0*w*]

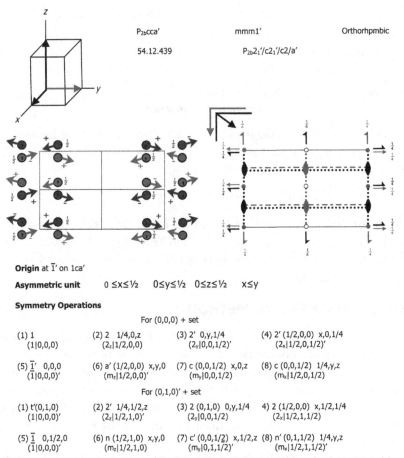

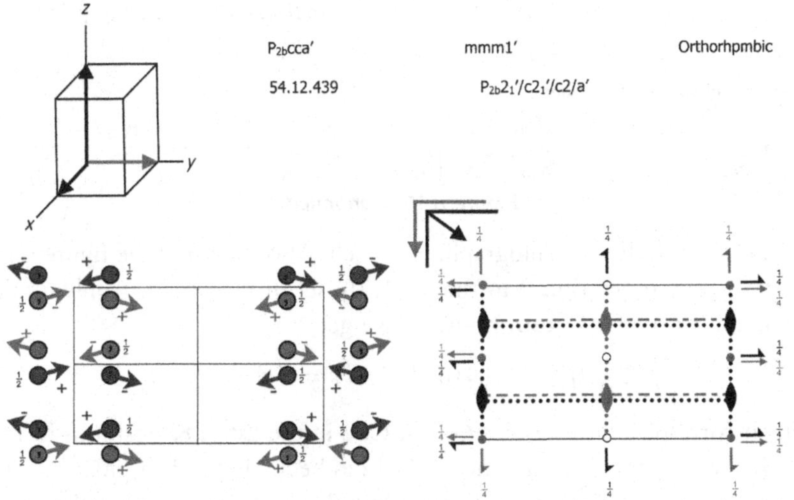

Origin at $\bar{1}'$ on 1ca'

Asymmetric unit 0 ≤x≤½ 0≤y≤½ 0≤z≤½ x≤y

Symmetry Operations

For (0,0,0) + set

(1) 1 (1|0,0,0)	(2) 2 1/4,0,z (2_z|1/2,0,0)	(3) 2′ 0,y,1/4 (2_y|0,0,1/2)′	(4) 2′ (1/2,0,0) x,0,1/4 (2_x|1/2,0,1/2)′
(5) $\bar{1}'$ 0,0,0 ($\bar{1}$|0,0,0)′	(6) a′ (1/2,0,0) x,y,0 (m_z|1/2,0,0)′	(7) c (0,0,1/2) x,0,z (m_y|0,0,1/2)	(8) c (0,0,1/2) 1/4,y,z (m_x|1/2,0,1/2)

For (0,1,0)′ + set

(1) t′(0,1,0) (1|0,0,0)′	(2) 2′ 1/4,1/2,z (2_z|1/2,1,0)′	(3) 2 (0,1,0) 0,y,1/4 (2_y|0,0,1/2)	4) 2 (1/2,0,0) x,1/2,1/4 (2_x|1/2,1,1/2)
(5) $\bar{1}$ 0,1/2,0 ($\bar{1}$|0,0,0)′	(6) n (1/2,1,0) x,y,0 (m_z|1/2,1,0)	(7) c′ (0,0,1/2) x,1/2,z (m_y|0,1,1/2)′	(8) n′ (0,1,1/2) 1/4,y,z (m_x|1/2,1,1/2)′

Figure 8.16 Space group $P_{2b}cca'$ (serial number 54.12.439) taken from Litvin's tables.

Generators selected (1): t(1,0,0); t(0,1,0); t(0,0,1); (2); (3); (5).

Positions

Multiplicity, Wyckoff letter, Site Symmetry			Coordinates			
16	f	1	(1) x,y,z [u,v,w]	(2) x+1/2,y,z [u,$\bar{v}$,w]	(3) $\bar{x}$,y,$\bar{z}$+1/2 [u,$\bar{v}$,w]	(4) x+1/2,$\bar{y}$,$\bar{z}$+1/2 [$\bar{u}$,v,w]
			(5) $\bar{x}$,$\bar{y}$,$\bar{z}$ [$\bar{u}$,$\bar{v}$,$\bar{w}$]	(6) x+1/2,y,$\bar{z}$ [u,v,$\bar{w}$]	(7) x,$\bar{y}$,z+1/2 [$\bar{u}$,v,$\bar{w}$]	(8) $\bar{x}$+1/2,y,z+1/2 [u,$\bar{v}$,$\bar{w}$]
8	e	..2′	1/4,1/2,z [u,v,0]	3/4,1/2,$\bar{z}$+1/2 [u,v,0]	3/4,1/2,$\bar{z}$ [u,v,0]	1/4,½,z+1/2 [u,$\bar{v}$,0]
8	d	..2	1/4,0,z [0,0,w]	3/4,0,$\bar{z}$+1/2 [0,0,w]	3/4,0,$\bar{z}$ [0,0,$\bar{w}$]	1/4,0,z+1/2 [0,0,$\bar{w}$]
8	c	.2′.	0,y,1/4 [u,0,w]	1/2,$\bar{y}$,1/4 [$\bar{u}$,0,w]	0,$\bar{y}$,3/4 [$\bar{u}$,0,$\bar{w}$]	1/2,y,3/4 [u,0,$\bar{w}$]
8	b	$\bar{1}$	0,1/2,0 [u,v,w]	1/2,1/2,0 [u,v,$\bar{w}$]	0,1/2,1/2 [u,$\bar{v}$,w]	1/2,1/2,1/2 [u,$\bar{v}$,$\bar{w}$]
8	a	$\bar{1}'$	0,0,0 [0,0,0]	1/2,0,0 [0,0,0]	0,0,1/2 [0,0,0]	1/2,0,1/2 [0,0,0]

Symmetry of Special Projections

Along [0,0,1] $p_{2a^*}2mm$	Along [1,0,0] $p_{c^*}2mm$	Along [0,1,0] p2mg1′
a* = **a** **b*** = **a**/2	**a*** = **b** **b*** = **c**/2	**a*** = -**a** **b*** = **c**/2
Origin at 0,0,z	Origin at x,1/2,1/4	Origin at 0,y,0

Figure 8.16 ***(Continued)*.**

with respect to the crystallographic unit cell. Also shown in the figure is the 2′ at ¼, ½, *z* (operation 2 in the (0, 1, 0)′ set in Fig. 8.16). Applying this operation to the Co1 atom position, we get

$$\{2[001]|½, 1, 0\}(0, 0.47, ¼) = (½, 0.53, ¼)$$

which corresponds to the position of Co2 in the list of 8*c* positions in Eqn [8.3] in which the spin direction $[u0w]$ has been changed to $[u0\overline{w}]$, that is the **a**-axis component is unchanged but the **c**-axis component has been reversed in direction.

8.8. REPRESENTATION METHOD

In some crystals, it is found that the arrangements of atoms occur with symmetries that are not contained in the set of operations specified by the space group. Such extra symmetry can lead to many varieties of structure, such as those seen in polytypes of SiC. In magnetic structures, such extra symmetry is very common, for example where long-period helical spin arrangements are found. In order to address this, E.F. Bertaut[4] developed an

[4] E.F. Bertaut, *Acta Cryst.*, **A24**, 217 (1968). Actually there was an earlier paper by S. Alexander (*Phys. Rev.*, **127**, 320, 1962) that first introduced the use of group representations to magnetic symmetry.

Table 8.1 Character table for point group 222(D_2). Bold symbols indicate non-distinct groups.

222	**1**	**2[100]**	**2[010]**	**2[001]**	
A_1	1	1	1	1	222
A_2	1	−1	−1	1	$2'2'2$
B_1	1	−1	1	−1	$\mathbf{2'22'}$
B_2	1	1	−1	−1	$\mathbf{22'2'}$

alternative approach in dealing with the transformation properties of magnetic structures under the operations of the normal 230 space groups, that is without the need to use magnetic space groups directly. This theory makes use of group representations to derive the possible spin vectors in general. That is not to say that the magnetic groups discussed earlier can be disregarded, especially when considering cases where magnetic and non-magnetic energy terms interact, such as in magnetoelectricity. However, for describing magnetic structures *per se*, the representation method has the advantages of simplicity, as well as the ability to describe complex spin patterns that the 1651 three-dimensional magnetic space groups cannot describe in full (although the use of superspace symmetry could be applied here). Nonetheless, it is worth noting that claims by Bertaut that this method of classifying magnetic structures is more general than classifying by magnetic groups have been challenged by Opechowski and Dreyfus.[5] We outline this method below, but the reader is encouraged to study Bertaut's original publication for complete details.

The representation approach makes use of the fact that magnetic groups are strictly isomorphous to classical groups and therefore, have the same irreducible representations. In particular, the magnetic groups **F**(**D**) are related to the one-dimensional representations of **F**.

As an example, Table 8.1 shows the usual character table for the point group 222(D_2) with the four irreducible representations A_1, A_2, B_1 and B_2. We now make use of the one-to-one correspondence with the magnetic classes: wherever the character is −1, we get the associated anti-operation. Thus, in the A_2 representation −1 under the 2[100] operation means that this operation is now replaced by $2[100]'$. We see therefore that the point group corresponding to A_2 is $2'2'2$. Note however, that the groups $2'2'2$, $2'22'$ and $22'2'$ are not distinct but are just different settings of the same group.

5 W. Opechowski & T. Dreyfus, *Acta Cryst.*, **A27**, 470 (1971).

Table 8.2 Character table for point group $2/m(C_{2h})$

2/m	1	2[001]	m[001]	$\bar{1}$	
A_g	1	1	1	1	$2/m$
B_g	1	−1	−1	1	$2'/m'$
A_u	1	1	−1	−1	$2/m'$
B_u	1	−1	1	−1	$2'/m$

Similarly, if we take point group $2/m(C_{2h})$, Table 8.2 gives rise to four distinct point groups: $2/m$, $2'/m'$, $2/m'$ and $2'/m$. Proceeding in this way with all 32 point groups one then obtains the 90 magnetic point groups.

It is evident from this simple treatment that it should be applicable also to the derivation of the 1651 space groups, provided that one again restricts consideration only to one-dimensional representations. For example, take the space group $Pbam(D_{2h}^9)$ and consider for the moment all cases where the magnetic spin arrangement has wave-vector $\mathbf{k} = 0$.

Table 8.3 shows the character table for space group *Pbam* at the center of the Brillouin zone ($\mathbf{k} = 0$). The coset representations for the operations are shown at the top in shorthand and should be obvious as to their meaning. We immediately see that this generates eight equi-translational space groups. Of these eight, there are six distinct space groups: for instance, *Pba′m′* and *Pb′am′* can be transformed into each other by interchanging the **a** and **b** axes, as well as the corresponding representations Γ_2 and Γ_3. The same is true for *Pb′am* and *Pba′m*.

Now, in order to extend the possible magnetic space groups based on *Pbam*, we need to consider the effect of setting $\mathbf{k} \neq \mathbf{0}$. The possible wave-vectors are (½, 0, 0), (0, ½, 0), (0, 0, ½), (0, ½, ½), (½, 0, ½), (½, ½, 0) and (½, ½, ½) in **k**-space. It turns out that only for $\mathbf{k} \equiv (0, 0, ½)$ are there

Table 8.3 Character table for space group *Pbam* at $\mathbf{k} = 0$. Bold symbols indicate non-distinct groups

Pbam	1	2[001]	2_1[010]	2_1[100]	$\bar{1}$	*m*[001]	*a*[010]	*b*[100]	
Γ_1	1	1	1	1	1	1	1	1	*Pbam*
Γ_2	1	−1	−1	1	1	−1	−1	1	*Pba′m′*
Γ_3	1	−1	1	−1	1	−1	1	−1	***Pb′am′***
Γ_4	1	1	−1	−1	1	1	−1	−1	*Pb′a′m*
Γ_5	1	1	1	1	−1	−1	−1	−1	*Pb′a′m′*
Γ_6	1	−1	−1	1	−1	1	1	−1	*Pb′am*
Γ_7	1	−1	1	−1	−1	1	−1	1	***Pba′m***
Γ_8	1	1	−1	−1	−1	−1	1	1	*Pbam′*

Table 8.4 Character table for space group *Pnma* at $\mathbf{k}=0$

Pnma	1	2_1[001]	2_1 [100]	2_1[010]	$\bar{1}$	*a*[001]	*n*[100]	*m*[010]	
$\Gamma_1=\Gamma_{1g}$	1	1	1	1	1	1	1	1	*Pnma*
$\Gamma_2=\Gamma_{2g}$	1	1	−1	−1	1	1	−1	−1	*Pn′m′a*
$\Gamma_3=\Gamma_{3g}$	1	−1	1	−1	1	−1	1	−1	*Pnm′a′*
$\Gamma_4=\Gamma_{4g}$	1	−1	−1	1	1	−1	−1	1	*Pn′ma′*
$\Gamma_5=\Gamma_{1u}$	1	1	1	1	−1	−1	−1	−1	*Pn′m′a′*
$\Gamma_6=\Gamma_{2u}$	1	1	−1	−1	−1	−1	1	1	*Pnma′*
$\Gamma_7=\Gamma_{3u}$	1	−1	1	−1	−1	1	−1	1	*Pn′ma*
$\Gamma_8=\Gamma_{4u}$	1	−1	−1	1	−1	1	1	−1	*Pnm′a*

one-dimensional representations[6], all other cases either being complex or two-dimensional. For this wave-vector, a phase factor of $\exp(2\pi i\mathbf{k}.\mathbf{c}) = -1$ is introduced showing that the magnetic cell is doubled along the **c**-axis direction. We thus find three space groups $P_{2c}bam$, $P_{2c}b'am$ and $P_{2c}b'a'm$, all others being related to these by changes of origin or setting. There is therefore a family of nine magnetic groups associated with *Pbam*.

Another example is that of space group P*nma*(D_{2h}^{16}) for which we can write the character Table 8.4 (note that Bertaut in his original publication uses the setting *Pbnm*: we have used the equivalent space group *Pnma* in order to make it easier to compare with the coordinates and directions of the setting chosen in the *ITA*). To save space, we use a condensed notation for the symmetry operations. In this case, we find eight distinct magnetic space groups for $\mathbf{k}=0$. For other values of the wave-vector it is found that no one-dimensional real representations exist and so these are the total number of magnetic space groups in the family derived from *Pnma*.

Now, consider Fig. 8.17 where the screw axes for space group *Pnma* are shown together with the special positions of the type 4*a*.

Let us place spins S_j onto these sites numbered $j=1$, 2, 3 and 4, respectively. These are at

(1) 0, 0, 0	(2) ½, 0, ½	(3) 0, ½, 0	(4) ½, ½, ½

Equivalently, we can locate spins at the 4*b* sites

(1) 0, 0, ½	(2) ½, 0, 0	(3) 0, ½, ½	(4) ½, ½, 0

6 You can check this with program REPRES at the Bilbao Crystallographic Server.

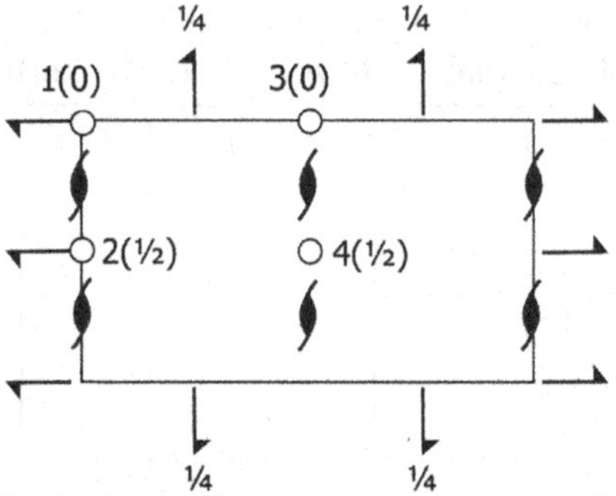

Figure 8.17 Screw axes in *Pnma*(D_{2h}^{16}). The circles 1, 2, 3 and 4 are the 4*a* Wyckoff positions (fractional heights given in parentheses).

The order of these coordinates is the same as in the *ITA*. By applying each of the first four operations in Table 8.4 in turn we derive the transformation of spins (Table 8.5).

We write the spins here with their directions in the form S_{jx}, S_{jy} and S_{jz} where in this case $x \equiv [100], y \equiv [010]$ and $z \equiv [001]$. Thus, for example, the screw operation $\{2[100]|½,½,½\}$ takes the point at (0, 0, 0) to (½, ½, ½), that is from position 1 to 4 and at the same time keeps the x-component of the spin unchanged, that is from $+S_{1x}$ to $+S_{4x}$. Note that applying the $\{\bar{1}|0,0,0\}$, $\{m[100]|½,½,½\}$, $\{m[010]|0,½,0\}$ and $\{m[001]|½,0,½\}$ operations one obtains the same results and so this adds nothing new.

In order to understand the spin arrangements we now use projection operator techniques. The projection operator is defined by

$$V^{\nu} \propto \sum_{R} \chi^{\nu}(R)R \qquad [8.4]$$

where ν refers to the particular representation and χ is the character of the νth representation. We can then write for the x-component

$$V_x^{\nu} = \chi^{\nu}(1)(S_{1x}) + \chi^{\nu}(2_1[001])\{2[001]|½,0,½\}S_{1x}$$
$$+ \chi^{\nu}(2_1[100])\{2[100]|½,½,½\}S_{1x} + \chi^{\nu}\ (2_1[010])\{2[010]|0,½,0\}S_{1x} \qquad [8.5]$$

Table 8.5 Multiplication table for identity plus screw operations acting on spin components for special position 1 on 4*a*/4*b* site. Seitz notation from common origin used

	{1|0,0,0}	{2[001]|½,0,½}	{2[100]|½,½,½}	{2[010]|0,½,0}
S_{1x}	$+S_{1x}$	$-S_{2x}$	$+S_{4x}$	$-S_{3x}$
S_{1y}	$+S_{1y}$	$-S_{2y}$	$-S_{4y}$	S_{3y}
S_{1z}	$+S_{1z}$	S_{2z}	$-S_{4z}$	$-S_{3z}$

Proceeding in this way for each representation, we obtain

$$
\begin{array}{ll}
\Gamma_{1g} & C_x = S_{1x} - S_{2x} - S_{3x} + S_{4x} \\
\Gamma_{2g} & G_x = S_{1x} - S_{2x} + S_{3x} - S_{4x} \\
\Gamma_{3g} & F_x = S_{1x} + S_{2x} + S_{3x} + S_{4x} \\
\Gamma_{4g} & A_x = S_{1x} + S_{2x} - S_{3x} - S_{4x}
\end{array}
\qquad [8.6]
$$

The G and F labels correspond to those used by Bertaut, whereas the C and A labels have been interchanged to reflect the order in which the coordinates have been chosen. Repeating this for the y and z components we can construct Table 8.6 for the 4*a* and 4*b* sites. The table also includes the transformation properties for an atom on a 4*c* site, in which the positions are in the order

$$(1)\ x, \tfrac{1}{2}, z \quad (2)\ \bar{x}+\tfrac{1}{2}, \tfrac{3}{4}, z\tfrac{1}{2} \quad (3)\ \bar{x}, \tfrac{3}{4}, \bar{z} \quad (4)\ x+\tfrac{1}{2}, \tfrac{1}{4}, \bar{z}+\tfrac{1}{2}$$

For 4*a* and 4*b* sites, $\Gamma_{1u} = \Gamma_{2u} = \Gamma_{3u} = \Gamma_{4u} = 0$. Note that the same pattern of spins can be read directly from Litvin's tables. The following is the list of positions and spin vector directions for the 4*a* sites:

Γ_{1g}	*Pnma*	0,0,0 $[u,v,w]$	½,0,½ $[\bar{u},\bar{v}\ w]$	0,½,0 $[\bar{u},v,\bar{w}]$	½,½,½ $[u,\bar{v},\bar{w}]$
Γ_{2g}	*Pn′m′a*	0,0,0 $[u,v,w]$	½,0,½ $[\bar{u},\bar{v},w]$	0,½,0 $[u,\bar{v},w]$	½,½,½ $[\bar{u},v,w]$
Γ_{3g}	*Pnm′d′*	0,0,0 $[u,v,w]$	½,0,½ $[u,v,\bar{w}]$	0,½,0 $[u,\bar{v},w]$	½,½,½ $[u,\bar{v},\bar{w}]$
Γ_{4g}	*Pn′md′*	0,0,0 $[u,v,w]$	½,0,½ $[u,v,\bar{w}]$	0,½,0 $[\bar{u},v,\bar{w}]$	½,½,½ $[\bar{u},v,w]$

For example, in *Pn′m′a* the *u*-directions follow the pattern + − + − corresponding to the G_x mode. The *v*-directions follow + − − + corresponding to C_y and the *w*-directions follow + + + + corresponding to the ferromagnetic mode F_z. In *Pnm′d′*, we get for the *u*-direction + + + + corresponding to F_x, for the *v*-direction + + − − corresponding to A_y, and

Table 8.6 Spin transformations in *Pnma*

Representation	**Atom on 4*a* or 4*b* site**			**Atom in 4*c* site**			**Magnetic group**
Γ_{1g}	C_x	G_y	A_z	—	G_y	—	*Pnma*
Γ_{2g}	G_x	C_y	F_z	G_x	—	F_z	*Pn′m′a*
Γ_{3g}	F_x	A_y	C_z	F_x	—	G_z	*Pnm′d′*
Γ_{4g}	A_x	F_y	G_z	—	F_y	—	*Pn′md′*
Γ_{1u}	—	—	—	C_x	—	A_z	*Pn′m′d′*
Γ_{2u}	—	—	—	—	C_y	—	*Pnmd′*
Γ_{3u}	—	—	—	—	A_y	—	*Pn′ma*
Γ_{4u}	—	—	—	A_x	—	C_z	*Pnm′a*

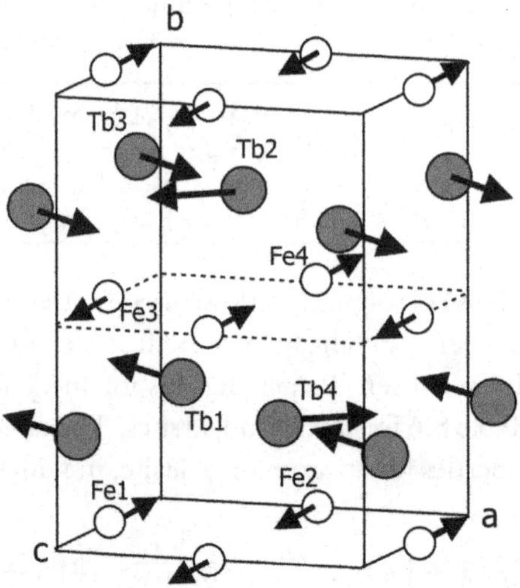

Figure 8.18 Magnetic structure of $TbFeO_3$ at 1.5 K. Tb is on 4*c* sites and Fe on 4*b*.[7]

for the *w*-direction + − + − corresponding to C_z. Finally, in $Pn'ma'$ the *u*-direction gives + + − − corresponding to A_x, the *v*-direction gives + + + + corresponding to F_y, and the *w*-direction gives + − − + corresponding to G_z.

An example of a magnetic structure in which the component modes are shown is that of $TbFeO_3$ measured at 1.5 K by neutron diffraction (Fig. 8.18). The atoms are labelled in the order given for the 4*b* and 4*c* sites as in the *ITA*. It was found that the Fe atoms in 4*b* sites have magnetic moments of 4.8 μ_B in the mode G_z in representation Γ_{4g} corresponding to the magnetic space group $Pn'ma'$. In addition, the Tb atoms, which are on 4*c* sites, have moments specified by the combination of two modes to form a non-collinear arrangement: C_zA_x in Γ_{4u} (space group $Pnm'a$) with $C_z = 6.51\ \mu_B$ and $A_x = 5.6\ \mu_B$, giving a total saturation moment of 8.6 μ_B.

Another interesting example is offered by the compound $CoAl_2O_4$, which crystallizes in the spinal structure in space group $Fd\bar{3}m(O_h^7)$. The cobalt atoms occupy 4*a* and 16*d* sites. We consider here only the Co atoms on 4*a* sites, which lie at the centers of oxygen tetrahedra.

Figure 8.19a shows the positions of these cobalt atoms in the unit cell together with the directions of magnetic moments as determined by neutron diffraction. Note that because all the moments are aligned along

7 E.F. Bertaut et al. *Solid State Comm.*, **5**, 293–298 (1967).

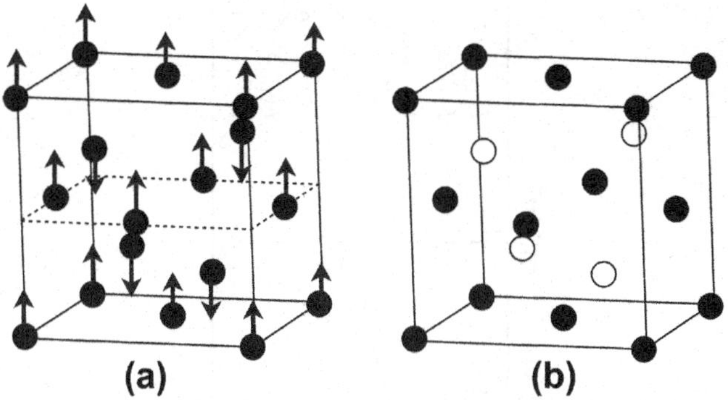

Figure 8.19 Antiferromagnetic structure of $CoAl_2O_4$: only Co on *4a* sites are shown (a) with spin vectors and (b) as black and white symmetry.

a single axis, the magnetic structure is no longer to be treated as cubic, but instead as tetragonal. From the *ITA*, we find that there is a Type I maximal subgroup $I4_1/amd(D_{4h}^{19})$. The character table for $4/mmm$ is given together with symbols for magnetic space groups based on $I4_1/amd$ (Table 8.7).

If we now consult Litvin's tables, we find that the arrangement of spins is given by magnetic group $I4_1'/a'md'$ (141.8.1220 in the Tables), that is representation Γ_{4u}, with the Co atoms on $4a$ sites:

$$0, 0, 0, [00w] \text{ and } 0, ½, ¼\ [00\overline{w}] + (½, ½, ½)$$

To emphasize the difference between magnetic and color symmetries, Fig. 8.19b shows the arrangement of Co atoms as black and white replacing the up and down spins, respectively, thus ignoring the effect of time-reversal symmetry on magnetic moment vectors. Check for yourself that this time, the symmetry remains cubic and is described in space group $Fd'\overline{3}'m$ with the atoms on $8a$ sites.

8.9. OG/BNS MAGNETIC GROUP SYMBOLS

The symbols that we have been using in this chapter are those used in Litvin's tables, which were introduced by W. Opechowski and R. Guccione[8] (OG symbols). However, N.V. Belov, N.N. Nerenova and T.S. Smirnova[9] adopted their own symbols (BNS symbols) for two-color groups, which in turn are sometimes used for three-dimensional magnetic group

[8] *Magnetism*, edited by G.T. Rado & H. Suhl, Vol. 2A, Ch. 3, New York: Academic Press (1965).
[9] *Sov. Phys. Crystallogr.* **1**, 487–488 (1955).

Table 8.7 Character table for $\mathbf{k}=0$ for $I4_1/amd$. Identity operation omitted

$I4_1/amd$	$4_1[001]$	2_1 [001]	2[010]	2[110]	$\bar{1}$	$\bar{4}$ [001]	*m*[001]	*m*[100]	m[110]	
Γ_{1g}	1	1	1	1	1	1	1	1	1	$I4_1/amd$
Γ_{2g}	1	1	−1	−1	1	1	1	−1	−1	$I4_1/am'd'$
Γ_{3g}	−1	1	1	−1	1	−1	1	1	−1	$I4_1'/amd'$
Γ_{4g}	−1	1	−1	1	1	−1	1	−1	1	$I4_1'/am'd$
Γ_{1u}	1	1	1	1	−1	−1	−1	−1	−1	$I4_1/d'm'd'$
Γ_{2u}	1	1	−1	−1	−1	−1	−1	1	1	$I4_1/d'm'd$
Γ_{3u}	−1	1	1	−1	−1	1	−1	−1	1	$I4_1'/d'm'd$
Γ_{4u}	−1	1	−1	1	−1	1	−1	1	−1	$I4_1'/d'md'$

types. As we saw earlier (Section 8.1), groups of the form **F**(**D**) were divided into $\mathbf{M_T}$ and $\mathbf{M_R}$ groups, where **D** is equitranslational or equi-class, respectively. The OG and BNS symbols are the same for the groups **F**, **F1′** and $\mathbf{M_T}$. However, in the case of $\mathbf{M_R}$ groups, the BNS symbol is based on the symbol for the **D** subgroup, whereas in OG it is based on the symbol for **F**.

For example, consider the space group with OG (Fig. 8.20) symbol $P_{2b}c'a'2_1$ (number 29.7.204 in Litvin).

The symmetry operations for this group are given by

$$\{1|0,0,0\} \quad \{m[100]|½,0,½\}' \quad \{m[010]|½,0,0\}' \quad \{2[001]|0,0,½\}$$

with the unprimed translations given by (**a**,2**b**,**c**). The group **F** is $Pca2_1$ whose operations are

$$\{1|0,0,0\} \quad \{m[100]|½,0,½\} \quad \{m[010]|½,0,0\} \quad \{2[001]|0,0,½\}$$

If we now consider the unprimed operations, the type IIb subgroup **D** is $Pna2_1$ (a non-magnetic subgroup of index 2) which has the operations

$$\{1|0,0,0\} \quad \{m[100]|½,1,½\} \quad \{m[010]|½,1,0\} \quad \{2[001]|0,0,½\}$$

The BNS symbol then is related to **D** as P_bna2_1. A full list of the OG and BNS symbols is given in Litvin's tables. Note that Grimmer[10] has published a critical comparison of the two systems of notation.

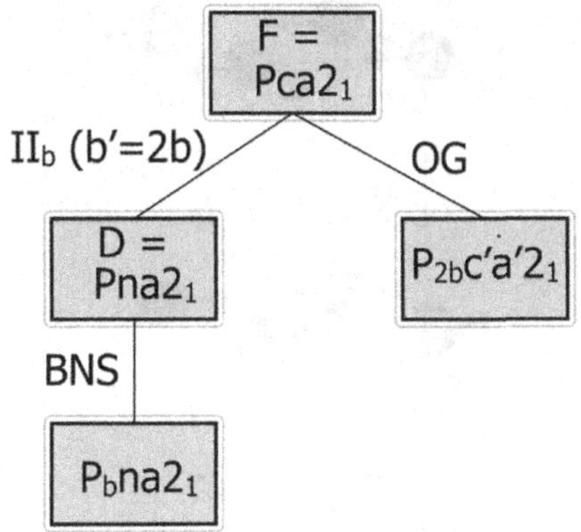

Figure 8.20 Flow chart for BNS and OG symbols.

[10] H. Grimmer. *Acta Cryst.*, **A65**, 145 (2009).

Problems

1. The diagrams below show some examples of Neolithic artwork. Identify the symbols for the black and white plane groups.

The black regions are now replaced by spin vectors pointing perpendicular to the planes (i.e. out of the page) in each case, while the white regions are replaced by spin vectors pointing in the opposite directions. What are the magnetic symmetries?

2. The heavy Fermion compound $CePt_3Si$ crystallizes in space group $P4mm(C_{4v}^1)$ with the atoms in the positions

Ce	1*a*	0, 0, 0
*Pt*1	2*c*	½, 0, 0.4036
*Pt*2	1*b*	½, ½, −0.1468
Si	1*b*	½, ½, 0.2650

It is found that at a certain temperature the magnetic structure is antiferromagnetic with the magnetic moments directed along ±[010] as in the following diagram:

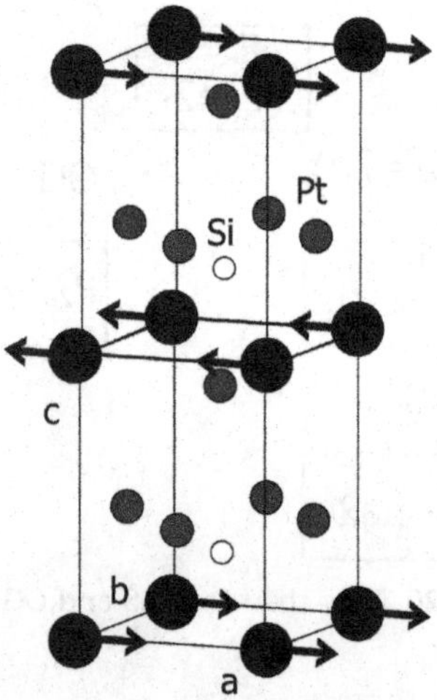

What is the space group symbol for this antiferromagnetic structure?

3. Use the group representation method to determine the distinct magnetic groups for the point group *6mm* and the space group *P2/c*.
4. A recent publication[11] of the magnetic space group for $CaFe_2As_2$ stated that the magnetic space group in BNS notation was B_Abcm, and this was illustrated by the following diagram. Identify the magnetic space group in OG notation.

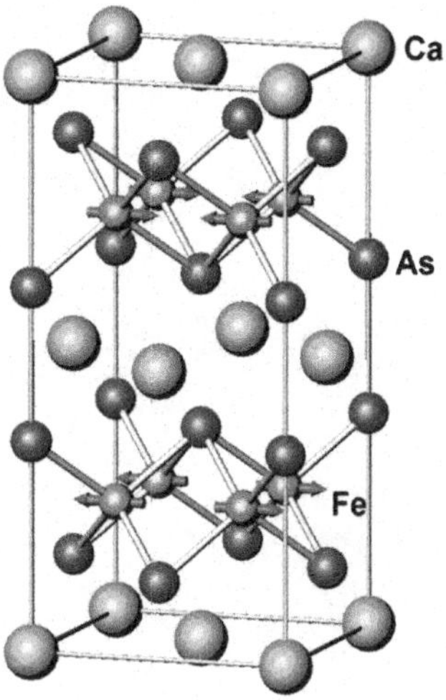

11 C.J. Howard & M.A. Carpenter (2012). *Acta Cryst.*, **B68**, 209.

APPENDIX 1

Matrices Representing the Symmetry Operations

The matrices given here are those for the active operation $\mathbf{r}' = R\mathbf{r}$. Thus multiplication of any coordinate position $\mathbf{r} \equiv (x, y, z)$ by the matrix, R, will give the symmetry-related coordinate. The matrices are grouped together according to the directions about which they operate. The directions used are those that are usually needed in problems related to crystal symmetry. The matrices applying *specifically* to the hexagonal (and trigonal) systems are marked with the letter H. The determinant of any of these matrices is always ±1, +1 for the proper operations and −1 for improper (or inversion) operations. This number taken together with the trace of each matrix (sum of diagonal elements) is characteristic of a particular point symmetry:

Operation	1	2	3	4	6	$\bar{1}$	m	$\bar{3}$	$\bar{4}$	$\bar{6}$
Trace	3	−1	0	1	2	−3	1	0	−1	−2
Determinant	1	1	1	1	1	−1	−1	−1	−1	−1

(Remember $\bar{2} = m$). If these two quantities are calculated for a particular matrix, it then makes it easier to find the operation in the following list.

Origin

$1(E)$	$\begin{bmatrix} 1 & 0 & 0 \\ 0 & 1 & 0 \\ 0 & 0 & 1 \end{bmatrix}$	$\bar{1}(i)$	$\begin{bmatrix} \bar{1} & 0 & 0 \\ 0 & \bar{1} & 0 \\ 0 & 0 & \bar{1} \end{bmatrix}$

Direction [1 0 0]

$2(C_2)$	$\begin{bmatrix} 1 & 0 & 0 \\ 0 & \bar{1} & 0 \\ 0 & 0 & \bar{1} \end{bmatrix}$	$\bar{2} \equiv m(\sigma)$	$\begin{bmatrix} \bar{1} & 0 & 0 \\ 0 & 1 & 0 \\ 0 & 0 & 1 \end{bmatrix}$

(Continued)

H $2(C_2)$	$\begin{bmatrix} 1 & \bar{1} & 0 \\ 0 & \bar{1} & 0 \\ 0 & 0 & \bar{1} \end{bmatrix}$	$\bar{2} \equiv m(\sigma)$	$\begin{bmatrix} \bar{1} & 1 & 0 \\ 0 & 1 & 0 \\ 0 & 0 & 1 \end{bmatrix}$
$4(C_4)$	$\begin{bmatrix} 1 & 0 & 0 \\ 0 & 0 & \bar{1} \\ 0 & 1 & 0 \end{bmatrix}$	$\bar{4}(S_4^3)$	$\begin{bmatrix} \bar{1} & 0 & 0 \\ 0 & 0 & 1 \\ 0 & \bar{1} & 0 \end{bmatrix}$
$4^2(C_4^2) = 2(C_2)$		$\bar{4}(S_4^2)^2 = 2(C_2)$	
$4^3(C_4^3)$	$\begin{bmatrix} 1 & 0 & 0 \\ 0 & 0 & 1 \\ 0 & \bar{1} & 0 \end{bmatrix}$	$\bar{4}^3(S_4^3)$	$\begin{bmatrix} \bar{1} & 0 & 0 \\ 0 & 0 & \bar{1} \\ 0 & 1 & 0 \end{bmatrix}$

Direction [0 1 0]

$2(C_2)$	$\begin{bmatrix} \bar{1} & 0 & 0 \\ 0 & 1 & 0 \\ 0 & 0 & \bar{1} \end{bmatrix}$	$\bar{2} \equiv m(\sigma)$	$\begin{bmatrix} 1 & 0 & 0 \\ 0 & \bar{1} & 0 \\ 0 & 0 & 1 \end{bmatrix}$
H $2(C_2)$	$\begin{bmatrix} \bar{1} & 0 & 0 \\ \bar{1} & 1 & 0 \\ 0 & 0 & \bar{1} \end{bmatrix}$	$\bar{2} \equiv m(\sigma)$	$\begin{bmatrix} 1 & 0 & 0 \\ 1 & \bar{1} & 0 \\ 0 & 0 & 1 \end{bmatrix}$
$4(C_4)$	$\begin{bmatrix} 0 & 0 & 1 \\ 0 & 1 & 0 \\ \bar{1} & 0 & 0 \end{bmatrix}$	$\bar{4}(S_4^3)$	$\begin{bmatrix} 0 & 0 & \bar{1} \\ 0 & \bar{1} & 0 \\ 1 & 0 & 0 \end{bmatrix}$
$4^2(C_4^2) = 2(C_2)$		$\bar{4}(S_4^2)^2 = 2(C_2)$	
$4^3(C_4^3)$	$\begin{bmatrix} 0 & 0 & \bar{1} \\ 0 & 1 & 0 \\ 1 & 0 & 0 \end{bmatrix}$	$\bar{4}^3(S_4^3)$	$\begin{bmatrix} 0 & 0 & 1 \\ 0 & \bar{1} & 0 \\ \bar{1} & 0 & 0 \end{bmatrix}$

Direction [0 0 1]

$2(C_2)$	$\begin{bmatrix} \bar{1} & 0 & 0 \\ 0 & \bar{1} & 0 \\ 0 & 0 & 1 \end{bmatrix}$	$\bar{2} \equiv m(\sigma)$	$\begin{bmatrix} 1 & 0 & 0 \\ 0 & 1 & 0 \\ 0 & 0 & \bar{1} \end{bmatrix}$

(Continued)

H $3(C_3)$	$\begin{bmatrix} 0 & \bar{1} & 0 \\ 1 & \bar{1} & 0 \\ 0 & 0 & 1 \end{bmatrix}$	$\bar{3}(S_6^5)$	$\begin{bmatrix} 0 & 1 & 0 \\ \bar{1} & 1 & 0 \\ 0 & 0 & \bar{1} \end{bmatrix}$
H $3^2(C_3^2)$	$\begin{bmatrix} \bar{1} & 1 & 0 \\ \bar{1} & 0 & 0 \\ 0 & 0 & 1 \end{bmatrix}$	$\bar{3}^2(S_6)$	$\begin{bmatrix} 1 & \bar{1} & 0 \\ 1 & 0 & 0 \\ 0 & 0 & \bar{1} \end{bmatrix}$
H		$\bar{3}^2(S_6^4) = 3^2(C_3^2)$	
H		$\bar{3}^3(S_6^3) = \bar{1}(i)$	
H		$\bar{3}^4(S_6^2) = 3\,(C_3)$	
$4(C_4)$	$\begin{bmatrix} 0 & \bar{1} & 0 \\ 1 & 0 & 0 \\ 0 & 0 & 1 \end{bmatrix}$	$\bar{4}(S_4^3)$	$\begin{bmatrix} 0 & 1 & 0 \\ \bar{1} & 0 & 0 \\ 0 & 0 & \bar{1} \end{bmatrix}$
$4^2(C_4^2) = 2(C_2)$		$\bar{4}^2(S_4^2) = 2(C_2)$	
$4^3(C_4^3)$	$\begin{bmatrix} 0 & 1 & 0 \\ \bar{1} & 0 & 0 \\ 0 & 0 & 1 \end{bmatrix}$	$\bar{4}^3(S_4^3)$	$\begin{bmatrix} 0 & \bar{1} & 0 \\ 1 & 0 & 0 \\ 0 & 0 & \bar{1} \end{bmatrix}$
H $6(C_6)$	$\begin{bmatrix} 1 & \bar{1} & 0 \\ 1 & 0 & 0 \\ 0 & 0 & 1 \end{bmatrix}$	$\bar{6}(S_3^5)$	$\begin{bmatrix} \bar{1} & 1 & 0 \\ \bar{1} & 0 & 0 \\ 0 & 0 & \bar{1} \end{bmatrix}$
H $6^2(C_6^2) = 3(C_3)$		$\bar{6}^2(S_3^4) = 3(C_3)$	
H $6^3(C_6^3) = 2(C_2)$		$\bar{6}(S_3^3) = m\,(\sigma)$	
H $6^4(C_6^4) = 3^2(C_3^2)$		$\bar{6}^4(S_3^2) = 3^2(C_3^2)$	
H $6^5(C_6^5)$	$\begin{bmatrix} 0 & 1 & 0 \\ \bar{1} & 1 & 0 \\ 0 & 0 & 1 \end{bmatrix}$	$\bar{6}^5(S_3)$	$\begin{bmatrix} 0 & \bar{1} & 0 \\ 1 & \bar{1} & 0 \\ 0 & 0 & \bar{1} \end{bmatrix}$

Direction [1 1 0]

$2(C_2)$	$\begin{bmatrix} 0 & 1 & 0 \\ 1 & 0 & 0 \\ 0 & 0 & \bar{1} \end{bmatrix}$	$\bar{2} \equiv m(\sigma)$	$\begin{bmatrix} 0 & \bar{1} & 0 \\ \bar{1} & 0 & 0 \\ 0 & 0 & 1 \end{bmatrix}$

Direction [1 0 1]

$2(C_2)$	$\begin{bmatrix} 0 & 0 & 1 \\ 0 & \bar{1} & 0 \\ 1 & 0 & 0 \end{bmatrix}$	$\bar{2} \equiv m(\sigma)$	$\begin{bmatrix} 0 & 0 & \bar{1} \\ 0 & 1 & 0 \\ \bar{1} & 0 & 0 \end{bmatrix}$

Direction [0 1 1]

$2(C_2)$	$\begin{bmatrix} \bar{1} & 0 & 0 \\ 0 & 0 & 1 \\ 0 & 1 & 0 \end{bmatrix}$	$\bar{2} \equiv m(\sigma)$	$\begin{bmatrix} 1 & 0 & 0 \\ 0 & 0 & \bar{1} \\ 0 & \bar{1} & 0 \end{bmatrix}$

Direction [1 $\bar{1}$ 0]

$2(C_2)$	$\begin{bmatrix} 0 & \bar{1} & 0 \\ \bar{1} & 0 & 0 \\ 0 & 0 & \bar{1} \end{bmatrix}$	$\bar{2} \equiv m(\sigma)$	$\begin{bmatrix} 0 & 1 & 0 \\ 1 & 0 & 0 \\ 0 & 0 & 1 \end{bmatrix}$

Direction [$\bar{1}$ 0 1]

$2(C_2)$	$\begin{bmatrix} 0 & 0 & \bar{1} \\ 0 & \bar{1} & 0 \\ \bar{1} & 0 & 0 \end{bmatrix}$	$\bar{2} \equiv m(\sigma)$	$\begin{bmatrix} 0 & 0 & 1 \\ 0 & 1 & 0 \\ 1 & 0 & 0 \end{bmatrix}$

Direction [0 1 $\bar{1}$]

$2(C_2)$	$\begin{bmatrix} \bar{1} & 0 & 0 \\ 0 & 0 & \bar{1} \\ 0 & \bar{1} & 0 \end{bmatrix}$	$\bar{2} \equiv m(\sigma)$	$\begin{bmatrix} 1 & 0 & 0 \\ 0 & 0 & 1 \\ 0 & 1 & 0 \end{bmatrix}$

Direction [1 1 1]

$3(C_3)$	$\begin{bmatrix} 0 & 0 & 1 \\ 1 & 0 & 0 \\ 0 & 1 & 0 \end{bmatrix}$	$\bar{3}(S_6^5)$	$\begin{bmatrix} 0 & 0 & \bar{1} \\ \bar{1} & 0 & 0 \\ 0 & \bar{1} & 0 \end{bmatrix}$
$3^2(C_3^2)$	$\begin{bmatrix} 0 & 0 & 1 \\ 1 & 0 & 0 \end{bmatrix}$	$\bar{3}^2(S_6)$	$\begin{bmatrix} 0 & 0 & \bar{1} \\ \bar{1} & 0 & 0 \end{bmatrix}$
		$\bar{3}^2(S_6^4) = 3^2(C_3^2)$	
		$\bar{3}^3(S_6^3) = \bar{1}(i)$	
		$\bar{3}^4(S_6^2) = 3(C_3)$	

Direction $[\bar{1}\ 1\ 1]$

$3(C_3)$	$\begin{bmatrix} 0 & \bar{1} & 0 \\ 0 & 0 & 1 \\ \bar{1} & 0 & 0 \end{bmatrix}$	$\bar{3}(S_6^5)$	$\begin{bmatrix} 0 & 1 & 0 \\ 0 & 0 & \bar{1} \\ 1 & 0 & 0 \end{bmatrix}$
$3^2(C_3^2)$	$\begin{bmatrix} 0 & 0 & \bar{1} \\ \bar{1} & 0 & 0 \\ 0 & 1 & 0 \end{bmatrix}$	$\bar{3}^2(S_6)$	$\begin{bmatrix} 0 & 0 & 1 \\ 1 & 0 & 0 \\ 0 & \bar{1} & 0 \end{bmatrix}$
		$\bar{3}^2(S_6^4) = 3^2(C_3^2)$	
		$\bar{3}^3(S_6^3) = \bar{1}(i)$	
		$\bar{3}^4(S_6^2) = 3(C_3)$	

Direction $[1\ \bar{1}\ 1]$

$3(C_3)$	$\begin{bmatrix} 0 & \bar{1} & 0 \\ 0 & 0 & \bar{1} \\ 1 & 0 & 0 \end{bmatrix}$	$\bar{3}(S_6^5)$	$\begin{bmatrix} 0 & 1 & 0 \\ 0 & 0 & 1 \\ \bar{1} & 0 & 0 \end{bmatrix}$

(*Continued*)

$3^2(C_3^2)$	$\begin{bmatrix} 0 & 0 & 1 \\ \bar{1} & 0 & 0 \\ 0 & \bar{1} & 0 \end{bmatrix}$	$\bar{3}^2(S_6)$	$\begin{bmatrix} 0 & 0 & \bar{1} \\ 1 & 0 & 0 \\ 0 & 1 & 0 \end{bmatrix}$

$\bar{3}^2(S_6^4) = 3^2(C_3^2)$
$\bar{3}^3(S_6^3) = \bar{1}(i)$
$\bar{3}^4(S_6^2) = 3(C_3)$

Direction [1 1 $\bar{1}$]

$3(C_3)$	$\begin{bmatrix} 0 & 1 & 0 \\ 0 & 0 & \bar{1} \\ \bar{1} & 0 & 0 \end{bmatrix}$	$\bar{3}(S_6^5)$	$\begin{bmatrix} 0 & \bar{1} & 0 \\ 0 & 0 & 1 \\ 1 & 0 & 0 \end{bmatrix}$
$3^2(C_3^2)$	$\begin{bmatrix} 0 & 0 & \bar{1} \\ 1 & 0 & 0 \\ 0 & \bar{1} & 0 \end{bmatrix}$	$\bar{3}^2(S_6)$	$\begin{bmatrix} 0 & 0 & 1 \\ \bar{1} & 0 & 0 \\ 0 & 1 & 0 \end{bmatrix}$

$\bar{3}^2(S_6^4) = 3^2(C_3^2)$
$\bar{3}^3(S_6^3) = \bar{1}(i)$
$\bar{3}^4(S_6^2) = 3(C_3)$

Direction [2 1 0]

H $2(C_2)$	$\begin{bmatrix} 1 & 0 & 0 \\ 1 & \bar{1} & 0 \\ 0 & 0 & \bar{1} \end{bmatrix}$	$\bar{2} \equiv m(\sigma)$	$\begin{bmatrix} \bar{1} & 0 & 0 \\ \bar{1} & 1 & 0 \\ 0 & 0 & 1 \end{bmatrix}$

Direction [1 2 0]

H $2(C_2)$	$\begin{bmatrix} \bar{1} & 1 & 0 \\ 0 & 1 & 0 \\ 0 & 0 & \bar{1} \end{bmatrix}$	$\bar{2} \equiv m(\sigma)$	$\begin{bmatrix} 1 & \bar{1} & 0 \\ 0 & \bar{1} & 0 \\ 0 & 0 & 1 \end{bmatrix}$

JONES' FAITHFUL REPRESENTATION SYMBOLS

The Jones symbol for the operator R is the vector $\mathbf{r}'$ formed by $R\mathbf{r}$, and is recognizable as one of the coordinates of the equivalent positions. It is easily obtained from any of the above matrices by multiplying with the vector $\mathbf{r} \equiv (x, y, z)$. For example, the inverse operator would be represented by $(-x, -y, -z)$, and 2[100] (C_2[100]) would be $(x, -y, -z)$ and so on. The Jones symbol is then simply an abbreviated form of the matrix operator. Multiplication of these symbols is similar to that for matrices. For example, if we wish to find $\{2[001]\}\{m[001]\}$ ($\{C_2[001]\}\{\sigma_h\}$) we can write the Jones' symbols:

$$2[001]m[001] \equiv (-x, -y, z)(x, y, -z) = (-x, -y, -z)$$

and hence $2[001]m[001] = \bar{1}$.

APPENDIX 2

Crystal Families, Systems, and Bravais Lattices

Crystal family	Symbol	Crystal system	Point groups*	Unit-cell restrictions	Bravais lattices
One dimension					
—	—	—	1, **m**	None	*p*
Two dimensions					
Oblique (monoclinic)	*m*	Oblique	1, **2**	None	*mp*
Rectangular (orthorhombic)	*o*	Rectangular	*m*, **2mm**	$\gamma = 90°$	*op* *oc*
Square (tetragonal)	*t*	Square	**4**, **4mm**	$a = b, \gamma = 90°$	*tp*
Hexagonal	*h*	Hexagonal	3, **6** 3*m*, **6mm**	$a = b, \gamma = 120°$	*hp*
Three dimensions					
Triclinic (anorthic)	*a*	Triclinic	1, $\mathbf{\bar{1}}$	None	*aP*
Monoclinic	*m*	Monoclinic	2, *m*, **2/m**	First setting $\alpha = \beta = 90°$ Second setting $\alpha = \gamma = 90°$	*mP* *mS (mA, mB, mI)* *mP* *mS (mA, mB, mI)*
Orthorhombic	*o*	Orthorhombic	222, *mm*2, **mmm**	$\alpha = \beta = \gamma = 90°$	*oP* *oS (oC, oA, oB)* *oI* oF
Tetragonal	*t*	Tetragonal	4, $\mathbf{\bar{4}}$, **4/m** 422, 4*mm*, $\bar{4}2m$, **4/mmm**	$a = b, \alpha = \beta = \gamma = 90°$	*tP* *tI*
Hexagonal	*h*	Trigonal	3, $\mathbf{\bar{3}}$ 32, 3*m*, $\mathbf{\bar{3}m}$	$a = b$, $\alpha = \beta = 90°$, $\gamma = 120°$	*hP*

(*Continued*)

Crystal family	Symbol	Crystal system	Point groups*	Unit-cell restrictions	Bravais lattices
				$a = b = c$, $\alpha = \beta = \gamma$ (rhombohedral axes, primitive cell) $a = b$, $\alpha = \beta = 90°$, $\gamma = 120°$ (hexagonal axes, triple obverse cell)	*hR*
		Hexagonal	6, $\mathbf{\bar{6}}$, $\mathbf{6/m}$ 622, 6*mm*, $\bar{6}2m$, $\mathbf{6/mmm}$	$a = b$, $\alpha = \beta = 90°$, $\gamma = 120°$	*hP*
Cubic	*c*	Cubic	23, $\mathbf{m\bar{3}}$ 432, $\bar{4}3m$, $\mathbf{m\bar{3}m}$	$a = b = c$, $\alpha = \beta = \gamma = 90°$	*cP* *cI* *cF*

*Symbols in bold are Laue groups.

APPENDIX 3

The 14 Bravais Lattices

These are the conventional unit cells of the 14 Bravais lattices. For the hexagonal lattice, the unit cell is outlined in black. A hexagonal prism is also shown, dotted; this is drawn only to help indicate the angles of the unit cell. For the monoclinic lattice, a *B*-centered cell is shown. This is for the first setting where the **c**-axis is taken as the unique 2-fold axis. For the second setting, the **b**-axis is the unique 2-fold axis and then the cell is C-centered.

24 WIGNER–SEITZ CELLS

In real space, the conventional parallepiped unit cells for the lattice are normally used in order to specify the coordinate positions of atoms making up the crystal structure. Although they may be multiply-primitive, these cells show, at a glance, the full symmetry of the lattice. However, as discussed in Section 3.4 (Fig. 3-10), the Wigner–Seitz construction is a useful way of defining a primitive unit cell which also shows the full symmetry of the lattice. The Wigner–Seitz cell can, of course, be used in direct space, although it is generally not a convenient shape with which to specify the positions of the atoms making up the crystal structure. However, it is more often used in reciprocal space to represent the first Brillouin zone, where the main purpose is to enable quantum states (typically electronic or vibrational) to be enumerated.

Corresponding to the 14 Bravais lattices, we expect to find 14 first Brillouin zones, one Wigner–Seitz cell for each lattice type. However, when topology is taken into account, it turns out that there are 24 distinct topological types of Wigner–Seitz cell. Figures 3-10c and d show this effect where the c/a ratio determines two topologically different Wigner–Seitz cells for a body-centered tetragonal lattice. The 24 topologically different Wigner–Seitz cells are shown below. For the five Wigner–Seitz, base-centered monoclinic cells, we have picked the first setting (**c** is unique) and have chosen *B*-centering (½, 0, ½). For this setting, $\gamma \neq 90^\circ$ is chosen; the other two unmarked angles are 90°.

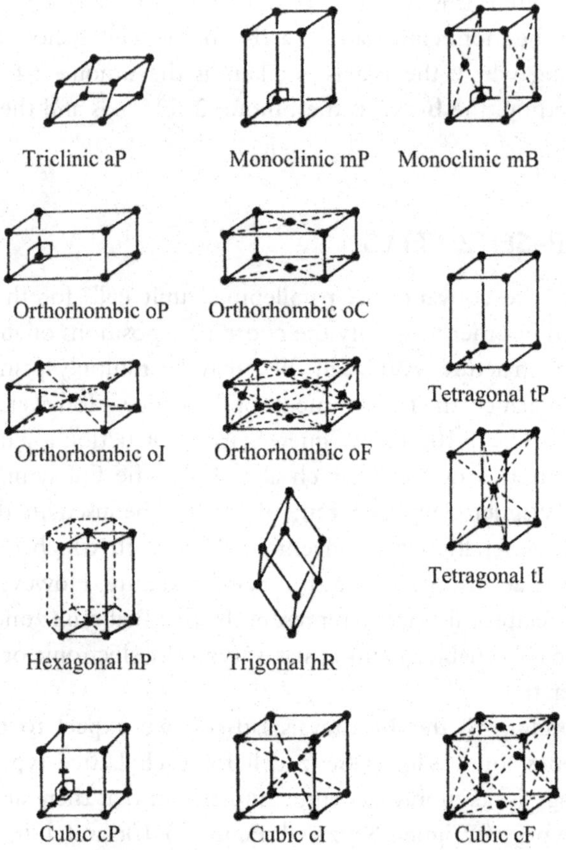
Triclinic aP
Monoclinic mP
Monoclinic mB
Orthorhombic oP
Orthorhombic oC
Tetragonal tP
Orthorhombic oI
Orthorhombic oF
Tetragonal tI
Hexagonal hP
Trigonal hR
Cubic cP
Cubic cI
Cubic cF

Wigner-Seitz cells

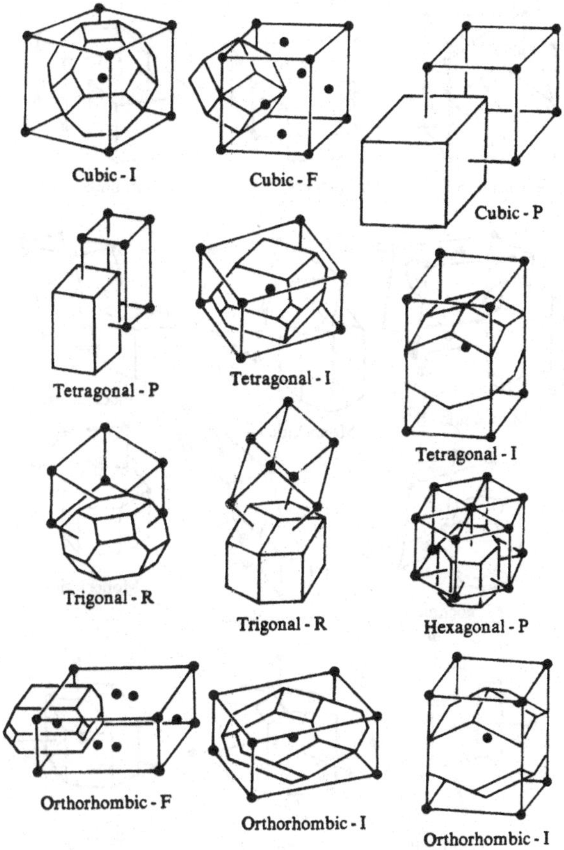

Wigner-Seitz cells

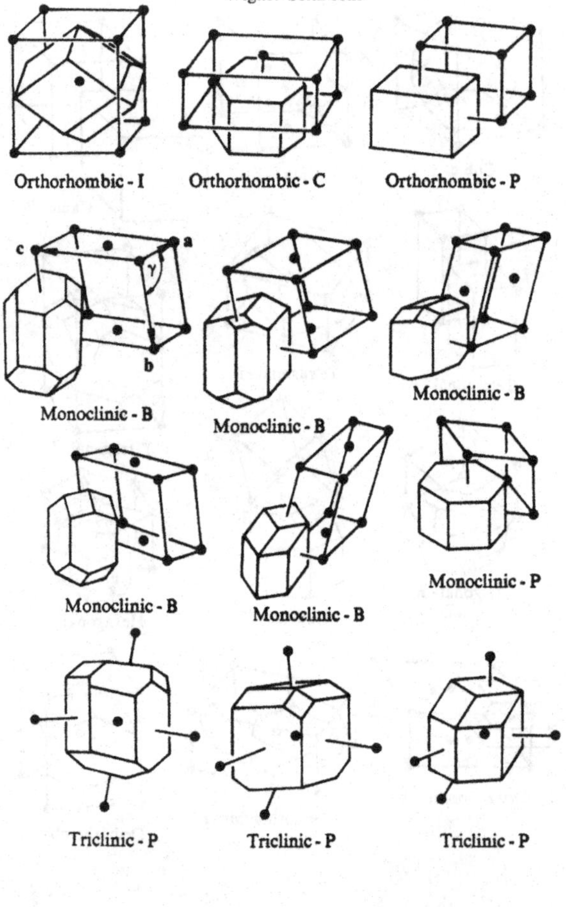

APPENDIX 4

The 32 Crystallographic Point Groups

The 32 crystallographic point groups are listed below using the Schoenflies, International and Full International notation. The symmetry operations, grouped by classes, as well as the space group numbers that correspond to each point group are shown.

Schoenflies	International	Full international symbol	Symmetry operations	Space group numbers
Triclinic				
C_1	1	1	E	1
$S_2(C_i)$	$\bar{1}$	$\bar{1}$	$E\ i$	2
Monoclinic				
C_2	2	2	$E\ C_2$	3–5
$C_{1h}(C_s)$	m	m	$E\ \sigma_h$	6–9
C_{2h}	$2/m$	$\frac{2}{m}$	$E\ C_2\ i\ \sigma_h$	10–15
Orthorhombic				
$D_2(V)$	222	222	$E\ C_2\ C_2'\ C_2''$	16–24
C_{2v}	$mm2$	$mm2$	$E\ C_2\ \sigma_v\ \sigma_v$	25–46
$D_{2h}(V_h)$	mmm	$\frac{2}{m}\frac{2}{m}\frac{2}{m}$	$E\ C_2\ C_2'C_2''\ i\ \sigma_h\ \sigma_v\ \sigma_v$	47–74
Tetragonal				
C_4	4	4	$E\ 2C_4\ C_2$	75–80
S_4	$\bar{4}$	$\bar{4}$	$E\ 2S_4\ C_2$	81–82
C_{4h}	$4/m$	$\frac{4}{m}$	$E\ 2C_4\ C_2\ i\ 2S_4\ \sigma_h$	83–88
D_4	422	422	$E\ 2C_4\ C_2\ 2C_2'\ 2C_2''$	89–98
C_{4v}	$4mm$	$4mm$	$E\ 2C_4\ C_2\ 2\sigma_v\ 2\sigma_d$	99–110
$D_{2d}(V_d)$	$\bar{4}2m$	$\bar{4}\ 2m$	$E\ C_2\ 2C_2'\ 2\sigma_d\ 2S_4$	111–122
D_{4h}	$4/mmm$	$\frac{4}{m}\frac{2}{m}\frac{2}{m}$	$E\ 2C_4\ C_2\ 2C_2'\ 2C_2''$ $i\ 2S_4\ \sigma_h\ 2\sigma_v\ 2\sigma_d$	123–142

(Continued)

Schoenflies	International	Full international symbol	Symmetry operations	Space group numbers
Trigonal (Rhombohedral)				
C_3	3	3	$E\ 2C_3$	143–146
$S_6(C_{3i})$	$\bar{3}$	$\bar{3}$	$E\ 2C_3\ i\ 2S_6$	147–148
D_3	32	32	$E\ 2C_3\ 3C_2$	149–155
C_{3v}	$3m$	$3m$	$E\ 2C_3\ 3\sigma_v$	156–161
D_{3d}	$\bar{3}m$	$\bar{3}\frac{2}{m}$	$E\ 2C_3\ 3C_2\ i\ 2S_6\ 3\sigma_v$	162–167
Hexagonal				
C_6	6	6	$E\ 2C_6\ 2C_3\ C_2$	168–173
C_{3h}	$\bar{6}$	$\bar{6}$	$E\ 2C_3\ \sigma_h\ 2S_3$	174
C_{6h}	$6/m$	$\frac{6}{m}$	$E\ 2C_6\ 2C_3\ C_2\ i\ 2S_3\ 2S_6\ \sigma_h$	175–176
D_6	622	622	$E\ 2C_6\ 2C_3\ C_2\ 3C_2'3C_2''$	177–182
C_{6v}	$6mm$	$6mm$	$E\ 2C_6\ 2C_3\ C_2\ 3\sigma_v\ 3\sigma_d$	183–186
D_{3h}	$\bar{6}m2$	$\bar{6}m2$	$E\ 2C_3\ 3C_2\ \sigma_h\ 2S_3\ 3\sigma_v$	187–190
D_{6h}	$6/mmm$	$\frac{6}{m}\frac{2}{m}\frac{2}{m}$	$E\ 2C_6\ 2C_3\ C_2\ 3C_2'\ 3C_2''$ $i\ 2S_3\ 2S_6\ \sigma_h\ 3\sigma_v\ 3\sigma_d$	191–194
Cubic				
T	23	23	$E\ 8C_3\ 3C_2$	195–199
T_h	$m\bar{3}$	$\frac{2}{m}\bar{3}$	$E\ 8C_3\ 3C_2\ i\ 8S_6\ 3\sigma_h$	200–206
O	432	432	$E\ 8C_3\ 3C_2\ 6C_2\ 6C_4$	207–241
T_d	$\bar{4}3m$	$\bar{4}3m$	$E\ 8C_3\ 3C_2\ 6\sigma_d\ 6S_4$	215–220
O_h	$m\bar{3}m$	$\frac{4}{m}\bar{3}\frac{2}{m}$	$E\ 8C_3\ 3C_2\ 6C_2\ 6C_4$ $i\ 8S_6\ 3\sigma_h\ 6\sigma_d\ 6S_4$	221–230

APPENDIX 5

Diagrams for the 32 Point Groups

STEREOGRAMS

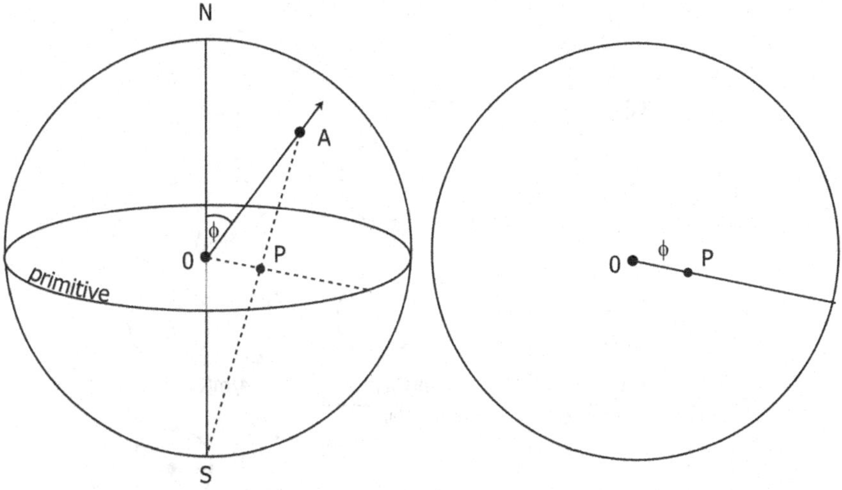

A stereogram is a useful way of visualizing the effects of symmetry operations on molecules and crystals. It is defined as follows. A unit sphere is drawn around the point O at its centre (left). A point in the northern hemisphere A is projected onto the equatorial plane (bounded by the so-called **primitive circle**) by drawing a line from A to the south pole S. This intersects the equatorial plane at P. This plane is then viewed down from the north pole (right) and the distance OP is a measure of the angle NOA $= \phi$ from the north pole. Similarly, points in the southern hemisphere are marked on the equatorial plane by projecting a line to the north pole (not shown in the diagram). The points marked on the equatorial plane are called **poles**. Thus, any two poles drawn on the stereogram are related by angles measured along great circles. In the stereograms that follow, a dot (•) represents a pole above the plane of projection while a circle (o) represents the one that is below the plane of projection. Their locations on the stereograms are determined by the symmetry operation shown on the right-hand side in each case.

SOME SHAPES ILLUSTRATING THE 32 POINT GROUPS

Also shown in this appendix is a collection of scroll shapes each having the point symmetry of one of the 32 crystallographic point groups (taken from Weinreich, p. 25).

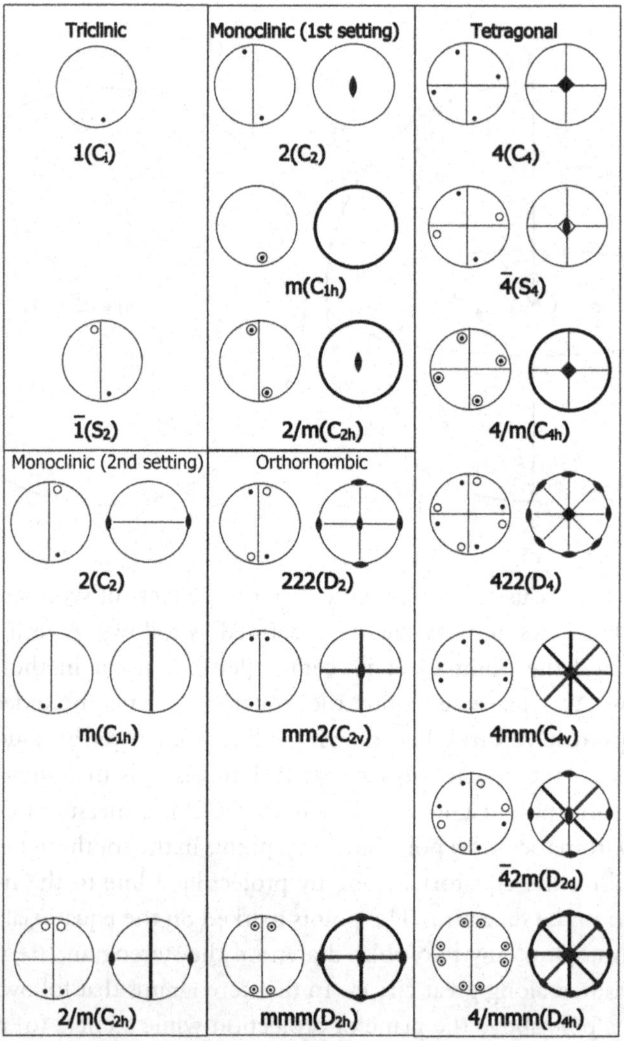

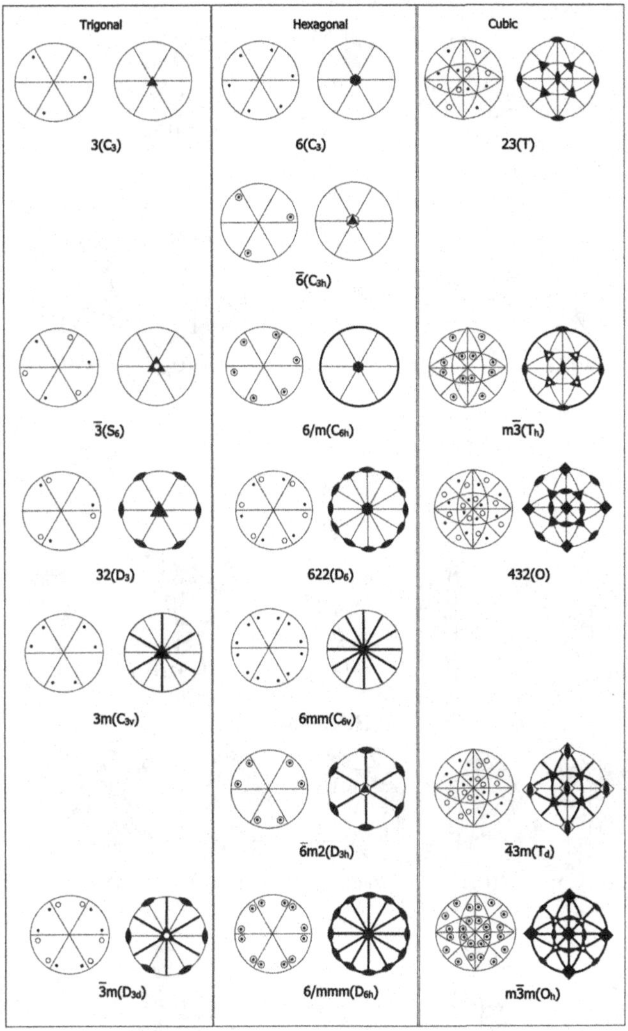

Trigonal
Hexagonal
Cubic
3(C3)
6(C3)
23(T)
6̄(C3h)
3̄(S6)
6/m(C6h)
m3̄(Th)
32(D3)
622(D6)
432(O)
3m(C3v)
6mm(C6v)
6̄m2(D3h)
4̄3m(Td)
3̄m(D3d)
6/mmm(D6h)
m3̄m(Oh)

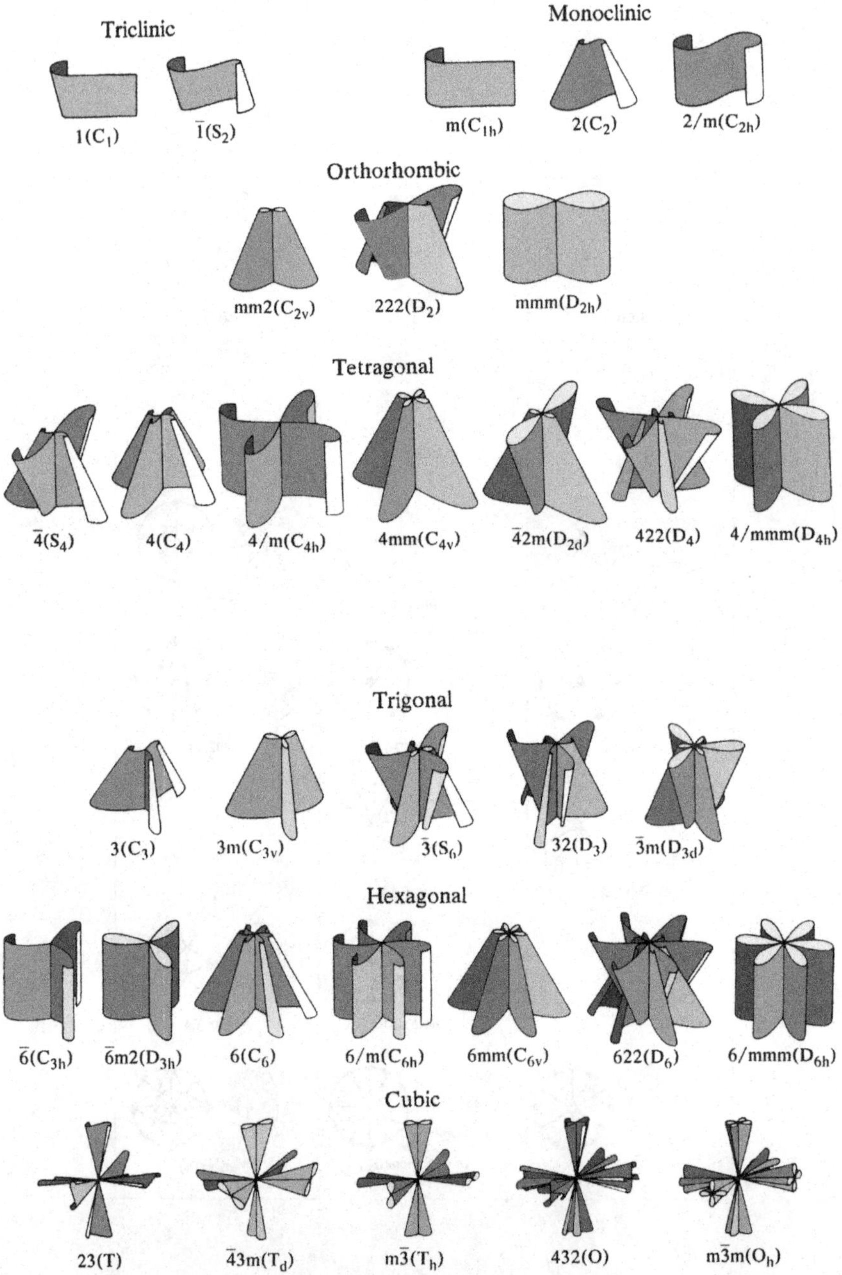
Triclinic
$1(C_1)$
$\bar{1}(S_2)$
Monoclinic
$m(C_{1h})$
$2(C_2)$
$2/m(C_{2h})$
Orthorhombic
$mm2(C_{2v})$
$222(D_2)$
$mmm(D_{2h})$
Tetragonal
$\bar{4}(S_4)$
$4(C_4)$
$4/m(C_{4h})$
$4mm(C_{4v})$
$\bar{4}2m(D_{2d})$
$422(D_4)$
$4/mmm(D_{4h})$
Trigonal
$3(C_3)$
$3m(C_{3v})$
$\bar{3}(S_6)$
$32(D_3)$
$\bar{3}m(D_{3d})$
Hexagonal
$\bar{6}(C_{3h})$
$\bar{6}m2(D_{3h})$
$6(C_6)$
$6/m(C_{6h})$
$6mm(C_{6v})$
$622(D_6)$
$6/mmm(D_{6h})$
Cubic
$23(T)$
$\bar{4}3m(T_d)$
$m\bar{3}(T_h)$
$432(O)$
$m\bar{3}m(O_h)$

APPENDIX 6

Symbols

SYMBOLS OF SYMMETRY PLANES

(a) Symmetry planes normal to the plane of projection (three dimensions) and symmetry lines in the plane of the figure (two dimensions)

Symmetry plane or symmetry line	Graphical symbol	Glide vector in units of lattice translation vectors parallel and normal to the projection plane	Printed symbol
Reflection plane, mirror plane Reflection line, mirror line (two dimensions)	———	None	*m*
'Axial' glide plane Glide line (two dimensions)	– – –	½ along line parallel to projection plane ½ along line in plane	*a, b* or *c* *g*
'Axial' glide plane	·········	½ normal to projection plane	*a, b* or *c*
'Diagonal' glide plane	-·-·-·-·-	½ along line parallel to projection plane, combined with ½ normal to projection plane	*n*
'Diamond' glide plane (pair of planes; in centred cells only)	–·–◄·– –·►–·–	¼ along line parallel to projection plane, combined with ¼ normal to projection plane (arrow indicates direction parallel to the projection plane for which the normal component is positive)	*d*
'Double' glide plane	··–··–··–··	Two glide vectors: ½ along line parallel to projection plane ½ normal to projection plane	*e*

For symbols used in cubic space groups see the *ITA* (Tables 1.4.3, 1.4.6 and 1.4.7).

(b) Symmetry planes parallel to the plane of projection

Symmetry plane	Graphical symbol	Glide vector in units of lattice translation vectors parallel to the projection plane	Printed symbol
Reflection plane, mirror plane		None	*m*
'Axial' glide plane	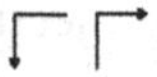	½ in the direction of the arrow	*a*, *b* or *c*
'Double' glide plane		½ in either of the directions of the two arrows	*e*
'Diagonal' glide plane		½ in the direction of the arrow	*n*
'Diamond' glide plane (pair of planes; in centred cells only)	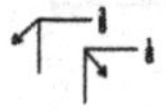	½ in the direction of the arrow; the glide vector is always half of a centring vector, i.e. one quarter of a diagonal of the conventional cell	*d*

SYMBOLS OF SYMMETRY AXES

Symmetry axes normal to the plane of projection (three dimensions) and symmetry points in the plane of the figure (two dimensions)

Symmetry axis or symmetry point	Graphical symbol	Screw vector of a right-handed screw rotation in units of the shortest lattice translation vector parallel to the axis	Printed symbol
Identity	None	None	1
Twofold rotation axis Twofold rotation point (two dimensions)		None	2
Twofold screw axis: '2 sub 1'		$^1/_2$	2_1
Threefold rotation axis Threefold rotation point (two dimensions)		None	3
Threefold screw axis: '3 sub 1'		$^1/_3$	3_1

(*Continued*)

Symmetry axis or symmetry point	Graphical symbol	Screw vector of a right-handed screw rotation in units of the shortest lattice translation vector parallel to the axis	Printed symbol
Threefold screw axis: '3 sub 2'		2/3	3_2
Fourfold rotation axis Fourfold rotation point (two dimensions)		None	4
Fourfold screw axis: '4 sub 1'		1/4	4_1
Fourfold screw axis: '4 sub 2'		1/2	4_2
Fourfold screw axis: '4 sub 3'		3/4	4_3
Sixfold rotation axis Sixfold rotation point (two dimensions)		None	6
Sixfold screw axis: '6 sub 1'		1/6	6_1
Sixfold screw axis: '6 sub 2'		1/3	6_2
Sixfold screw axis: '6 sub 3'		1/2	6_3
Sixfold screw axis: '6 sub 4'		2/3	6_4
Sixfold screw axis: '6 sub 5'		5/6	6_5
Centre of symmetry, inversion centre: '1 bar' Reflection point, mirror point (one dimension)		None	$\bar{1}$
Inversion axis: '3 bar'		None	$\bar{3}$
Inversion axis: '4 bar'		None	$\bar{4}$
Inversion axis: '6 bar'		None	$\bar{6}$
Twofold rotation axis with centre of symmetry		None	$2/m$

(*Continued*)

Symmetry axis or symmetry point	Graphical symbol	Screw vector of a right-handed screw rotation in units of the shortest lattice translation vector parallel to the axis	Printed symbol
Twofold screw axis with centre of symmetry		1/2	$2_1/m$
Fourfold rotation axis with centre of symmetry		None	$4/m$
'4 sub 2' screw axis with centre of symmetry		1/2	$4_2/m$
Sixfold rotation axis with centre of symmetry		None	$6/m$
'6 sub 3' screw axis with centre of symmetry		1/2	$6_3/m$

ORDER OF SYMBOLS

This table summarizes the order of positions in the three-dimensional space group or point group symbols. Thus, for example, in the orthorhombic system, *mm*2 refers to mirror planes *perpendicular* to **a** and **b**, and a 2-fold axis *along* **c**. Similarly, in the hexagonal system $\bar{6}m2$ means a $\bar{6}$ axis along **c**, mirror planes *perpendicular* to **a**, **b** and **a** + **b** ([110]), and twofold axes *perpendicular* to **a**, **b** and **a** + **b**. Finally, in the cubic system, $\bar{4}3\mathrm{m}$ means $\bar{4}$ axes *along* the three cube axes, **a**, **b** and **c**, four 3-fold axes *along* the body diagonals <111> and diagonal mirror planes *perpendicular* to <110>. In $m\bar{3}$ the first position tells us that there are mirror planes *perpendicular* to the cube axes and the second position tells us that there are 3-fold axes *along* the body diagonals. There is nothing in the third position, as in this case there is no symmetry operation acting about the <110> directions, except, of course, for the trivial identity operation.

THREE-DIMENSIONAL LATTICES

Lattice	Position in International Symbol		
	1	2	3
Triclinic	Only one symbol used		
Monoclinic	First setting: **c**-axis unique Second setting: **b**-axis unique		
Orthorhombic	2 or $\bar{2}$ along **a**	2 or $\bar{2}$ along **b**	2 or $\bar{2}$ along **c**
Tetragonal	4 or $\bar{4}$ along **c**	2 or $\bar{2}$ along **a** and **b**	2 or $\bar{2}$ along [110] and $[1\bar{1}0]$
Trigonal	3 or $\bar{3}$ along **c**	2 or $\bar{2}$ along **a**, **b** and [110]	2 and $\bar{2}$ perpendicular to **a**, **b** and [110]
Hexagonal	6 or $\bar{6}$ along **c**	2 or $\bar{2}$ along **a**, **b** and [110]	2 or $\bar{2}$ perpendicular to **a**, **b** and [110]
Cubic	4, $\bar{4}$, 2 or $\bar{2}$ along **a**, **b** and **c**	3 or $\bar{3}$ along <111>	2 or $\bar{2}$ along <110>

Remember: $\bar{2}$ is equivalent to a mirror plane *m* perpendicular to $\bar{2}$ axis.

TWO-DIMENSIONAL LATTICES

Lattice	Position in International Symbol		
	1	2	3
Oblique	Rotation point in plane		
Rectangular	"	[10]	[01]
Square	"	[10] [01]	$[1\bar{1}]$ [11]
Hexagonal	"	[10] [01] $[\bar{1}\bar{1}]$	$[1\bar{1}]$ [12] $[\bar{2}\bar{1}]$

APPENDIX 7

The Space Groups

11 ENANTIOMORPHIC SPACE GROUP PAIRS

The 11 enantiomorphic space group pairs are listed in the International and Schoenflies notation. (Naturally, these space groups cannot contain a mirror plane).

$P3_1$	$P3_2$	C_3^2	C_3^3
$P4_1$	$P4_3$	C_4^2	C_4^4
$P6_1$	$P6_5$	C_6^2	C_6^3
$P6_2$	$P6_4$	C_6^4	C_6^5
$P3_112$	$P3_212$	D_3^3	D_3^5
$P3_121$	$P3_221$	D_3^4	D_3^6
$P4_122$	$P4_322$	D_3^4	D_4^7
$P4_12_12$	$P4_32_12$	D_4^4	D_4^8
$P6_122$	$P6_522$	D_6^2	D_6^3
$P6_422$	$P6_222$	D_6^5	D_6^6
$P4_332$	$P4_132$	O^6	O^7

THE 230 SPACE GROUPS

The 230 space groups are listed with a number and the Schoenflies, international standard short and international standard full symbol. A line under the number indicates that the space group is symmorphic.

TRICLINIC SYSTEM

No. of space group	Schoenflies symbol	
$\underline{1}$	C_1^1	$P1$
$\underline{2}$	C_i^1	$P\bar{1}$

MONOCLINIC SYSTEM

No. of space group	Schoenflies symbol	*c*-axis unique (first setting)		*b*-axis unique (second setting)	
		Short symbol	Full symbol	Short symbol	Full symbol
3	C_2^1	$P2$	$P112$	$P2$	$P121$
4	C_2^2	$P2_1$	$P112_1$	$P2_1$	$P12_11$
5	C_2^3	$B2$	$B112$	$C2$	$C121$
6	C_s^1	Pm	$P11m$	Pm	$P1m1$
7	C_s^2	Pb	$P11b$	Pc	$P1c1$
8	C_s^3	Bm	$B11m$	Cm	$C1m1$
9	C_s^4	Bb	$B11b$	Cc	$C1c1$
10	C_{2h}^1	$P2/m$	$P11\frac{2}{m}$	$P2/m$	$P1\frac{2}{m}1$
11	C_{2h}^2	$P2_1/m$	$P11\frac{2_1}{m}$	$P2_1/m$	$P1\frac{2_1}{m}1$
12	C_{2h}^3	$B2/m$	$B11\frac{2}{m}$	$C2/m$	$C1\frac{2}{m}1$
13	C_{2h}^4	$P2/b$	$P11\frac{2}{b}$	$P2/c$	$P1\frac{2}{c}1$
14	C_{2h}^5	$P2_1/b$	$P11\frac{2_1}{b}$	$P2_1/c$	$P1\frac{2_1}{c}1$
15	C_{2h}^6	$B2/b$	$B11\frac{2}{b}$	$C2/c$	$C1\frac{2}{c}1$

ORTHORHOMBIC SYSTEM

No. of space group	Schoenflies symbol	Short symbol	Full symbol	No of space group	Schoenflies symbol	Short symbol	Full symbol
16	$D_2^1 = V^1$	$P222$	$P222$	36	C_{2v}^{12}	$Cmc2_1$	$Cmc2_1$
17	$D_2^2 = V^2$	$P222_1$	$P222_1$	37	C_{2v}^{13}	$Ccc2$	$Ccc2$
18	$D_2^3 = V^3$	$P2_12_12$	$P2_12_12$	38	C_{2v}^{14}	$Amm2$	$Amm2$
19	$D_2^4 = V^4$	$P2_12_12_1$	$P2_12_12_1$	39	C_{2v}^{15}	$Aem2$ ($Abm2$)	$Aem2$ ($Abm2$)
20	$D_2^5 = V^5$	$C222_1$	$C222_1$	40	C_{2v}^{16}	$Ama2$	$Ama2$
21	$D_2^6 = V^6$	$C222$	$C222$	41	C_{2v}^{17}	$Aea2$ ($Aba2$)	$Aea2$ ($Aba2$)
22	$D_2^7 = V^7$	$F222$	$F222$	42	C_{2v}^{18}	$Fmm2$	$Fmm2$

(*Continued*)

No. of space group	Schoenflies symbol	Short symbol	Full symbol	No of space group	Schoenflies symbol	Short symbol	Full symbol
23	$D_2^8 = V^8$	$I222$	$I222$	43	C_{2v}^{19}	$Fdd2$	$Fdd2$
24	$D_2^9 = V^9$	$I2_12_12_1$	$I2_12_12_1$	44	C_{2v}^{20}	$Imm2$	$Imm2$
25	C_{2v}^1	$Pmm2$	$Pmm2$	45	C_{2v}^{21}	$Iba2$	$Iba2$
26	C_{2v}^2	$Pmc2_1$	$Pmc2_1$	46	C_{2v}^{22}	$Ima2$	$Ima2$
27	C_{2v}^3	$Pcc2$	$Pcc2$	47	$D_{2h}^1 = V_h^1$	$Pmmm$	$P\frac{2}{m}\frac{2}{m}\frac{2}{m}$
28	C_{2v}^4	$Pma2$	$Pma2$	48	$D_{2h}^2 = V_h^2$	$Pnnn$	$P\frac{2}{n}\frac{2}{n}\frac{2}{n}$
29	C_{2v}^5	$Pca2_1$	$Pca2_1$	49	$D_{2h}^3 = V_h^3$	$Pccm$	$P\frac{2}{c}\frac{2}{c}\frac{2}{m}$
30	C_{2v}^6	$Pnc2$	$Pnc2$	50	$D_{2h}^4 = V_h^4$	$Pban$	$P\frac{2}{b}\frac{2}{a}\frac{2}{n}$
31	C_{2v}^7	$Pmn2_1$	$Pmn2_1$	51	$D_{2h}^5 = V_h^5$	$Pmma$	$P\frac{2_1}{m}\frac{2}{m}\frac{2}{a}$
32	C_{2v}^8	$Pba2$	$Pba2$				
33	C_{2v}^9	$Pna2_1$	$Pna2_1$				
34	C_{2v}^{10}	$Pnn2$	$Pnn2$				
35	C_{2v}^{11}	$Cmm2$	$Cmm2$				

ORTHORHOMBIC SYSTEM

No. of space group	Schoenflies symbol	Short symbol	Full symbol	No. of space group	Schoenflies symbol	Short symbol	Full symbol
52	$D_{2h}^6 = V_h^6$	$Pnna$	$P\frac{2}{n}\frac{2_1}{n}\frac{2}{a}$	64	$D_{2h}^{18} = V_h^{18}$	$Cmce$ ($Cmca$)	$C\frac{2}{m}\frac{2}{c}\frac{2_1}{a}$
53	$D_{2h}^7 = V_h^7$	$Pmna$	$P\frac{2}{m}\frac{2}{n}\frac{2_1}{a}$	65	$D_{2h}^{19} = V_h^{19}$	$Cmmm$	$C\frac{2}{m}\frac{2}{m}\frac{2}{m}$
54	$D_{2h}^8 = V_h^8$	$Pcca$	$P\frac{2_1}{c}\frac{2}{c}\frac{2}{a}$	66	$D_{2h}^{20} = V_h^{20}$	$Cccm$	$C\frac{2}{c}\frac{2}{c}\frac{2}{m}$
55	$D_{2h}^9 = V_h^9$	$Pbam$	$P\frac{2_1}{b}\frac{2_1}{a}\frac{2}{m}$	67	$D_{2h}^{21} = V_h^{21}$	$Cmme$ ($Cmma$)	$C\frac{2}{m}\frac{2}{m}\frac{2}{a}$
56	$D_{2h}^{10} = V_h^{10}$	$Pccn$	$P\frac{2_1}{c}\frac{2_1}{c}\frac{2}{n}$	68	$D_{2h}^{22} = V_h^{22}$	$Ccce$ ($Cmma$)	$C\frac{2}{c}\frac{2}{c}\frac{2}{a}$

(Continued)

No. of space group	Schoenflies symbol	Short symbol	Full symbol	No. of space group	Schoenflies symbol	Short symbol	Full symbol
57	$D_{2h}^{11} = V_h^{11}$	*Pbcm*	$P\frac{2}{b}\frac{2_1}{c}\frac{2_1}{m}$	69	$D_{2h}^{23} = V_h^{23}$	*Fmmm*	$F\frac{2}{m}\frac{2}{m}\frac{2}{m}$
58	$D_{2h}^{12} = V_h^{12}$	*Pnnm*	$P\frac{2_1}{n}\frac{2_1}{n}\frac{2}{m}$	70	$D_{2h}^{24} = V_h^{24}$	*Fddd*	$F\frac{2}{d}\frac{2}{d}\frac{2}{d}$
59	$D_{2h}^{13} = V_h^{13}$	*Pmmn*	$P\frac{2_1}{m}\frac{2_1}{m}\frac{2}{n}$	71	$D_{2h}^{25} = V_h^{25}$	*Immm*	$I\frac{2}{m}\frac{2}{m}\frac{2}{m}$
60	$D_{2h}^{14} = V_h^{14}$	*Pbcn*	$P\frac{2_1}{b}\frac{2}{c}\frac{2_1}{n}$	72	$D_{2h}^{26} = V_h^{26}$	*Ibam*	$I\frac{2}{b}\frac{2}{a}\frac{2}{m}$
61	$D_{2h}^{15} = V_h^{15}$	*Pbca*	$P\frac{2_1}{b}\frac{2_1}{c}\frac{2_1}{a}$	73	$D_{2h}^{27} = V_h^{27}$	*Ibca*	$I\frac{2}{b}\frac{2}{c}\frac{2}{a}$
62	$D_{2h}^{16} = V_h^{16}$	*Pnma*	$P\frac{2_1}{n}\frac{2_1}{m}\frac{2_1}{a}$	74	$D_{2h}^{28} = V_h^{28}$	*Imma*	$I\frac{2}{m}\frac{2}{m}\frac{2}{a}$
63	$D_{2h}^{17} = V_h^{17}$	*Cmcm*	$C\frac{2}{m}\frac{2}{c}\frac{2_1}{m}$				

TETRAGONAL SYSTEM

No. of space group	Schoenflies symbol	Symbols: a b c (standard)	Symbols: (a±b) (b∓a) c	No. of space group	Schoenflies symbol	Symbols: a b c (standard	Symbols: (a±b) (b∓a) c
75	C_4^1	$P4$	$C4$	99	C_{4v}^1	*P4mm*	*C4mm*
76	C_4^2	$P4_1$	$C4_1$	100	C_{4v}^2	*P4bm*	*C4mb*
77	C_4^3	$P4_2$	$C4_2$	101	C_{4v}^3	$P4_2cm$	$C4_2mc$
78	C_4^4	$P4_3$	$C4_3$	102	C_{4v}^4	$P4_2nm$	$C4_2mn$
79	C_4^5	$I4$	$F4$	103	C_{4v}^5	*P4cc*	*C4cc*
80	C_4^6	$I4_1$	$F4_1$	104	C_{4v}^6	*P4nc*	*C4cn*
81	S_4^1	$P\overline{4}$	$C\overline{4}$	105	C_{4v}^7	$P4_2mc$	$C4_2cm$
82	S_4^2	$I\overline{4}$	$F\overline{4}$	106	C_{4v}^8	$P4_2bc$	$C4_2cb$

(Continued)

No. of space group	Schoenflies symbol	Symbols: a b c (standard)	Symbols: (a±b) (b∓a) c	No. of space group	Schoenflies symbol	Symbols: a b c (standard	Symbols: (a±b) (b∓a) c
83	C_{4h}^{1}	$P4/m$	$C4/m$	107	C_{4v}^{9}	$I4mm$	$F4mm$
84	C_{4h}^{2}	$P4_2/m$	$C4_2/m$	108	C_{4v}^{10}	$I4cm$	$F4mc$
85	C_{4h}^{3}	$P4/n$	$C4/n$	109	C_{4v}^{11}	$I4_1md$	$F4_1dm$
86	C_{4h}^{4}	$P4_2/n$	$C4_2/n$	110	C_{4v}^{12}	$I4_1cd$	$F4_1dc$
87	C_{4h}^{5}	$I4/m$	$F4/m$	111	$D_{2d}^{1} = V_d^1$	$P\bar{4}2m$	$C\bar{4}m2$
88	C_{4h}^{6}	$I4_1/a$	$F4_1/a$	112	$D_{2d}^{2} = V_d^2$	$P\bar{4}2c$	$C\bar{4}c2$
89	D_4^1	$P422$	$C422$	113	$D_{2d}^{3} = V_d^3$	$P\bar{4}2_1m$	$C\bar{4}m2_1$
90	D_4^2	$P42_12$	$C422_1$	114	$D_{2d}^{4} = V_d^4$	$P\bar{4}2_1c$	$C\bar{4}c2_1$
91	D_4^3	$P4_122$	$C4_122$	115	$D_{2d}^{5} = V_d^5$	$P\bar{4}m2$	$C\bar{4}2m$
92	D_4^4	$P4_12_12$	$C4_122_1$	116	$D_{2d}^{6} = V_d^6$	$P\bar{4}c2$	$C\bar{4}2c$
93	D_4^5	$P4_222$	$C4_222$	117	$D_{2d}^{7} = V_d^7$	$P\bar{4}b2$	$C\bar{4}2b$
94	D_4^6	$P4_22_12$	$C4_222_1$	118	$D_{2d}^{8} = V_d^8$	$P\bar{4}n2$	$C\bar{4}2\mathrm{n}$
95	D_4^7	$P4_322$	$C4_322$	119	$D_{2d}^{9} = V_d^9$	$I\bar{4}m2$	$F\bar{4}2m$
96	D_4^8	$P4_32_12$	$C4_322_1$	120	$D_{2d}^{10} = V_d^{10}$	$I\bar{4}c2$	$F\bar{4}2c$
97	D_4^9	$I422$	$F422$	121	$D_{2d}^{11} = V_d^{11}$	$I\bar{4}2m$	$F\bar{4}m2$
98	D_4^{10}	$I4_122$	$F4_122$	122	$D_{2d}^{12} = V_d^{12}$	$I\bar{4}2d$	$F\bar{4}d2$

TETRAGONAL SYSTEM

No. of space group	Schoenflies symbol	Short symbols: a b c (standard)	Short symbols: (a±b) (b∓a) c	Full symbols: a b c	Full symbols: (a±b) (b∓at) c
123	D_{4h}^{1}	$P4/mmm$	$C4/mmm$	$P\frac{4}{m}\frac{2}{m}\frac{2}{m}$	$C\frac{4}{m}\frac{2}{m}\frac{2}{m}$
124	D_{4h}^{2}	$P4/mcc$	$C4/mcc$	$P\frac{4}{m}\frac{2}{c}\frac{2}{c}$	$C\frac{4}{m}\frac{2}{c}\frac{2}{c}$

(Continued)

No. of space group	Schoenflies symbol	Short symbols		Full symbols	
		a b c (standard)	**(a±b) (b∓a) c**	**a b c**	**(a±b) (b∓at) c**
125	D_{4h}^{3}	*P*4/*nbm*	*C*4/*nmb*	$P\frac{4}{n}\frac{2}{b}\frac{2}{m}$	$C\frac{4}{n}\frac{2}{m}\frac{2}{b}$
126	D_{4h}^{4}	*P*4/*nnc*	*C*4/*ncn*	$P\frac{4}{n}\frac{2}{n}\frac{2}{c}$	$C\frac{4}{n}\frac{2}{c}\frac{2}{n}$
127	D_{4h}^{5}	*P*4/*mbm*	*C*4/*mmb*	$P\frac{4}{m}\frac{2_1}{b}\frac{2}{m}$	$C\frac{4}{m}\frac{2}{m}\frac{2_1}{b}$
128	D_{4h}^{6}	*P*4/*mnc*	*C*4/*mcn*	$P\frac{4}{m}\frac{2_1}{n}\frac{2}{c}$	$C\frac{4}{m}\frac{2}{c}\frac{2_1}{n}$
129	D_{4h}^{7}	*P*4/*nmm*	*C*4/*nmm*	$P\frac{4}{n}\frac{2_1}{m}\frac{2}{m}$	$C\frac{4}{n}\frac{2}{m}\frac{2_1}{m}$
130	D_{4h}^{8}	*P*4/*ncc*	*C*4/*ncc*	$P\frac{4}{n}\frac{2_1}{c}\frac{2}{c}$	$C\frac{4}{n}\frac{2}{c}\frac{2_1}{c}$
131	D_{4h}^{9}	$P4_2/mmc$	$C4_2/mcm$	$P\frac{4_2}{m}\frac{2}{m}\frac{2}{c}$	$C\frac{4_2}{m}\frac{2}{c}\frac{2}{m}$
132	D_{4h}^{10}	$P4_2/mcm$	$C4_2/mmc$	$P\frac{4_2}{m}\frac{2}{c}\frac{2}{m}$	$C\frac{4_2}{m}\frac{2}{m}\frac{2}{c}$
133	D_{4h}^{11}	$P4_2/nbc$	$C4_2/ncb$	$P\frac{4_2}{n}\frac{2}{b}\frac{2}{c}$	$C\frac{4_2}{n}\frac{2}{c}\frac{2}{b}$
134	D_{4h}^{12}	$P4_2/nnm$	$C4_2/nmn$	$P\frac{4_2}{n}\frac{2}{n}\frac{2}{m}$	$C\frac{4_2}{n}\frac{2}{m}\frac{2}{n}$
135	D_{4h}^{13}	$P4_2/mbc$	$C4_2/mcb$	$P\frac{4_2}{m}\frac{2_1}{b}\frac{2}{c}$	$C\frac{4_2}{m}\frac{2}{c}\frac{2_1}{b}$
136	D_{4h}^{14}	$P4_2/mnm$	$C4_2/mmn$	$P\frac{4_2}{m}\frac{2_1}{n}\frac{2}{m}$	$C\frac{4_2}{m}\frac{2}{m}\frac{2_1}{n}$
137	D_{4h}^{15}	$P4_2/nmc$	$C4_2/ncm$	$P\frac{4_2}{n}\frac{2_1}{m}\frac{2}{c}$	$C\frac{4_2}{n}\frac{2}{c}\frac{2_1}{m}$
138	D_{4h}^{16}	$P4_2/ncm$	$C4_2/nmc$	$P\frac{4_2}{n}\frac{2_1}{c}\frac{2}{m}$	$C\frac{4_2}{n}\frac{2}{m}\frac{2_1}{c}$
139	D_{4h}^{17}	*I*4/*mmm*	*F*4/*mmm*	$I\frac{4}{m}\frac{2}{m}\frac{2}{m}$	$F\frac{4}{m}\frac{2}{m}\frac{2}{m}$
140	D_{4h}^{18}	*I*4/*mcm*	*F*4/*mmc*	$I\frac{4}{m}\frac{2}{c}\frac{2}{m}$	$F\frac{4}{m}\frac{2}{m}\frac{2}{c}$

(Continued)

No. of space group	Schoenflies symbol	Short symbols		Full symbols	
		a b c (standard)	(a±b) (b∓a) c	a b c	(a±b) (b∓at) c
141	D_{4h}^{19}	$I4_1/amd$	$F4_1/adm$	$I\frac{4_1}{a}\frac{2}{m}\frac{2}{d}$	$F\frac{4_1}{a}\frac{2}{d}\frac{2}{m}$
142	D_{4h}^{20}	$I4_1/acd$	$F4_1/adc$	$I\frac{4_1}{a}\frac{2}{c}\frac{2}{d}$	$F\frac{4_1}{a}\frac{2}{d}\frac{2}{c}$

TRIGONAL SYSTEM HEXAGONAL SYSTEM

No. of space group	Schoenflies symbol	Short symbol	Full symbol	No. of space group	Schoenflies symbol	Short symbol	Full symbol
143	C_3^1	$P3$		168	C_6^1	$P6$	
144	C_3^2	$P3_1$		169	C_6^2	$P6_1$	
145	C_3^3	$P3_2$		170	C_6^3	$P6_5$	
146	C_3^4	$R3$		171	C_6^4	$P6_2$	
147	C_{3i}^1	$P\overline{3}$		172	C_6^5	$P6_4$	
148	C_{3i}^2	$R\overline{3}$		173	C_6^6	$P6_3$	
149	D_3^1	$P312$		174	C_{3h}^1	$P\overline{6}$	
150	D_3^2	$P321$		175	C_{3h}^1	$P6/m$	
151	D_3^3	$P3_112$		176	C_{3h}^2	$P6_3/m$	
152	D_3^4	$P3_121$		177	D_6^1	$P622$	
153	D_3^5	$P3_212$		178	D_6^2	$P6_122$	
154	D_3^6	$P3_221$		179	D_6^3	$P6_522$	
155	D_3^7	$R32$		180	D_6^4	$P6_222$	
156	C_{3v}^1	$P3m1$		181	D_6^5	$P6_422$	
157	C_{3v}^2	$P31m$		182	D_6^6	$P6_322$	
158	C_{3v}^3	$P3c1$		183	C_{6v}^1	$P6mm$	
159	C_{3v}^4	$P31c$		184	C_{6v}^2	$P6cc$	
160	C_{3v}^5	$R3m$		185	C_{6v}^3	$P6_3cm$	
161	C_{3v}^6	$R3c$		186	C_{6v}^4	$P6_3mc$	

(Continued)

No. of space group	Schoenflies symbol	Short symbol	Full symbol	No. of space group	Schoenflies symbol	Short symbol	Full symbol
162	D_{3d}^{1}	$P\bar{3}1m$	$P\bar{3}1\frac{2}{m}$	187	D_{3h}^{1}	$P\bar{6}m2$	
163	D_{3d}^{2}	$P\bar{3}1c$	$P\bar{3}1\frac{2}{c}$	188	D_{3h}^{2}	$P\bar{6}c2$	
164	D_{3d}^{3}	$P\bar{3}m1$	$P\bar{3}\frac{2}{m}1$	189	D_{3h}^{3}	$P\bar{6}2m$	
165	D_{3d}^{4}	$P\bar{3}c1$	$P\bar{3}\frac{2}{c}1$	190	D_{3h}^{4}	$P\bar{6}2c$	
166	D_{3d}^{5}	$R\bar{3}m$	$R\bar{3}\frac{2}{m}$	191	D_{6h}^{1}	$P6/mmm$	$P\frac{6}{m}\frac{2}{m}\frac{2}{m}$
167	D_{3d}^{6}	$R\bar{3}c$	$R\bar{3}\frac{2}{c}$	192	D_{6h}^{2}	$P6/mcc$	$P\frac{6}{m}\frac{2}{c}\frac{2}{c}$
				193	D_{6h}^{3}	$P6_3/mcm$	$P\frac{6_3}{m}\frac{2}{c}\frac{2}{m}$
				194	D_{6h}^{4}	$P6_3/mmc$	$P\frac{6_3}{m}\frac{2}{m}\frac{2}{c}$

CUBIC SYSTEM

No. of space group	Schoenflies symbol	Short symbol	Full symbol	No. of space group	Schoenflies symbol	Short symbol	Full symbol
195	T^{1}	$P23$		213	O^{7}	$P4_132$	
196	T^{2}	$F23$		214	O^{8}	$I4_132$	
197	T^{3}	$I23$		215	T_d^{1}	$P\bar{4}3m$	
198	T^{4}	$P2_13$		216	T_d^{2}	$F\bar{4}3m$	
199	T^{5}	$I2_13$		217	T_d^{3}	$I\bar{4}3m$	
200	T_h^{1}	$Pm\bar{3}$	$P\frac{2}{m}\bar{3}$	218	T_d^{4}	$P\bar{4}3n$	
201	T_h^{2}	$Pn\bar{3}$	$P\frac{2}{n}\bar{3}$	219	T_d^{5}	$F\bar{4}3c$	
202	T_h^{3}	$Fm\bar{3}$	$F\frac{2}{m}\bar{3}$	220	T_d^{6}	$I\bar{4}3d$	

(Continued)

No. of space group	Schoenflies symbol	Short symbol	Full symbol	No. of space group	Schoenflies symbol	Short symbol	Full symbol
203	T_h^4	$Fd\bar{3}$	$F\frac{2}{d}\bar{3}$	221	O_h^1	$Pm\bar{3}m$	$P\frac{4}{m}\bar{3}\frac{2}{m}$
204	T_h^5	$Im\bar{3}$	$I\frac{2}{m}\bar{3}$	222	O_h^2	$Pn\bar{3}n$	$P\frac{4}{n}\bar{3}\frac{2}{n}$
205	T_h^6	$Pa\bar{3}$	$P\frac{2}{a}\bar{3}$	223	O_h^3	$Pm\bar{3}n$	$P\frac{4}{m}\bar{3}\frac{2}{n}$
206	T_h^7	$Ia\bar{3}$	$I\frac{2}{a}\bar{3}$	224	O_h^4	$Pn\bar{3}m$	$P\frac{4}{n}\bar{3}\frac{2}{m}$
207	O^1	$P432$		225	O_h^5	$Fm\bar{3}m$	$F\frac{4}{n}\bar{3}\frac{2}{m}$
208	O^2	$P4_232$		226	O_h^6	$Fm\bar{3}c$	$F\frac{4}{m}\bar{3}\frac{2}{c}$
209	O^3	$F432$		227	O_h^7	$Fd\bar{3}m$	$F\frac{4}{d}\bar{3}\frac{2}{m}$
210	O^4	$F4_132$		228	O_h^8	$Fd\bar{3}c$	$F\frac{4}{d}\bar{3}\frac{2}{c}$
211	O^5	$I432$		229	O_h^9	$Im\bar{3}m$	$I\frac{4}{m}\bar{3}\frac{2}{m}$
212	O^6	$P4_332$		230	O_h^{10}	$Ia\bar{3}d$	$I\frac{4}{a}\bar{3}\frac{2}{d}$

APPENDIX 8

The Reciprocal Lattice and Diffraction

A diffraction pattern of a crystal may be obtained using X-rays, electrons or neutrons. However, independent of the way it is obtained, the fundamental question is in what way does the translational symmetry of the crystal affect the diffraction pattern? Recall that we can describe a crystal, $C(\mathbf{r})$, in terms of a convolution of a lattice function, $L(\mathbf{r})$, with a basis function, $B(\mathbf{r})$:

$$C(\mathbf{r}) = L(\mathbf{r}) * B(\mathbf{r}) \qquad [A8.1]$$

where the vector $\mathbf{r}$ is a general position vector in the crystal (i.e. it is in direct space). Now, to obtain the diffraction amplitudes, it is necessary to calculate the Fourier transform of the scattering object, which in this case is the crystal. To do this, we use the so-called **convolution theorem**, which states that the Fourier transform of the convolution of two functions is equal to the product of the Fourier transforms of the individual functions (this also works the other way round). Applying this to the definition of the crystal, we can write

$$C(\mathbf{s}) = L(\mathbf{s}) \times B(\mathbf{s}) \qquad [A8.2]$$

where $C(\mathbf{s})$ is the amplitude of scattering in the diffraction pattern, with scattering vector $\mathbf{s}$; $L(\mathbf{s})$ is the Fourier transform of the lattice; and $B(\mathbf{s})$ is the Fourier transform of the basis.

We now ask what does $L(\mathbf{s})$ look like? One way of viewing this is to decompose the original lattice function $L(\mathbf{r})$ into the convolution of three functions, each in different directions and each representing a one-dimensional row of lattice points in direct space:

$$L(\mathbf{r}) = L(\mathbf{r}_1) * L(\mathbf{r}_2) * L(\mathbf{r}_3) \qquad [A8.3]$$

Because these functions are convoluted together, each row of lattice points will appear in sympathy with the lattice points in a different row to produce a three-dimensional lattice, as shown schematically in Figs A8.la and A8.lb. Now, let us use the convolution theorem again to obtain the Fourier transform of the lattice:

$$L(\mathbf{s}) = L(\mathbf{s}_1) \times L(\mathbf{s}_2) \times L(\mathbf{s}_3) \qquad [A8.4]$$

The Fourier transform of a single row of points, for example $L(\mathbf{r}_1)$, is a set of infinite, parallel planes perpendicular to the row of points (Fig. A8.1c) with a repeat spacing inversely proportional to the spacing of the points in direct space. Because $L(\mathbf{s})$ consists of the product of three sets of such planes in different orientations, there will be cancellation everywhere except where these planes intersect, that is, along the lines shown in Fig. A8.1d, for the two-dimensional case and at points for the three-dimensional case. Thus, the result is that $L(\mathbf{s})$ is itself a lattice with dimensions reciprocally related to the dimensions of the original lattice. The two lattices are called the **reciprocal lattice** and the **direct lattice** (or **real lattice**), respectively, with the vector spaces called reciprocal space and direct or real space. Thus, the reciprocal lattice is the Fourier transform of the real lattice, and the real lattice is the Fourier transform of the reciprocal lattice. It can be further

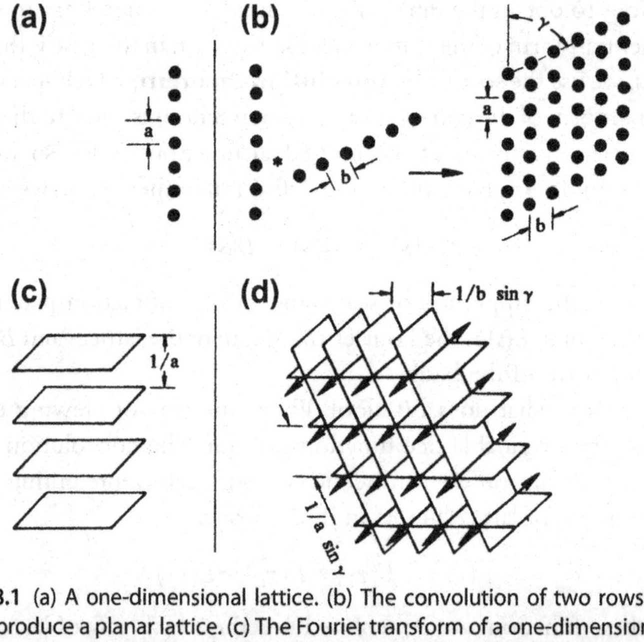

Figure A8.1 (a) A one-dimensional lattice. (b) The convolution of two rows of lattice points to produce a planar lattice. (c) The Fourier transform of a one-dimensional lattice. The planes shown are infinite in extent. (d) The Fourier transform of a two-dimensional lattice (shown slightly inclined) derived as the product of the Fourier transforms of the original single lattice rows. In two dimensions, the multiplication of two sets of planes produces zero everywhere, except on the lines where they intersect (marked by arrows), separated, in this example, by the distances $l/(a \sin \gamma)$ and $l/(b \sin \gamma)$. In three dimensions, there would be three sets of intersecting planes to produce points of intersection (i.e. the reciprocal lattice).

shown that there is a 1:1 correspondence between the symmetries of the two lattices, in the following way:

$$P \text{ (primitive)} \Leftrightarrow P \text{ (primitive)}$$
$$F \text{ (face-centered)} \Leftrightarrow I \text{ (body-centered)} \qquad [A8.5]$$
$$C \text{ (one-face-centered)} \Leftrightarrow C \text{ (one-face-centered)}$$

Corresponding to the real lattice vectors **a, b** and **c**, we can also define reciprocal lattice vectors $\mathbf{a}^*$, $\mathbf{b}^*$ and $\mathbf{c}^*$, given by

$$\mathbf{a}^* = 2\pi\frac{\mathbf{b}\times\mathbf{c}}{V_\mathrm{p}}, \quad \mathbf{b}^* = 2\pi\frac{\mathbf{c}\times\mathbf{a}}{V_\mathrm{p}}, \quad \mathbf{c}^* = 2\pi\frac{\mathbf{a}\times\mathbf{b}}{V_\mathrm{p}} \qquad [A8.6]$$

where $V_\mathrm{p} = \mathbf{a}\cdot(\mathbf{b}\times\mathbf{c})$ is the volume of the unit cell in real space. Note that crystallographers normally define the reciprocal lattice with the factor of 2π replaced by unity. It can be seen from this that when the units of direct space are taken as Å, then the units of reciprocal space are Å^{-1}. *This means that if by some means the lattice is doubled in real space, the corresponding lengths are halved in reciprocal space.*

Now, a diffraction pattern from a crystal consists of 'spots' corresponding to the points, or nodes, of the reciprocal lattice and a reciprocal lattice vector from some origin to any of these nodes can be defined as

$$\mathbf{Q} = h\mathbf{a}^* + k\mathbf{b}^* + l\mathbf{c}^* \qquad [A8.7]$$

where h, k and l are integers indexing the particular node. These integers are in fact the so-called Miller indices defined for planes in direct space by the intercepts (a/h, b/k, c/l) that a plane makes on the axes (Fig. A8.2). These were first used to denote the forms of macroscopic crystals, but they are now also used to symbolize the planes of atoms and molecules in a crystal structure. Note that crystallographers refer to planes by (hkl) in parentheses (and $\{hkl\}$ for the set of symmetry-related planes). It follows from the definition of reciprocal axes that the reciprocal vector **Q** is perpendicular to the (hkl) plane.

The effect of multiplying the reciprocal lattice function by the Fourier transform of the basis Eqn [A8.2] is to make the amplitudes different at each node. The intensity of diffraction is given by the amplitude multiplied by its complex conjugate; thus, at each node of the reciprocal lattice, the scattered wave will have different intensities. This is what we observe as **the diffraction pattern of a crystal.**

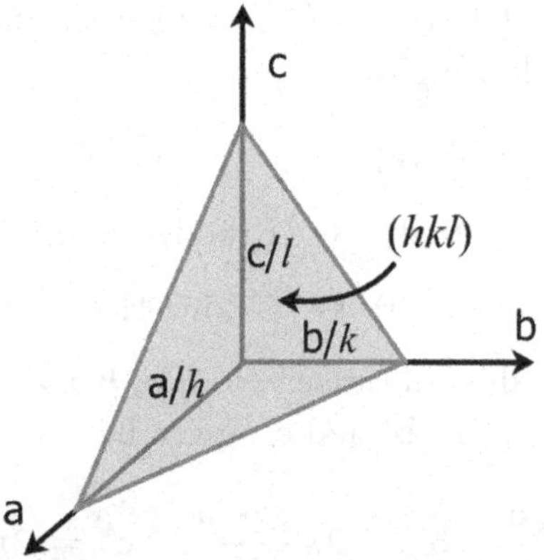

Figure A8.2 Definition of Miller index of a plane.

Crystals of finite size can be modelled by multiplying Eqn [A8.1] by $S(\mathbf{r})$, a function with value unity within a certain volume and zero elsewhere. On Fourier transforming, Eqn [A8.2] is convoluted by $S(\mathbf{s})$, the Fourier transform of $S(\mathbf{r})$. The effect of this changes the width of *each* diffraction peak – the smaller the crystal the broader the diffraction peak.

Superstructures of a crystal structure are formed when one or more of the real space lattice dimensions is multiplied by an integer (2, 3, 4, ...). The result is new reciprocal lattice points (nodes) that produce a lattice spacing reciprocal to that in the real space lattice. Thus, extra reciprocal lattice points appear (i.e. extra 'spots' in the diffraction pattern). For example, in $SrTiO_3$ there is a phase transition from a high-temperature cubic structure with a unit cell of roughly $4 \times 4 \times 4$ Å^3 to a tetragonal, low-temperature structure with the unit cell doubled in all three directions to give a new cell roughly $8 \times 8 \times 8$ Å^3. This doubling of the direct space unit-cell axes gives rise to extra diffraction spots, which have half-integral indices when indexed with respect to the original high-temperature reciprocal lattice. When referred to the appropriate reciprocal lattice corresponding to the $8 \times 8 \times 8$ Å^3 real space lattice, the extra spots have indices containing odd integers, with the original spots having all indices with even integers.

Incommensurate (modulated) crystals (Section 7.21) have structures closely related to that of a superstructure. The convolution approach can be used also to understand the diffraction effects from such crystals. First, let us describe the incommensurate structure $C_{inc}(\mathbf{r})$ as follows:

$$C_{inc}(\mathbf{r}) = [L(\mathbf{r}) \times M(\mathbf{r})] * B(\mathbf{r}) \qquad \text{[A8.8]}$$

where $L(\mathbf{r})$ is a regular lattice and $M(\mathbf{r})$ is some sort of modulation function. Consider the one-dimensional lattice in Fig. A8.3a. On multiplying by a modulation function consisting of a simple sinusoidal function (Fig. A8.3b), the lattice is distorted to form a series of points distributed

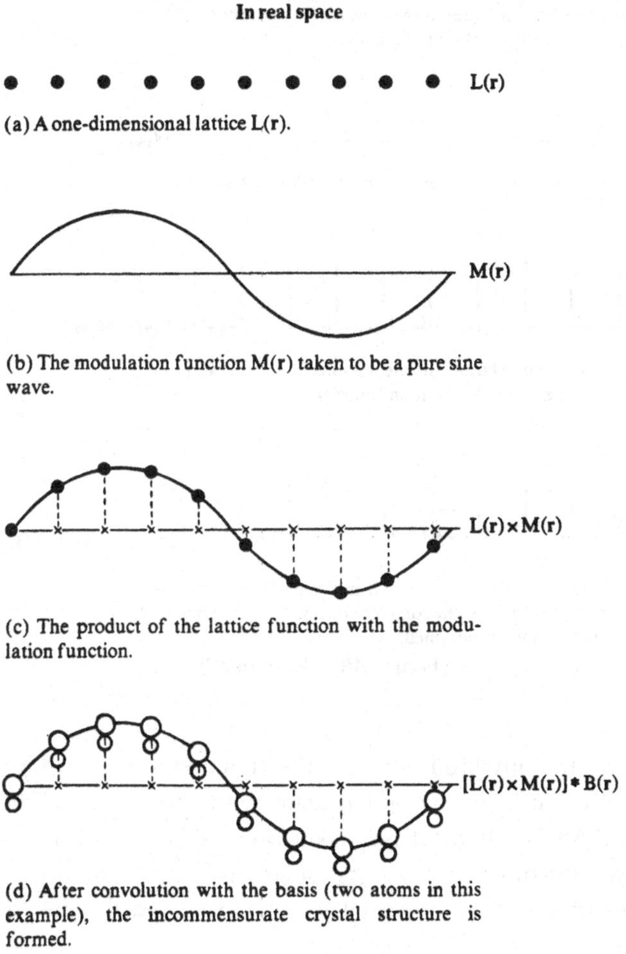
In real space
L(r)
(a) A one-dimensional lattice L(r).
M(r)
(b) The modulation function M(r) taken to be a pure sine wave.
L(r)×M(r)
(c) The product of the lattice function with the modulation function.
[L(r)×M(r)]∗B(r)
(d) After convolution with the basis (two atoms in this example), the incommensurate crystal structure is formed.

Figure A8.3 Diffraction from an incommensurate crystal. (a) A one-dimensional lattice $L(\mathbf{r})$. (b) The modulation function $M(\mathbf{r})$ taken to be a pure sine wave. (c) The product of the lattice function with the modulation function. (d) After convolution with the basis (two atoms in this example), the incommensurate crystal structure is formed. (e) The Fourier transform of the real lattice, shown here as a repeating set of delta functions. (f) The Fourier transform of the modulation function. (g) The convolution of the reciprocal lattice and the transform of the modulation function. (h) After multiplying by the transform of the basis and then multiplying by the complex conjugate, the intensities of diffraction are obtained.

In reciprocal space

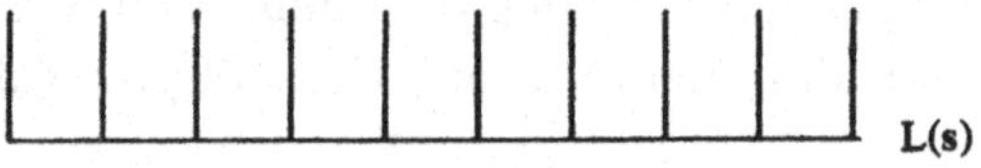

(e) The Fourier Transform of the real lattice, shown here as a repeating set of delta-functions.

(f) The Fourier Transform of the modulation function.

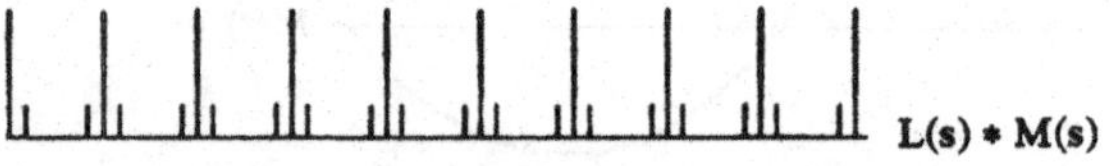

(g) The convolution of the reciprocal lattice and the transform of the modulation function.

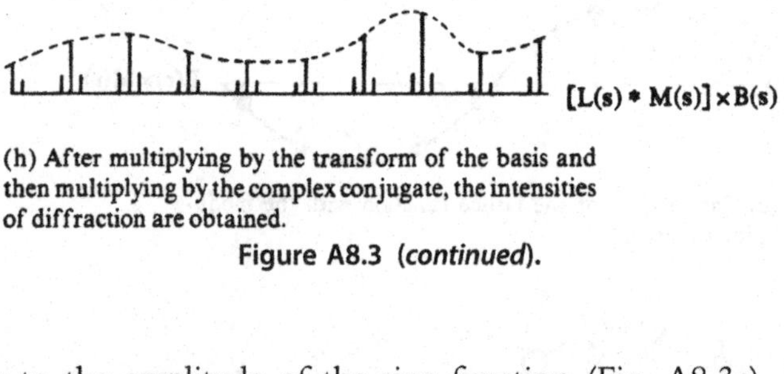

(h) After multiplying by the transform of the basis and then multiplying by the complex conjugate, the intensities of diffraction are obtained.

Figure A8.3 ***(continued)*****.**

according to the amplitude of the sine function (Fig. A8.3c), but with a wavelength that is not commensurate with the repeat of the original lattice (Fig. A8.3a). If the basis (taken here as a pair of atoms) is now convoluted with this pattern, a crystal structure is obtained in which the atoms are successively displaced in sympathy with the sine wave (Fig. 8.3d).

Now consider the Fourier transform of this crystal. From the convolution theorem, we can immediately write

$$C_{\text{inc}}(\mathbf{s}) = [L(\mathbf{s}) * M(\mathbf{s})] \times B(\mathbf{s}) \qquad \text{[A8.9]}$$

$L(\mathbf{s})$ is the usual reciprocal lattice (i.e., a regular repeating pattern of points, Fig. A8.3e), and $M(\mathbf{s})$ is the Fourier transform of the modulation function. If $M(\mathbf{r})$ is a sine function, its Fourier transform is two delta functions on either side of $|\mathbf{s}| = 0$ (Fig. A8.3f), with a distance between them proportional to the reciprocal of the sine period. Since $M(\mathbf{s})$ is convoluted with the reciprocal lattice, the effect is to produce extra diffraction spots on either sides of each of the original reciprocal lattice points (i.e., those from the undistorted structure) as shown in Fig. A8.3g. On multiplying by the transform of the basis, all the amplitudes, and hence the intensities, at each node are changed to give the observed diffraction pattern (Fig. A8.3h).

In reality, whether dealing with incommensurate modulations or superstructures, the modulation function can never be a perfect sine wave, since it is determined by *discrete* atomic positions. Thus the Fourier transform will contain higher harmonics. For example, in $NaNbO_3$, phase R (Section 7.21.1), the 6-fold repeat is caused by discrete Nb displacements resulting in diffraction intensity in reciprocal space at $\pm\,\mathbf{c}^*/6$ and $\pm\,2\mathbf{c}^*/6$, the latter reflection being generally weaker than the first. Because of cancellation effects, there is no peak due to Nb displacements at $3\mathbf{c}^*/6$ (the oxygens, however, do contribute a small amount of intensity to this position).

The main reflections, those due to the commensurate part of the structure, give information about the average structure of the crystal; the extra reflections tell us about the incommensurate modulation of the structure. We can see from this that a superstructure diffraction pattern is a special case of the incommensurate diffraction pattern; that is, it is obtained when the incommensurate points are moved to positions in reciprocal space where they represent rational fractional spacings of the main repeat.

SCATTERING FROM DISORDERED STRUCTURES

In practice, not all crystal structures are perfectly ordered. Instead, they have various amounts and kinds of disorder. When X-rays, neutrons or electrons are diffracted by such imperfect materials, extra scattering effects can often be observed. In order to understand how this arises, consider a crystal structure which is made up from two sorts of molecular basis as indicated by the white and black triangles in Fig. A8.4; these are placed at lattice sites in a random arrangement (Fig. A8.4a). Such a structure is not disordered in the way that a gas would be, since all the molecules are spaced apart regularly. On the other hand, because the black and white objects

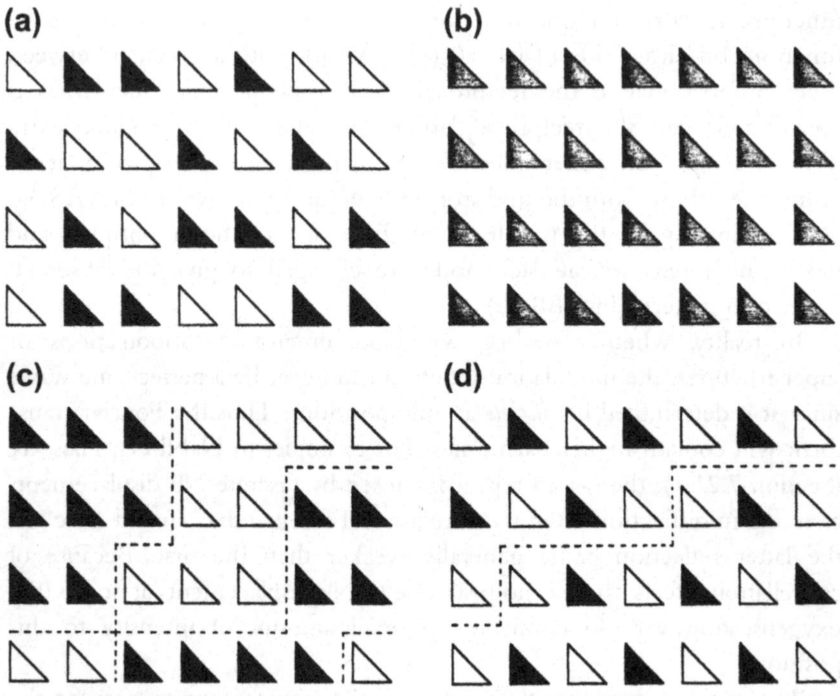

Figure A8.4 (a) A random arrangement of two types of basis on a regular lattice. (b) The average crystal structure. (c) Short-range order in which domains of similar bases appear. (d) Short-range order in which domains of alternating bases appear. In (c) and (d), the dashed lines are guides to the eye to make the domain arrangement clearer.

have been set down randomly on the regular sites, there is no short-range order. Let

$$\mathbf{F}_{\mathrm{w}} = \text{Fourier transform of a white molecule}$$

$$\mathbf{F}_{\mathrm{b}} = \text{Fourier transform of a black molecule}$$

Now, suppose that there are equal numbers of the two types of molecule, and that N is the total number of molecules in the crystal. The expression for the diffracted intensity can be shown to be

$$I = 0.5N^2(\mathbf{F}_{\mathrm{w}}{}^2 + \mathbf{F}_{\mathrm{b}}{}^2) + 0.25N^2(\mathbf{F}_{\mathrm{w}} - \mathbf{F}_{\mathrm{b}})^2$$

$$= N^2\overline{\mathbf{F}^2} + 0.25N^2(\mathbf{F}_{\mathrm{w}} - \mathbf{F}_{\mathrm{b}})^2 \qquad \text{[A8.10]}$$

This expression contains two terms, the first one being due to the average of the square of the Fourier transform. This corresponds to the diffraction intensity from a crystal structure consisting of an average of the white and black molecules, symbolized by the grey units in Fig. A8.4b. Since this appears to be effectively a fully ordered structure *on average*, the result is a set of sharp diffraction spots (Bragg scattering), just like that from a perfect structure. The second term depends on the difference in scattering between a white and a black molecule; it gives rise to a broad diffuse scattering superimposed on the Bragg spots.

However, in general the disordering may not be random, but instead may have a degree of short-range order. Figures A8.4c and A8.4d show two such possibilities; in one (Fig. A8.4c), there is a tendency for white molecules to be close to one another, and similarly for black molecules. In the other (Fig. A8.4d), there is a tendency for black and white molecules to alternate. The result in both cases is to produce small ordered domains separated by irregular and random boundaries. Clearly, this short-range ordering is describable in terms of probabilities for 'mistakes' in the ordering to occur. The probability distribution function P then modifies the above expression for the diffracted intensity, thus

$$I = N^2\overline{\mathbf{F}^2} + 0.25\ PN^2(\mathbf{F}_\mathrm{W} - \mathbf{F}_\mathrm{b})^2 \qquad [\mathrm{A8.11}]$$

and when P is such as to give short-range order like that in Figs A8.4c and A8.4d, this causes the diffuse scattering to be patterned, rather than to be uniform and smooth. The structure in Fig. A8.4c causes the diffuse scattering to peak on Bragg spots, whereas the alternating structure of Fig. A8.4d gives rise to diffuse maxima halfway between the Bragg spots (if the alternating order were perfect, this would mean that the unit-cell edge lengths would be precisely doubled, leading to sharp Bragg spots halfway between the original Bragg spots, i.e. a superstructure).

It is possible to imagine many sorts of short-range order or disorder, which can be determined by the nature of the probability function P, and these could be determined by diffraction techniques. P can generally be written as a series of probability functions:

$$P = P_{ij} + P_{ijk} + P_{ijkl} + \ldots\ldots \qquad [\mathrm{A8.12}]$$

P_{ij} is a probability function that depends on pair-wise, (i.e. two-body, interactions), P_{ijk} is one that depends on three-body interactions, and so on.

As an example of the effect of some disorder in an otherwise ordered crystal, we consider the perovskite $BaTiO_3$ which was discussed in Section 7.11.2. This shows the averaged position of the atoms in all four crystallographic structures. In very careful X-ray scattering work, diffuse scattering has been observed in all of the phases except for the

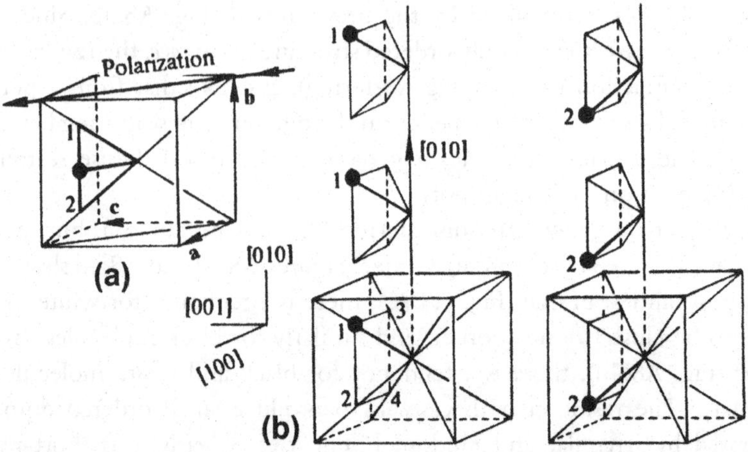

Figure A8.5 (a) $BaTiO_3$ in the orthorhombic phase. The Ti displacement may be described as the average of two component displacements along [111] and $[1\bar{1}1]$ as in (b).[1]

lowest-temperature (rhombohedral) phase; in this phase the structure was considered to be fully ordered with the Ti atoms all displaced from the centeres of the oxygen octahedra along [111].

Now, consider the orthorhombic phase taking x, y and z along the Ti–O bond directions. When the ferroelectric polarization is in the [101] direction (Fig. A8.5a), planes of diffuse scattering are observed perpendicular to the [010] direction (y-axis). A model along the following lines can be used to explain the diffuse scattering. Figure A8.5a shows the average crystal structure of $BaTiO_3$ in the orthorhombic phase; just the Ti atomic positions are shown. Suppose that this is averaged from chains of off-center Ti atoms; the Ti atoms have small displacements $\pm\delta$ along the y-axis and the chains are orientated along the [010] axis as indicated in Fig. A8.5b. The labelled positions 1 or 2 correspond to two equally probable positions for finding the Ti atoms in the orthorhombic cell. Now, if the chains are such that along the length of any one chain all the Ti atoms are in the same position, but that neighbouring chains are completely uncorrelated with one another, the result will be extended planes of diffuse scattering perpendicular to the chain direction.

This model has been further extended to the tetragonal phase where the actual Ti positions are proposed to be at positions 1, 2, 3 or 4 in Fig. A8.5b,

1 R. Comes, M. Lambert, and A. Guinier, *Solid State Commun.* **6**, 715 (1968).

in chains along the [100] and [010] directions. This leads to an averaged position of the Ti atom along [001], the direction of the polarization in this ferroelectric phase. Allowing neighbouring chains to be uncorrelated leads to perpendicular sets of planar scattering in agreement with what is observed. This was extended even to the cubic phase of $BaTiO_3$ where three sets of perpendicular chains were used. Note that when the correlation between the chains is long, the diffuse planes are narrow. For shorter correlation lengths (ℓ), the diffuse planes will have a width $\propto 1/\ell$.

APPENDIX 9

Some Interesting Structures

Below we give a brief summary of some crystal structures that are of interest to various groups of solid-state scientists.

A9-1 CeM_2Si_2

There are many intermetallic compounds whose formulae can be written as CeM_2Si_2 with M = Au, Ag, Pd, or Cr, Cu, ..., and also with Si replaced by Ge. These materials have the space group $I4/mmm(D_{4h}^{17})$ with atomic positions given by

Ce :	$2a$	$(0, 0, 0)$
Si :	$4d$	$(0, ½, ¼); (½, 0, ¼)$
M :	$4e$	$\pm(0, 0, z)$ with $z \approx 0.38$

Typically, a ≈ 4 Å and c ≈ 10 Å. The structure is shown in Fig. A9.1. Compounds with this structure often show antiferromagnetic ordering with the spin arrangement shown in Fig. A9.1b, for example, when M = Au. The corner Ce atoms have spin up while that at the center of the unit cell have spin down. Thus, if the magnetic structure is considered, the unit cell is primitive tetragonal.

For $CePd_2Si_2$, the spins are in the (001) planes pointing along $[\bar{1}10]$, as shown in Fig. A9.1c. Clearly, the magnetic structure has a unit cell that is double that of the structural unit cell along **a** and **b**. This is shown in Fig. A9.1d. Note that the arrangement of spins means that the magnetic symmetry belongs to the orthorhombic system with a unit cell given by $\mathbf{a}_o$, $\mathbf{b}_o$ and $\mathbf{c}_o$, as outlined in the figure. This corresponds to the magnetic group $F_Cm'm'm$. In Fig. A9.1e, the symmetry elements of this magnetic group are shown for the (0,0,0) set of translation vectors. See Litvin's tables (69.9.613) for the remaining symmetry elements.

This structure type includes materials that show superconductivity, such as $CeCu_2Si_2$, cooperative magnetism and normal metallic behaviour (as in $LaCu_2Si_2$), all of which are examples of **heavy fermion**

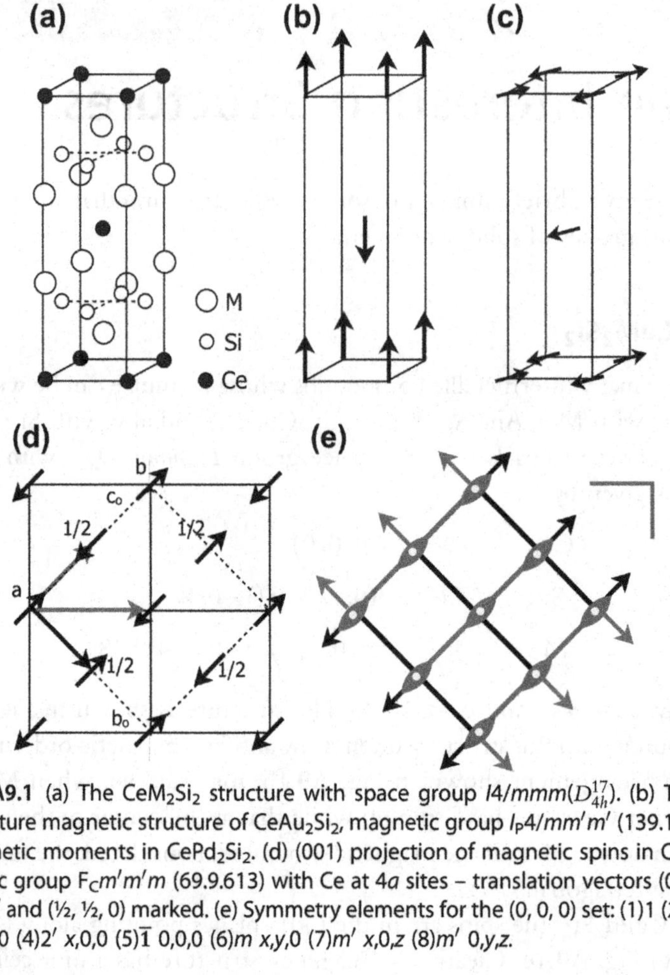

Figure A9.1 (a) The CeM_2Si_2 structure with space group $I4/mmm(D_{4h}^{17})$. (b) The low-temperature magnetic structure of $CeAu_2Si_2$, magnetic group $I_P4/mm'm'$ (139.15.1193). (c) Magnetic moments in $CePd_2Si_2$. (d) (001) projection of magnetic spins in $CePd_2Si_2$: magnetic group $F_Cm'm'm$ (69.9.613) with Ce at 4*a* sites – translation vectors (0, ½, ½)′, (½, 0, ½)′ and (½, ½, 0) marked. (e) Symmetry elements for the (0, 0, 0) set: (1)1 (2) 2 0,0,*z* (3)2′ 0,*y*,0 (4)2′ *x*,0,0 (5)$\bar{1}$ 0,0,0 (6)*m x*,*y*,0 (7)*m*′ *x*,0,*z* (8)*m*′ 0,*y*,*z*.

metals[1]. $CeCu_2Si_2$ was the first (1979) heavy fermion superconductor to be discovered.

A9-2 RUTILE

A common structure with octahedral coordination is that of rutile (TiO_2) or **cassiterite** (SnO_2). This has the space group $P4/mmm(D_{4h}^{14})$ with the atoms at

$$\text{Ti}: \quad 2a \quad (0,0,0); (½,½,½)$$

$$\text{O}: \quad 4f \quad \pm(u,u,0); (u+½, ½-u, ½)$$

[1] G.R. Stewart, *Rev. Mod. Phys.*, **56**, 755 (1984)

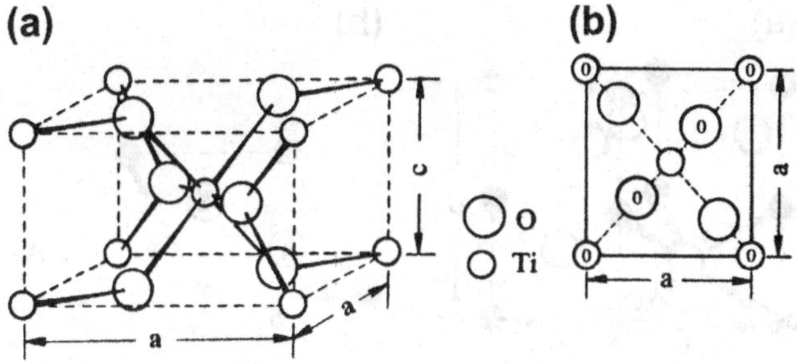

Figure A9.2 The rutile structure (TiO_2) in (a) perspective and (b) projected onto (001).

and so there are two formula units per cell. The structure is shown in Fig. A9.2. The coordination of Ti is 6 and that of oxygen is 3 (6:3). Typical unit cell dimensions for rutile and a large number of other oxides that adopt this structure type (e.g. Ti can be replaced by Cr, Mn, Nd, Mo, Ta, W, ... etc.) are c ≈ 4.6 Å, a ≈ 3.0 Å , with $u \approx 0.305$. Some fluorides, such as NiF_2 also crystallize in this structure: Ni can be replaced by other transition metals or by Mg or Pd. GeO_2 can be grown in the rutile form as well as in the α-quartz structure with coordination (4:2). In fact, TiO_2 itself can be found in two other modifications: **anatase**, space group $I4/amd(D_{4h}^{19})$ and **brookite,** space group $Pbca(D_{2h}^{15})$. In the latter structure, all of the atoms are in general positions.

$CaCl_2$, $CaBr_2$ and $CrCl_2$ all have a structure that is similar to rutile, but with the orthorhombic space group $Pnnm(D_{2h}^{12})$, and a and b differing by about 5%.

A9-3 NICKEL ARSENIDE

Many AB compounds, where A is a transition metal and B is an intermetallic element, crystallize in this structure type. The space group is $P6_3mc(C_{6v}^4)$ with the atoms in the following positions:

$$\text{A}: \quad 2a \quad (0,0,0); (0,0,½)$$

$$\text{B}: \quad 2b \quad (1/3,2/3,u);(2/3,1/3,u+ ½)$$

so that there are two formula units per cell. When $u = ¼$, the B atom lies at the center of a trigonal prism of A atoms (Fig. A9.3), otherwise it is slightly displaced along [001]. The coordination is (6:6). Typical examples of

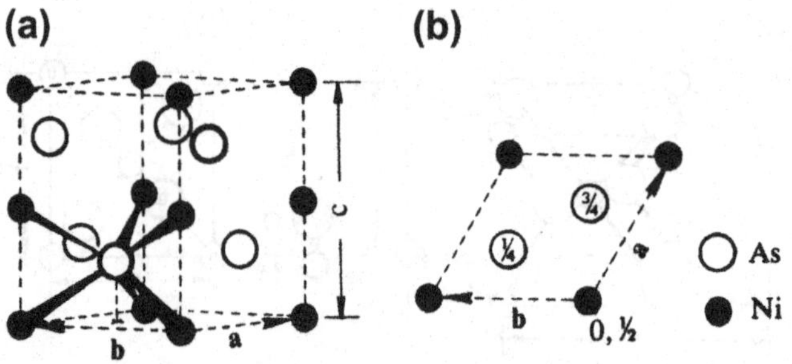

Figure A9.3 The nickel arsenide (NiAs) structure in (a) perspective and (b) projected onto (001).

materials with this structure are FeS, NiS, NiSn, CrS and CoTe. In the compound IrSb, the parameter u is very close to ¼. In general, the c/a ratio lies between the limits 1.2–1.7. There are a few cases where the A and B atoms occupy the opposite sites, such as PtB and NbN. In addition, in many cases, superstructures with very large cell dimensions are formed from this simple structure.

A9-4 CUPRITE

Cuprite, Cu_2O, adopts a cubic structure, space group $Pn\overline{3}(T_h^2)$, with atoms in special positions:

$$\text{Cu}: \quad 4b \quad (\tfrac{1}{4},\tfrac{1}{4},\tfrac{1}{4});(3/4,3/4,\tfrac{1}{4});(3/4,\tfrac{1}{4},3/4);(\tfrac{1}{4},3/4,3/4)$$

$$\text{O}: \quad 2a \quad (0,0,0);(\tfrac{1}{2},\tfrac{1}{2},\tfrac{1}{2})$$

and so there are two formula units per cell. In this structure, the metal atom is coordinated by two oxygens and each oxygen is surrounded by a tetrahedron of metal atoms (Fig. A9.4) for coordination (2:4). This structure is also found in compounds such as As_2O and Pb_2O.

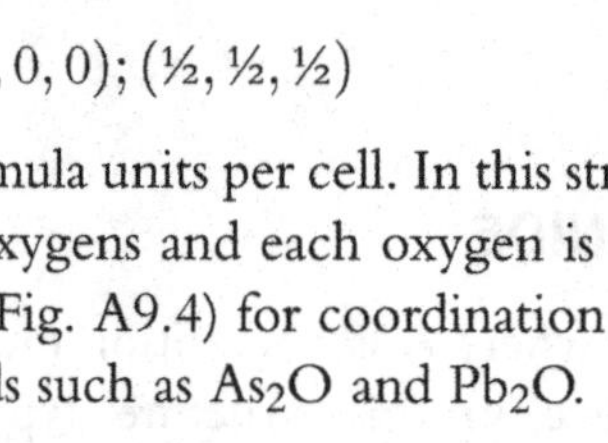

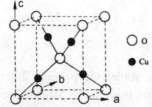

Figure A9.4 The cuprite (Cu_2O) structure.

A9-5 Nb_3Sn

Nb_3Sn is an important superconductor material. It crystallizes in a cubic structure with space group $Pm\bar{3}n(O_h^3)$ and atoms in the special positions:

$$\begin{aligned} \text{Nb}: \quad 6c \quad & (0, ½, ¼); (¼, 0, ½); (½, ¼, 0); \\ & (0, ½, 3/4); (3/4, 0, ½); (½, 3/4, 0) \\ \text{Sn}: \quad 2a \quad & (0, 0, 0); (½, ½, ½) \end{aligned}$$

and so there are two formula units per unit cell. This structure (Fig. A9.5) is found also in some intermetallic compounds such as $SiCr_3$ and SiV_3. In addition, one of the forms of tungsten (β-W) adopts this structure, with W atoms occupying both the Nb and the Sn positions. This structure is also often known as the A15 structure.

A9-6 PEROVSKITES AND THEIR SUPERSTRUCTURES

The perovskite prototype structure (Section 7.11.2) consists of a formula unit of ABX_3 in a cubic unit cell of edge length a ≈ 4 Å and space group $Pm\bar{3}m(O_h^1)$. The structure with the B atom at the origin or with the A atom at the origin is shown in Figs 7.5 and 7.14a, respectively. Although this is a simple structure, it is capable of considerable variation leading to complicated forms. We have already seen some of these in Section 7.11.2, where two modifications that lead to the formation of superstructures were described.

Inverse perovskites: The ABX_3 formula usually applies to materials such as $BaTiO_3$ or $NaMnF_3$, where X is an anion. However, there are examples of so-called inverse perovskites where X is a cation. Mn_3XC,

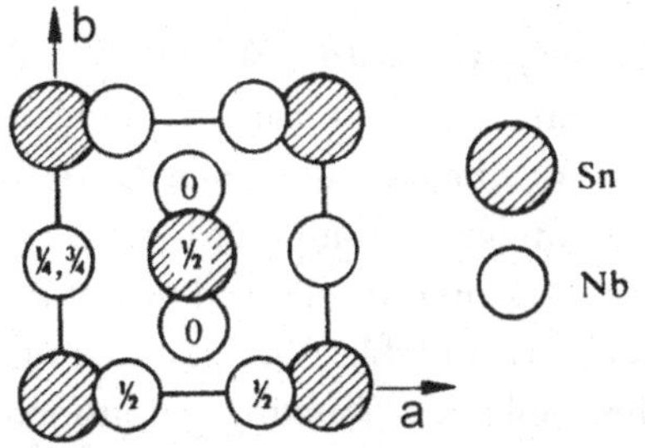

Figure A9.5 The (001)-projection of the Nb_3Sn structure.

where X = Zn, Ga, In or Sn, is a typical example with carbon at the center of the Mn octahedron (the B-site) and N in the interstices between the octahedra. These particular compounds exhibit a variety of magnetic orderings as well as structural phase transitions.

Cation ordering in perovskites: Suppose that at the B-site of a perovskite two different atoms are substituted, that is, $A(B'_{0.5}\ B''_{0.5})X_3$ such as $Pb_2(Sc,Ta)O_6$. Note that these B-site ions have a charge of 3+ and 5+. Then it is possible to order them in a regular fashion along one or more directions in the crystal. Suppose that the B′ and B″ atoms are arranged to alternate from one unit cell to the next in all directions. The result will be that the true unit-cell repeat is now double that of the original 4 Å pseudocubic perovskite cell with the atoms at the following positions of the space group $Fm\bar{3}m(O_h^5)$:

$$
\begin{array}{lll}
\text{Sc}: & 4a & (0,0,0);+ \\
\text{Ta}: & 4b & (\tfrac{1}{2},\tfrac{1}{2},\tfrac{1}{2});+ \\
\text{Pb}: & 8c & \pm(\tfrac{1}{4},\tfrac{1}{4},\tfrac{1}{4})+ \\
\text{O}: & 24e & \pm(x,0,0);(0,x,0);(0,0,x);+
\end{array}
$$

where + means adding on the face centering conditions. Ordering will lead to extra 'superlattice' points in reciprocal space, and therefore, to extra reflections in the diffraction pattern with indices h, k and l that are half integral. If we index these points with respect to the doubled cell, rather than with respect to the original cell, these indices will be odd-integral and the main reflections will be even-integral. These extra reflections are called superstructure, superlattice or difference reflections. They, in general, will be considerably weaker in intensity than the main reflections, and are sensitive to the difference in scattering powers of the B′ and B″ atoms. They are therefore of importance in determining the details of the particular ordering of the perovskite.

Tilted octahedra in perovskites: When one of the anion octahedra in the perovskite structure is tilted in some particular way, it causes tilting of the neighbouring octahedra. In general, the resulting structural arrangements can be complex and difficult to envisage. However, a notation and a method of visualizing the allowed tilts of the octahedra has been established and is widely used (Table A9.1). A total of 23 distinct structural types are possible when only cell-doubling is considered (the vast majority of cases).

Table A9.1 The 23 tilt systems for perovskites

Serial No.	Tilt system[2,3]	Space group	Sp. Gp. No.	Howard and Stokes[4]	Gopalan and Litvin[5]
(1)	$\mathbf{a}^+\mathbf{b}^+\mathbf{c}^+$	*Immm*	71	√	√
(2)	$\mathbf{a}^+\mathbf{b}^+\mathbf{b}^+$	*Immm*	71		√
(3)	$\mathbf{a}^+\mathbf{a}^+\mathbf{a}^+$	$Im\bar{3}$	204	√	√
(4)	$\mathbf{a}^+\mathbf{b}^+\mathbf{c}^-$	*Pmmn*	59		
(5)	$\mathbf{a}^+\mathbf{a}^+\mathbf{c}^-$	$P4_2/nmc$	137	√	√
(6)	$\mathbf{a}^+\mathbf{b}^+\mathbf{b}^-$	*Pmmn*	59		√
(7)	$\mathbf{a}^+\mathbf{a}^+\mathbf{a}^-$	$P4_2/nmc$	137		
(8)	$\mathbf{a}^+\mathbf{b}^-\mathbf{c}^-$	$A2_1/m11$	11	√	√
(9)	$\mathbf{a}^+\mathbf{a}^-\mathbf{c}^-$	$A2_1/m11$	11		
(10)	$\mathbf{a}^+\mathbf{b}^-\mathbf{b}^-$	*Pmnb*	62	√	√
(11)	$\mathbf{a}^+\mathbf{a}^-\mathbf{a}^-$	*Pmnb*	62		
(12)	$\mathbf{a}^-\mathbf{b}^-\mathbf{c}^-$	$F\bar{1}$	2	√	√
(13)	$\mathbf{a}^-\mathbf{b}^-\mathbf{b}^-$	*I2/a*	15	√	√
(14)	$\mathbf{a}^-\mathbf{a}^-\mathbf{a}^-$	$R\bar{3}c$	167	√	√
(15)	$\mathbf{a}^0\mathbf{b}^+\mathbf{c}^+$	*Immm*	71		√
(16)	$\mathbf{a}^0\mathbf{b}^+\mathbf{b}^+$	*I4/mmm*	139	√	√
(17)	$\mathbf{a}^0\mathbf{b}^+\mathbf{c}^-$	*Bmmb*	63	√	√
(18)	$\mathbf{a}^0\mathbf{b}^+\mathbf{b}^-$	*Bmmb*	63		
(19)	$\mathbf{a}^0\mathbf{b}^-\mathbf{c}^-$	*F2/m11*	12	√	√
(20)	$\mathbf{a}^0\mathbf{b}^-\mathbf{b}^-$	*Imcm*	74	√	√
(21)	$\mathbf{a}^0\mathbf{a}^0\mathbf{c}^+$	*C4/mmb*	127	√	√
(22)	$\mathbf{a}^0\mathbf{a}^0\mathbf{c}^-$	*F4/mmc*	140	√	√
(23)	$\mathbf{a}^0\mathbf{a}^0\mathbf{a}^0$	$Pm\bar{3}m$	221	√	√

[2] A.M. Glazer, *Acta Cryst.*, **B28**, 3384 (1972)
[3] K. Aleksandrov, *Kristallografiya*, **21**, 249 (1976); *Ferroelectrics*, **14**, 801 (1976)
[4] C.J. Howard & H.T. Stokes, *Acta Cryst.*, **B54**, 782–789 (1998)
[5] V. Gopalan & D.B. Litvin, *Nat. Mat.*,**10**, 376–381 (2011)

To understand this scheme, consider any particular tilt in terms of component tilts about the three tetrad axes of the octahedron. For small angles of tilt (as occur in practice) the component tilts can be taken about the pseudocubic axes of the untilted perovskite. The magnitudes of the angles of tilt about the three unit-cell axes are denoted by the letter code **abc** referring to tilts about [100], [010], [001], respectively. Equality of tilts is represented by repeating one of the letters, that is, **aac** means equal tilts about [100] and [010] with a different tilt about [001]. In addition to the magnitude of the tilt, it is also necessary to consider the sign of the tilt. If a particular octahedron is tilted *about* an axis, then the next octahedron *along* this axis can be tilted in-phase or out of phase with it. If the tilts are in-phase along a certain axis, a superscript + is used; if they are out of phase, the superscript is −. If the successive octahedra along the axis are not tilted, the superscript 0 is used. Thus, for example, the symbol $\mathbf{a}^0\mathbf{a}^0\mathbf{c}^-$ means that octahedra along the [001] direction are tilted about that direction alternately in opposite directions, but that there are no tilts about [100] and [010].

The result is a tetragonal structure with space group $F4/mmc(D_{4h}^{18})$. The unit cell is doubled in all three directions, so that superlattice reflections are found in reciprocal space at half-integral positions in each direction. That is, half-integer $h\ k\ l$ reflections appear in addition to the main integer $h\ k\ l$ reflections. The low-temperature structure of $SrTiO_3$ is probably the most famous example of this structure type and is shown in Fig. 7.16. Other perovskites with this tilt system are $KMnF_3$ and $RbCaF_3$.

The tilt system $\mathbf{a}^0\mathbf{a}^0\mathbf{c}^+$ has the octahedra along [001] tilted in-phase about this axis. The result is again a tetragonal structure, but with space group $C4/mmb(D_{4h}^5)$ and the unit-cell axes doubled only along **a** and **b**. Thus, in the diffraction pattern, half-integral reflections are found for h and k, but with l integral. Examples of crystals with this tilt system are $NaNbO_3$ (T_2 phase), $CsCaCl_3$ and $CsPbCl_3$.

Another interesting tilt system is $\mathbf{a}^+\mathbf{a}^+\mathbf{a}^+$, which consists of equal in-phase tilts about all three axes. This results in a body-centered cubic structure with space group $Im\overline{3}(T_h^5)$ and all axes doubled. Finally, the most common tilt system in perovskites is $\mathbf{a}^+\mathbf{b}^-\mathbf{b}^-$, the tilt system of $CaTiO_3$, orthorhombic, with space group $Pmnb(D_{2h}^{16})$, the mineral perovskite itself. Fig. A9.6 shows this tilt system in $NaMnF_3$.

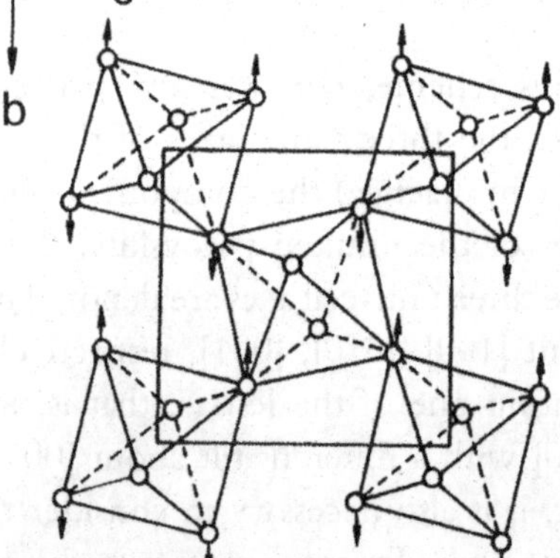

Figure A9.6 $NaMnF_3$ showing a layer of the $\mathbf{a}^+\ \mathbf{b}^-\ \mathbf{b}^-$ tilt system.

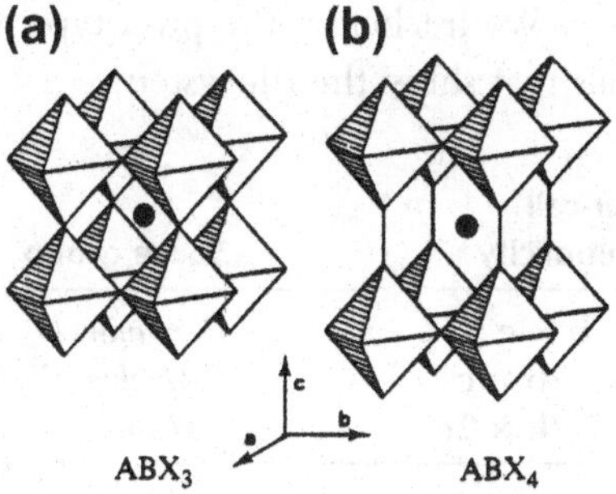

Figure A9.7 (a) The undistorted perovskite ABX_3 and (b) perovskite-like ABX_4.

$LiNbO_3$ can be described by $\mathbf{a^-a^-a^-}$ which gives a rhombohedral structure. However, in addition to the tilts, the positive and negatively charged atoms are displaced with respect to each other. This results in $LiNbO_3$ being ferroelectric. We list all of the 23 tilt systems together with their space groups in Table A9.1. More recently, Howard and Stokes applied group theory to these tilt systems and showed that 15 of the 23 possibilities were unique, while even more recently, Gopalan and Litvin developed a new type of roto-symmetry and this classification suggested 18 unique tilt arrangements. The tick marks in the table indicate the tilt systems found in each case.

A9-7 PEROVSKITE-LIKE PHASES

The relationship between the ABX_3 perovskite structure and the 'perovskite-like' ABX_4 structure can be seen by comparison of Fig. A9.7a and A9.7b. Many fluorides with the ABX_4[6] formula have this structure, such as $RbFeF_4$, $RbAlF_4$ and $TlAlF_4$. The tetragonal space group is $P4/mmm(D^1_{4h})$ with typically c ≈ 6.3 Å and a ≈ 3.6 Å for the fluorides. In the ABX_4 structure, because of the corner linkage, the BX_6 octahedra form two-dimensional networks (Fig. A9.7b) while in the perovskites they are three-dimensional (Fig. A9.7a). As in the ABX_3 perovskite, the ABX_4 structures undergo phase transitions associated with many different soft-phonon modes yielding a large variety of low-temperature space groups.

Simply by considering symmetry-reducing operations due to tilting of the BX_6 octahedra, it is found that there are 75 possible tilting systems

6 R. Deblieck, G. Van Tendeloo, J. Van Landuyt & S. Amelinckx, *Acta Cryst.*, **B41**, 319 (1985).

(75 possible structures). We list below the prototype and two examples of tilts along with crystals that show the tilt system.

Symbol	Unit-cell multiplicity	Space group	Material
$\mathbf{a^0a^0c^0}$	$\mathbf{a \times b \times c}$	$P4/mmm(D^1_{4h})$	$RbAlF_4$
$\mathbf{a^0a^0c^+}$	$\mathbf{2a \times 2b \times c}$	$C4/mbm(D^5_{4h})$	$KAlF_4$
$\mathbf{a^0a^0c^-}$	$\mathbf{2a \times 2b \times 2c}$	$F4/mmc(D^{18}_{4h})$	NH_4AlF_4

The International space group symbols listed for the tilted systems are non-standard (the standard ones are listed in the *ITA*). The ones listed here enable an easy comparison with the high-temperature prototype structure.

There are other 'perovskite-like' compounds where consideration of BX_6 octahedra tilts leads to an understanding of a variety of structural modifications. The unusual transition in $La_{1.9}Ba_{0.1}CuO_4$ (Section 7.13.1) from *Cmce* to $P4_2/ncm$ can also be described by tilts. First, in the orthorhombic phase, each layer has the tilt system $\mathbf{a^-a^-c^0}$. There are two layers of octahedra in the structure with the same tilt system, and so we can describe this as $\mathbf{a^-a^-c^0; a^-a^-c^0}$. Second, in the tetragonal phase, one of the layers perpendicular to [001] has the tilt system $\mathbf{a^-b^0c^0}$. The next layer is rotated through 90° and therefore has the tilt $\mathbf{b^0a^-c^0}$ so that the combined tilt system is $\mathbf{a^-b^0c^0; b^0a^-c^0}$.

There are a large number of crystals with the general formula A_2BX_6 typified by K_2PtCl_6 in the cubic space group $Fm\bar{3}m(O^5_h)$ with four formula units per unit cell and with a ≈ 10 Å for the chlorides. The BX_6 octahedra in these compounds are mechanically isolated from each other, only connected by forces via intermediate atoms. These crystals also display a large variety of tilts.

A9-8 STRUKTURBERICHT NOTATION

The Strukturbericht symbol is a fairly arbitrary notation that describes a crystal structure type. It is not much used today, except perhaps by metallurgists. Some examples are given below. The basic idea of the notation is to use a capital letter followed by a number to indicate the structure type. Of the more than 100 symbols, below are just a few examples.

*A*1 − fcc	*A*2 − bcc	*A*3 − hcp	*A*4 − diamond
*B*1 − NaCl	*B*2 − CsCl	*B*3 − ZnSi	*B*4 − wurtzite
*C*1 − CaF_2	*C*2 − FeS_2	*C*3 − Cu_2O	*C*4 − TiO_2
$E2_1$ − $CaTiO_3$		$H1_1$ − $MgAl_2O_4$	
$L1_0$ − AuCu	$L1_2$ − Cu_3Au		

The A15-Nb_3Sn structure was discussed in Section A9.5.

APPENDIX 10

Translational Subgroups of Magnetic Space Groups

Triclinic

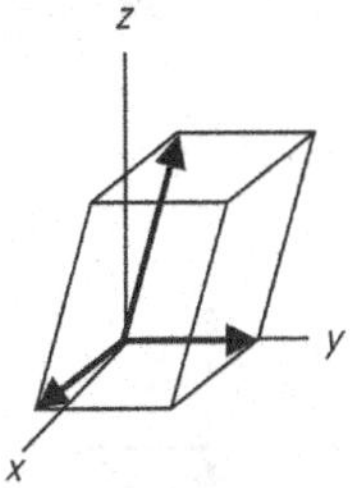

$P = P_{a,b,c}$

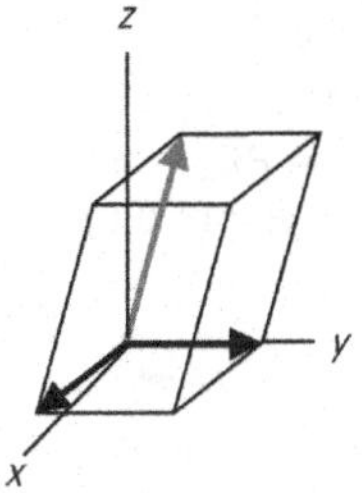

$P_{2a} = P_{a,b,2c}$
$\mathbf{t}_\alpha = c = (0,0,1)$

Monoclinic (2-fold axis along y)

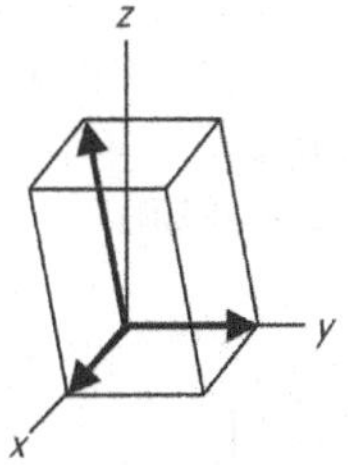

$P = P_{a,b,c}$

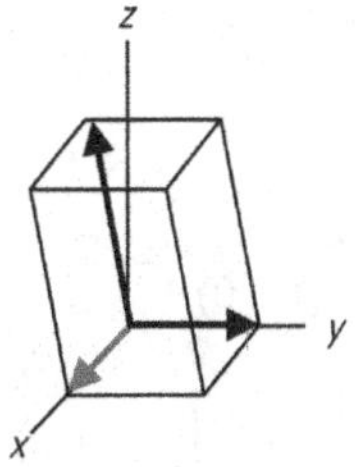

$P_{2a} = P_{2a,b,c}$
$\mathbf{t}_\alpha = a = (1,0,0)$

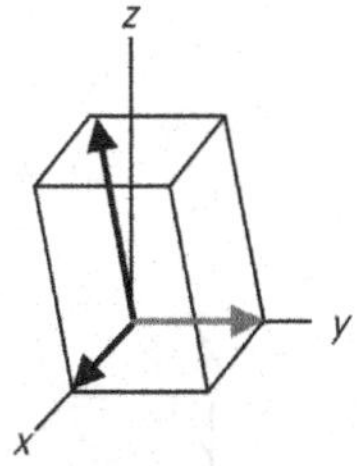

$P_{2b} = P_{a,2b,c}$
$\mathbf{t}_\alpha = b = (0,1,0)$

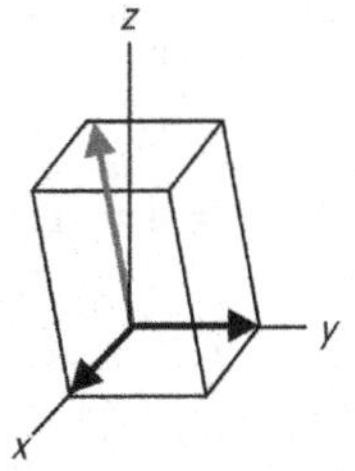

$P_{2c} = P_{a,b,2c}$
$\mathbf{t}_\alpha = c = (0,0,1)$

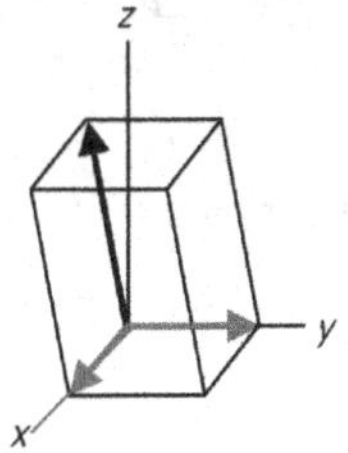

$P_C = P_{2a,a+b,c} = P_{a-b,a+b,c}$
$\mathbf{t}_\alpha = a = (1,0,0)$

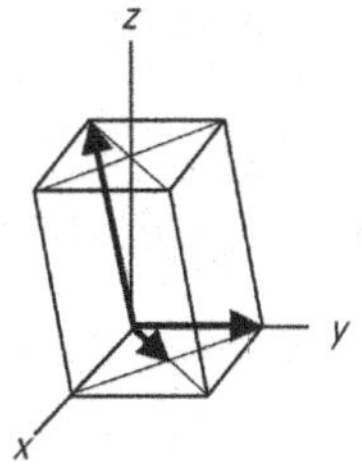

$C = C_{\frac{1}{2}(a+b),b,c}$

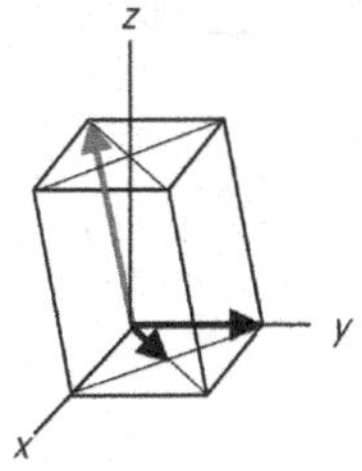

$C_{2c} = C_{\frac{1}{2}(a+b),b,2c}$
$\mathbf{t}_\alpha = c = (0,0,1)$

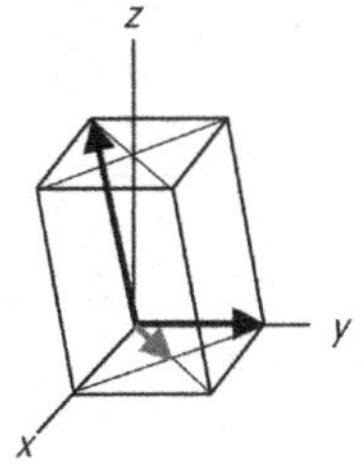

$C_P = C_{a+b,b,2,c} = C_{a,b,c}$
$\mathbf{t}_\alpha = \frac{1}{2}(a+b) = (\frac{1}{2},\frac{1}{2},0)$

Orthorhombic

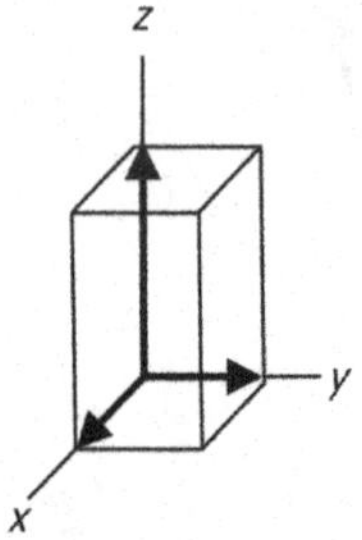

$P = P_{a,b,c}$

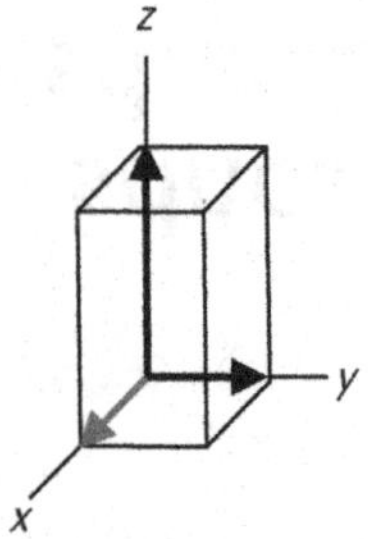

$P_{2a} = P_{2a,b,c}$
$\mathbf{t}_\alpha = a = (1,0,0)$

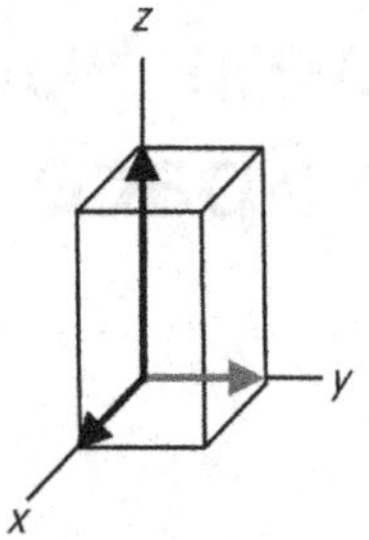

$P_{2b} = P_{a,2b,c}$
$\mathbf{t}_\alpha = b = (0,1,0)$

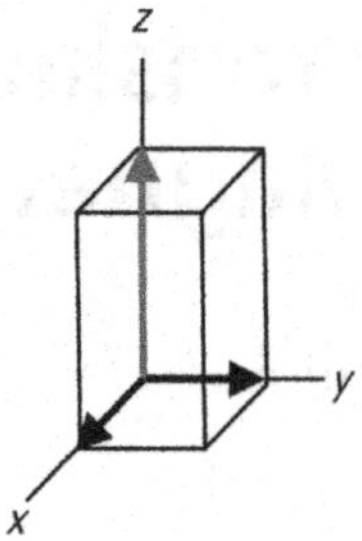

$P_{2c} = P_{a,b,2c}$
$\mathbf{t}_\alpha = c = (0,0,1)$

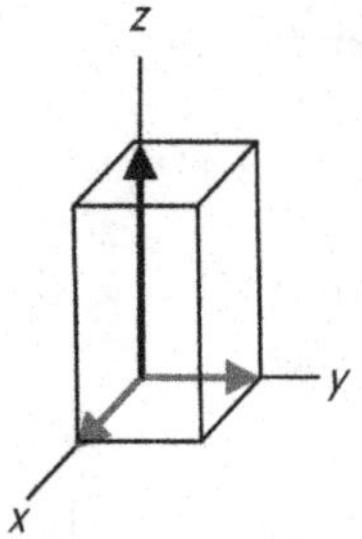

$P_C = P_{2a,a+b,c}$
$\mathbf{t}_\alpha = a = (1,0,0)$

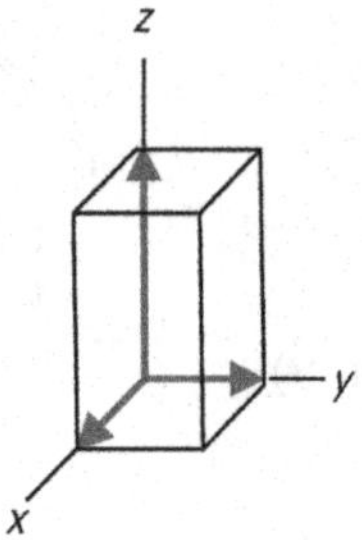

$P_F = P_{2a,a+b,a+c}$
$= P_{a+b,b+c,a+c}$
$\mathbf{t}_\alpha = a = (1,0,0)$

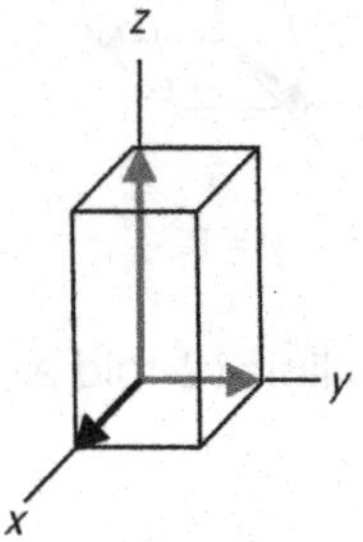

$P_A = P_{a,2b,b+c}$
$\mathbf{t}_\alpha = a = (0,1,0)$

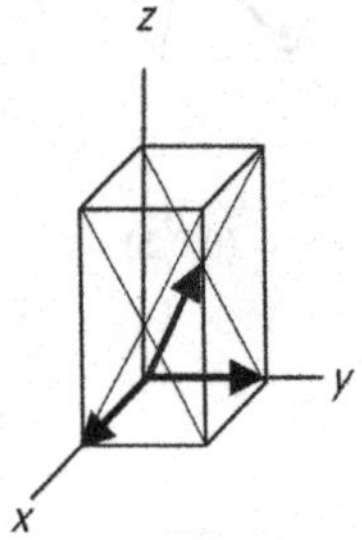

$A = A_{a,b,½(b+c)}$

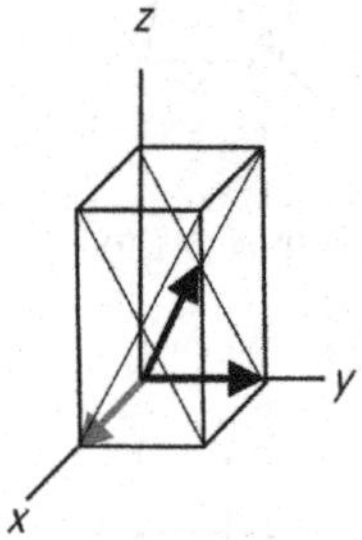

$A_{2a} = A_{2a,b,b+c}$
$\mathbf{t}_\alpha = a = (1,0,0)$

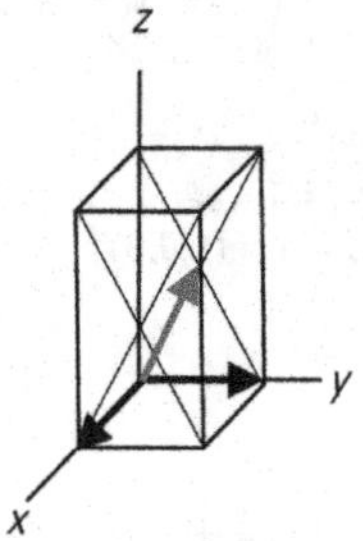

$A_P = A_{a,b,c}$
$\mathbf{t}_\alpha = ½(b+c) = (0,½,½)$

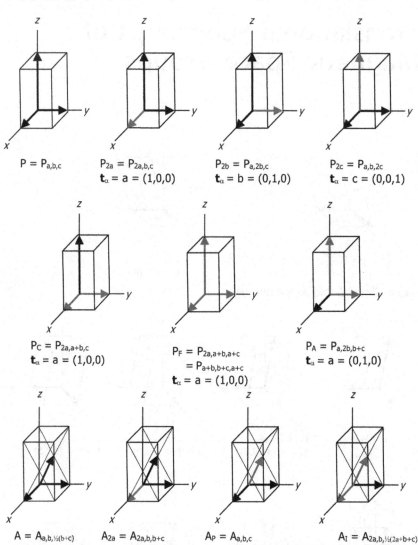

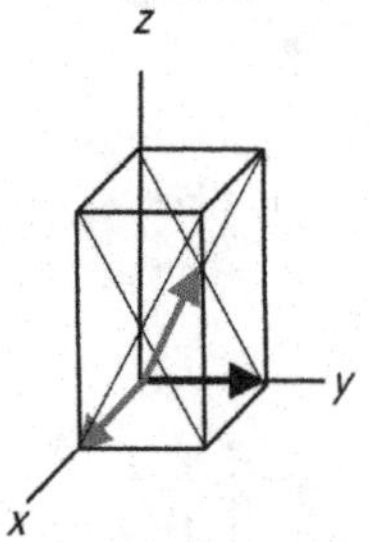

$A_I = A_{2a,b,½(2a+b+c)}$
$\mathbf{t}_\alpha = a = (1,0,0)$

Orthorhombic (cont.)

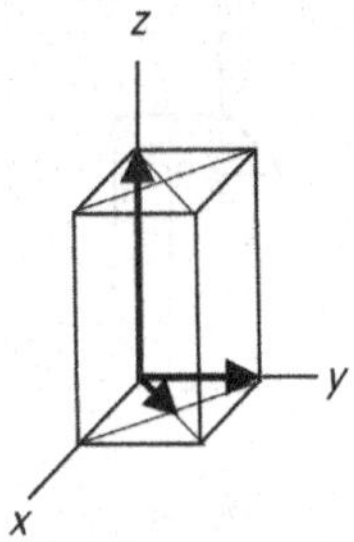

$C = C_{\frac{1}{2}(a+b),b,c}$

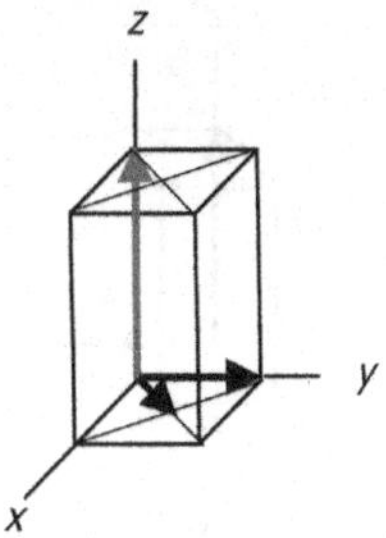

$C_{2c} = C_{\frac{1}{2}(a+b),b,2c}$

$\mathbf{t}_\alpha = c = (0,0,1)$

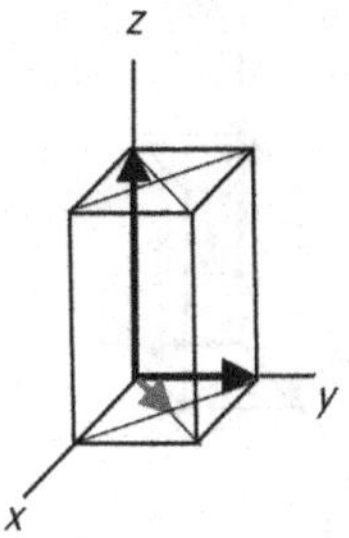

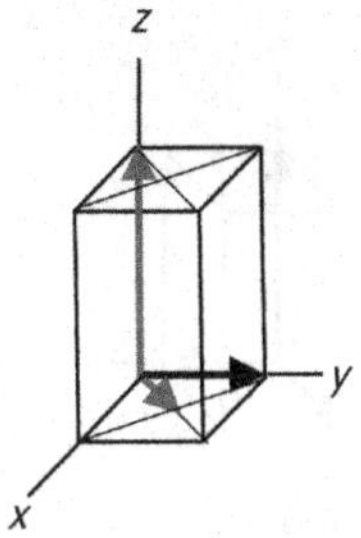

$C_I = C_{a,b,\frac{1}{2}(a+b+2c)}$

$\mathbf{t}_\alpha = c = (0,0,1)$

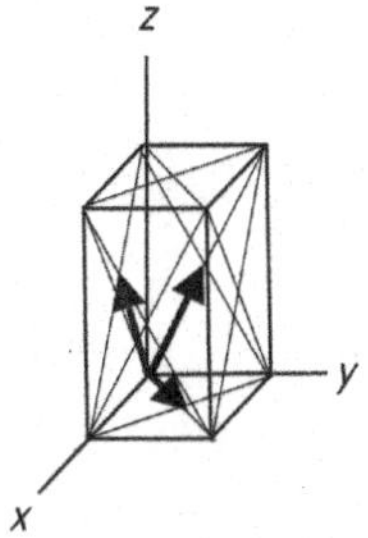

$F = F_{\frac{1}{2}(a+b),\frac{1}{2}(b+c),\frac{1}{2}(a+c)}$

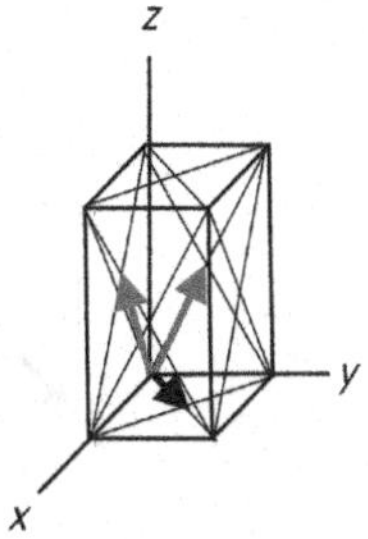

$F_C = F_{\frac{1}{2}(a+b),b,c}$

$\mathbf{t}_\alpha = \frac{1}{2}(a+c) = (\frac{1}{2},0,\frac{1}{2})$

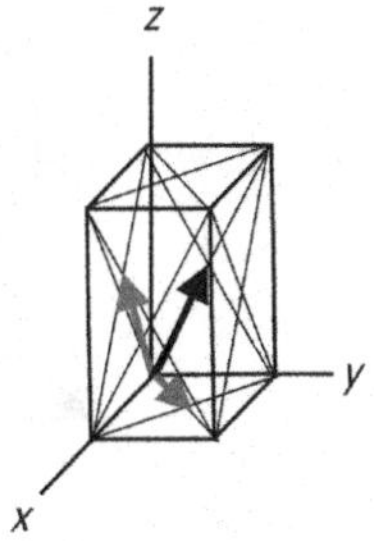

$F_A = F_{\frac{1}{2}(b+c),c,a}$

$\mathbf{t}_\alpha = \frac{1}{2}(a+b) = (\frac{1}{2},\frac{1}{2},0)$

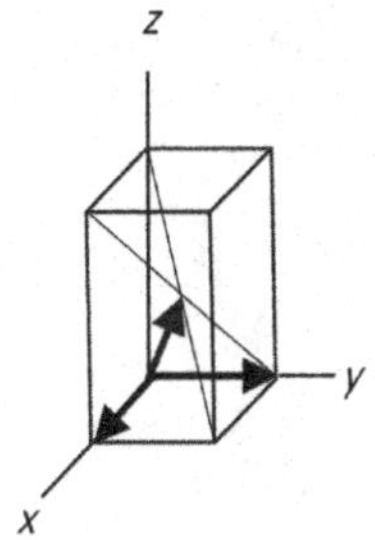

$I = I_{a,b,\frac{1}{2}(a+b+c)}$

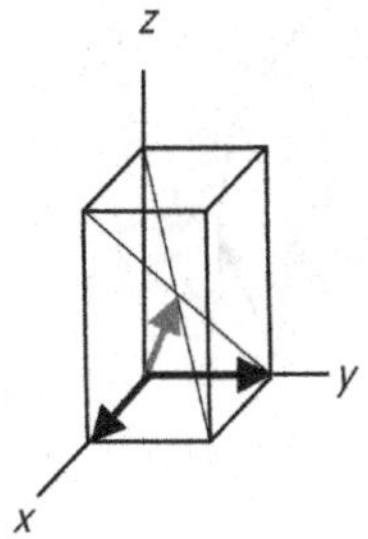

$I_P = I_{a,b,c}$

$\mathbf{t}_\alpha = \frac{1}{2}(a+b+c) = (\frac{1}{2},\frac{1}{2},\frac{1}{2})$

Tetragonal

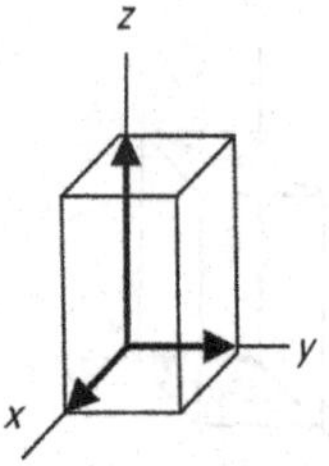

$P = P_{a,b,c}$

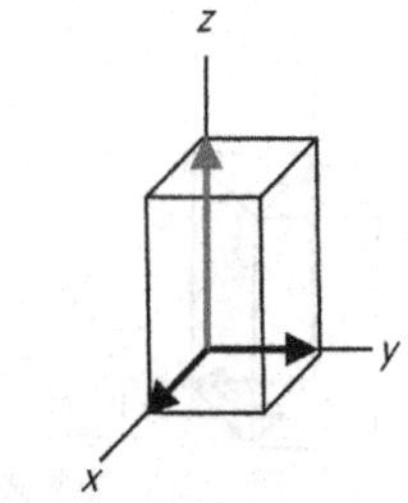

$P_{2c} = P_{a,b,2c}$
$\mathbf{t}_\alpha = c = (0,0,1)$

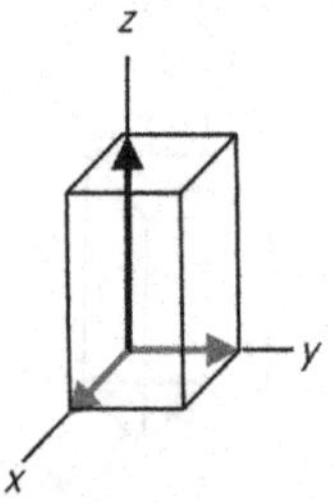

$P_{a-b,a+b}(P_C) = P_{a-b,a+b,c}$
$= P_P = P_{2a,a+b,c}$
$\mathbf{t}_\alpha = a = (1,0,0)$

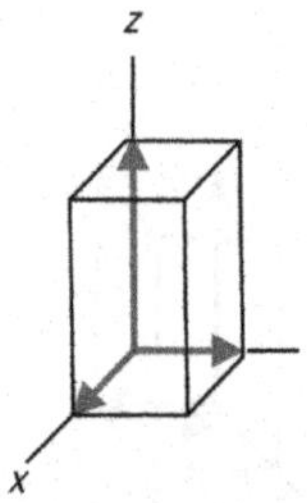

$P_I = P_{a-b,a+b,a+c}$
$\mathbf{t}_\alpha = a = (1,0,0)$

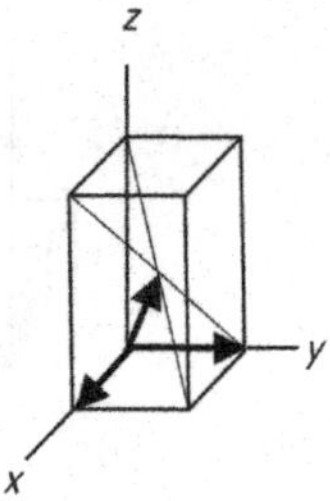

$I = I_{a,b,\frac{1}{2}(a+b+c)}$

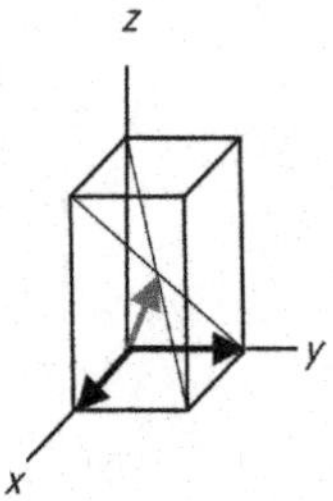

$I_P = I_{a,b,c}$
$\mathbf{t}_\alpha = \frac{1}{2}(a+b+c) = (\frac{1}{2},\frac{1}{2},\frac{1}{2})$

Trigonal

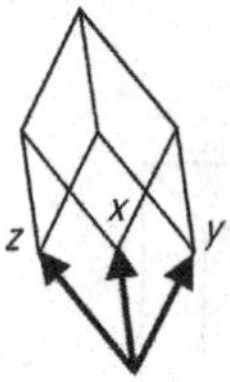

$R = R_{a,b,c}$

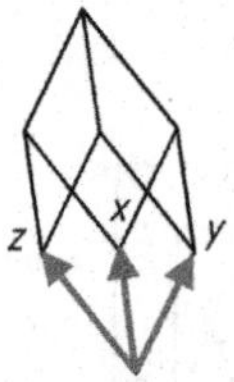

$R_{2a,a+b,a+c} = R_R = R_{a+b,b+c,a+c}$
$\mathbf{t}_\alpha = a = (1,0,0)$

Trigonal (Hexagonal Axes)

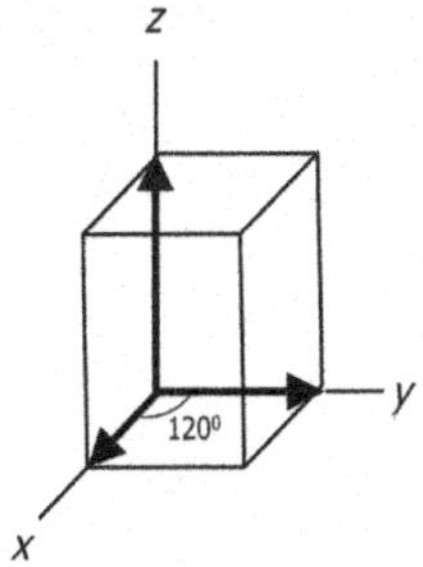

$P = P_{a,b,c}$

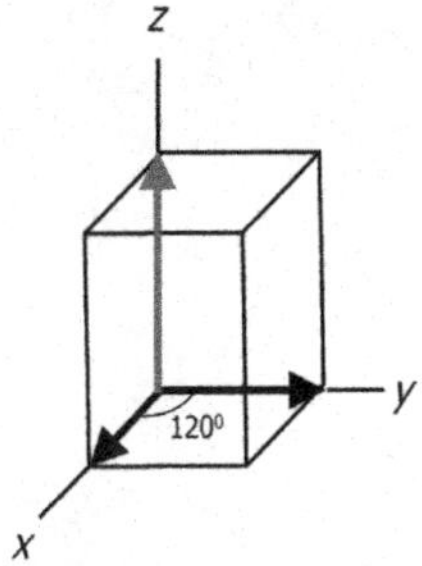

$P_{2c} = P_{a,b,2c}$
$\mathbf{t}_{\alpha} = c = (0,0,1)$

Cubic

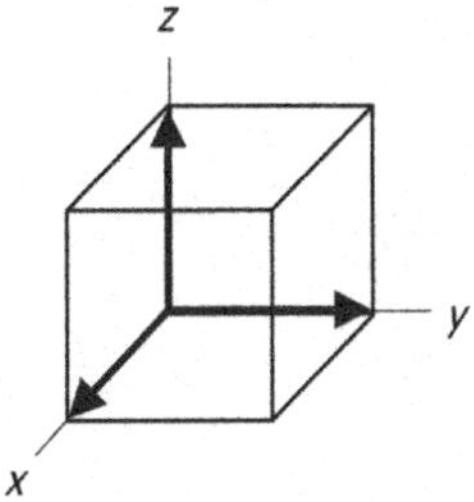

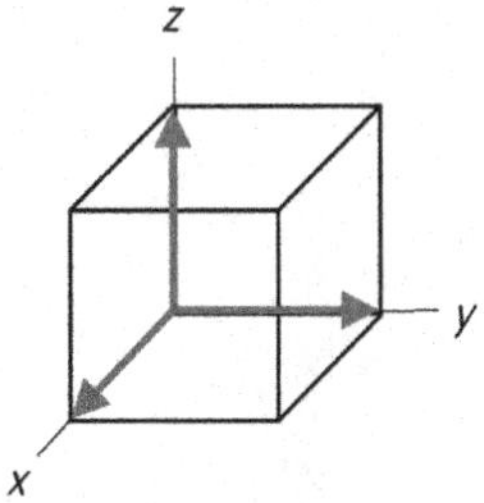

$P_F = P_{2a,a+b,a+c}$
$\mathbf{t}_{\alpha} = a = (1,0,0)$

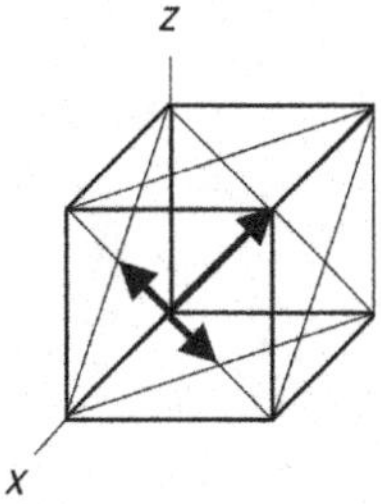

$F = F_{\frac{1}{2}(a+b),\frac{1}{2}(b+}$

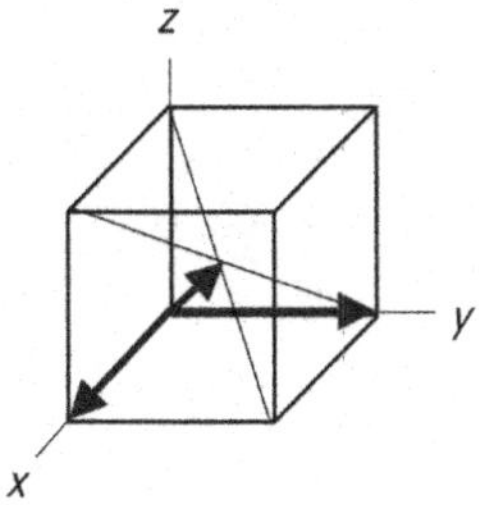

$I = I_{a,b,\frac{1}{2}(a+b+c)}$

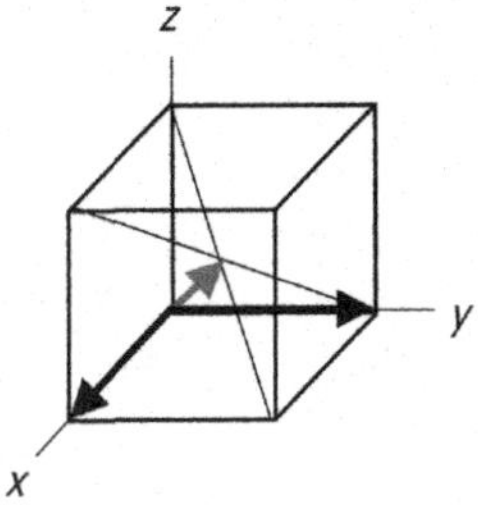

$I_P = I_{a,b,c}$
$\mathbf{t}_{\alpha} = \frac{1}{2}(a+b+c) = (\frac{1}{2},\frac{1}{2},\frac{1}{2})$

APPENDIX 11

Cubic Space Group Diagrams

The *ITA* includes diagrams for the cubic space groups which need a little clarification. In particular, 3-fold type axes are shown that are inclined to the direction of view, and at first sight look like they are not related by the other symmetry operations in the space group. This arises because the inclined axes are drawn where they intersect the plane of projection. For instance, consider the diagram (top) for space group $P2_13(T^4)$.

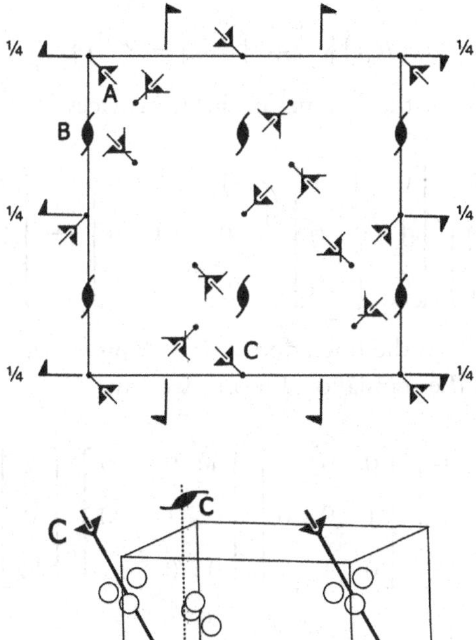

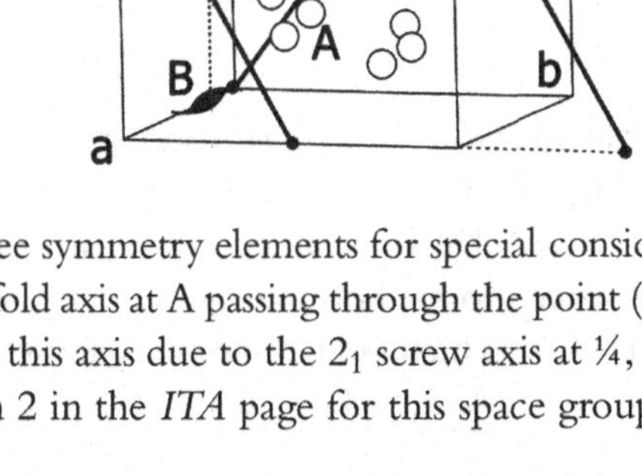

We have marked three symmetry elements for special consideration. Starting from the inclined 3-fold axis at A passing through the point (0, 0, 0), let us ask what is the effect on this axis due to the 2_1 screw axis at ¼, 0, z marked at B (symmetry operation 2 in the *ITA* page for this space group)?

In order to answer this, we need to perform a similarity transformation that relates symmetry operations in the same class. In other words, if there are symmetry operations X, E and F where we can write

$$D = XFX^{-1}$$

then operations D and F are in the same class. Thus, in the present case, we need to consider the following similarity transformation

$$\{2[001]|(1/2, 0, 1/2)\}\{3[111]|0, 0, 0\}\{2[001]^{-1}| - 2[001]^{-1}(1/2, 0, 1/2)\}$$

where we have used the identity

$$\{R|\mathbf{t})\}^{-1} = \{R^{-1}| - R^{-1}\mathbf{t}\}$$

Then we have for the product of the R matrices

$$\begin{bmatrix} -1 & 0 & 0 \\ 0 & -1 & 0 \\ 0 & 0 & 1 \end{bmatrix} \begin{bmatrix} 0 & 0 & 1 \\ 1 & 0 & 0 \\ 0 & 1 & 0 \end{bmatrix} \begin{bmatrix} -1 & 0 & 0 \\ 0 & -1 & 0 \\ 0 & 0 & 1 \end{bmatrix} = \begin{bmatrix} 0 & 0 & -1 \\ 1 & 0 & 0 \\ 0 & -1 & 0 \end{bmatrix}$$

which corresponds to the operation $3^2(C_3^2)$ along $[11\bar{1}]$.

In the case of the translational terms, we have

$$\begin{bmatrix} -1 & 0 & 0 \\ 0 & -1 & 0 \\ 0 & 0 & 1 \end{bmatrix} \begin{bmatrix} 0 & 0 & 1 \\ 1 & 0 & 0 \\ 0 & 1 & 0 \end{bmatrix} \begin{bmatrix} 1 & 0 & 0 \\ 0 & 1 & 0 \\ 0 & 0 & -1 \end{bmatrix} \begin{bmatrix} 1/2 \\ 0 \\ 1/2 \end{bmatrix} + \begin{bmatrix} 1/2 \\ 0 \\ 1/2 \end{bmatrix}$$

$$= \begin{bmatrix} 1 \\ -1/2 \\ 1/2 \end{bmatrix}$$

This is equivalent (see the Bilbao server program SYMMETRY OPERATIONS http://www.cryst.ehu.es/cryst/matrices.html) to a 3-fold axis through $\bar{x}+1$, $\bar{x}+$ ½, x, number (8) in the *ITA* list of symmetry operations in this space group. This intersects the plane of projection given by $x=0$ i.e. at 1, ½, 0 which is at position C shown in the space group diagram.

In order to aid in visualizing this, the lower diagram shows this effect in perspective. Three atoms at A are related to each other by the 3-fold axis through the origin of the unit cell. The 2-fold screw axis at B rotates these three atoms about the **c**-axis and translates them through **c**/2 to the position marked C. It is then easy to see that the 3-fold axis relating these atoms intersects the **ab**-plane of the unit cell at 1, ½, 0.

APPENDIX 12

Pitfalls

It is common to find incorrect or misleading statements about crystallographic symmetry. Here, we list some of these crystallographic horrors, and hope that you will not make such mistakes yourself!

Don't	use the term 'lattice structure'. This is a misleading term since it confuses the concept of a crystal structure with that of a lattice. Use the term 'lattice' when you mean the infinite regular array of points and use the term 'crystal structure' when talking about the regular array of atoms.
Don't	say 'the diamond lattice' when you mean the diamond crystal structure. The diamond structure can be described by a basis of two atoms convoluted with a face-centered cubic lattice. NaCl, Cu, GaAs, Si and probably 10,000 other structures all have an fcc lattice but with different atomic arrangements and hence, different structures.
Don't	mix up atoms with lattice points. Lattice points are infinitesimal points in space. Atoms are physical objects. Of course, some elements have simple crystal structures with only one atom per lattice point. For example, Cu has an fcc structure with one Cu atom per lattice point. Similarly, iron crystallizes in a body-centered cubic structure with one Fe atom per lattice point. Thus, for such simple structures, the drawings of the lattices closely resemble those of the crystal structures.
Don't	fall into the trap of describing, for example, the CsCl structure as bcc. It is primitive, since the atom at the center of the unit cell cannot be transformed by a symmetry operation into the one at the corner. No symmetry operation that we know of will transform Cs into Cl! It is important to understand this difference, for example, in looking at the electronic properties of such a crystal, in which case the correct choice of Brillouin zone needs to be made.
Don't	describe any crystal structure in terms of 'interpenetrating fcc lattices', 'interpenetrating bcc lattices' etc. You will only confuse others and perhaps even yourself.

(Continued)

Don't	define a cubic crystal in terms of the relationships between its unit-cell axes. Remember that it is the symmetry (in this case the presence of four 3-fold axes) that decides whether a particular crystal belongs to the cubic crystal system, and not the relationships between cell edges, which may appear to be equal within the accuracy of the measurements. Similar considerations apply to all of the crystal systems.
Don't	think that all cubic crystals must contain 4-fold axes. For instance, crystals in cubic point groups 23 or $m\bar{3}$ contain neither proper nor improper 4-fold axes.
Don't	*define* a Bravais lattice as an array of points given by the formula $$\mathbf{t} = n_1\mathbf{a} + n_2\mathbf{b} + n_3\mathbf{c}$$ This is just a definition of any lattice. The term 'Bravais' refers to the uniqueness of all possible lattices in the vector space being considered (there are 14 in three-dimensional space).
Don't	talk about a 'monatomic crystal'. This is a meaningless phrase.
Don't	confuse 'symmetry element' with 'symmetry operation'.
Don't	*define* a Brillouin zone by the Wigner–Seitz cell. The Wigner–Seitz construction produces a primitive unit cell and this can be made in real or reciprocal space. A Brillouin zone is given by a unit cell in reciprocal space. The Wigner–Seitz cell is only one possible unit cell out of an infinite set of choices. However, the Wigner–Seitz cell is the most useful (especially for crystals of high symmetry) unit cell when describing a Brillouin zone.
Don't	panic!

BIBLIOGRAPHY

TABLES

Bradley, C.J., Cracknell, A.P., 1972. The Mathematical Theory of Symmetry in Solids. Clarendon, Oxford (republished 2010).
Faddeyev, D.K., 1964. Tables of the Principal Unitary Representations of Fedorov Groups. Pergamon, Oxford.
Hahn, T. (Ed.), 2006. International Tables for Crystallography, vol. A. Space Group Symmetry. Published for the International Union of Crystallography by Wiley.
Henry, N.F.M., Lonsdale, K. (Eds.), 1952, 1965, 1969. International Tables for X-Ray Crystallography, vol. 1. Kynoch, Birmingham.
Koptsik, V.A., 1966. Shubnikov Groups. Handbook on the Symmetry and Physical Properties of Crystal Structures. University Press, Moscow (in Russian).
Kovalev, O.V., 1965. Irreducible Representations of the Space Groups. Gordon and Breach, New York.
Stokes, H.T., Hatch, D.M., 1988. Isotropy Subgroups of the 230 Crystallographic Space Groups. World Scientific, New Jersey.
Zak, J., Casher, A., Glück, M., Gur, Y., 1969. The Irreducible Representations of Space Groups. Benajmin, New York.

DESCRIPTION OF SPACE GROUPS

Barlow, W., 1894. On the geometric properties of homogeneous rigid structures and their application to crystals. Z. Krist. 23, 1 (in German).
Buerger, M.J., 1956, 1963. Elementary Crystallography, an Introduction to the Fundamental Geometrical Features of Crystals. Wiley, New York.
Buerger, M.J., 1971. Introduction to Crystal Geometry. McGraw-Hill, New York.
Burckhardt, J.J., 1966. The Motion Groups of Crystallography. Birkhauser, Basel (in German).
Burckhardt, J.J., 1967. The history of the discovery of the 230 space groups. Arch. His. Exact Sci. 4, 235 (in German).
Burns, G., 1985. Solid State Physics. Academic Press, New York.
Dauter, Z., Jaskolski, M., 2010. How to read (and understand) Volume A of International Tables for Crystallography: an introduction for nonspecialists. J. Appl. Cryst. 43, 1155–1171.
Fedorov, E.S., 1891. Symmetry of regular systems of figures. Vseross. min. Obshch., Zap. (II) 28, 1 (in Russian). A full translation, by D. and K. Harker, is given in the American Crystallographic Association Monograph No. 7.
Hilton, H., 1903. Mathematical Crystallography and the Theory of Groups of Movements. Clarendon, Oxford (reprinted by Dover, New York, 1963).
Schoenflies, A., 1891. Theory of Crystal Structure. Teubner, Leipzig (in German).
Sohncke, L., 1879. Development of a Theory of Crystal Structure. Teubner, Leipzig (in German).
Terpstra, P., 1955. Introduction to the Space Groups. Walters, Groningen.
Wyckoff, R.W.G., 1930. The Analytical Expression of the Results of the Theory of Space Groups. Carnegie Inst., Washington.

GENERAL CRYSTALLOGRAPHY, SYMMETRY AND STRUCTURES

Acta Crystallographica., IUCr, 1948. Munksgaard. Copenhagen
Bragg, W.L., 1962. The Crystalline State. Bell, London.
Galasso, F.S., 1969. Structure, Properties and Preparation of Perovskite-Type Compounds. Pergamon Press, Oxford.
Giacovazzo, C., Monaco, H.L., Artioli, G., Viterbo, D., Milaneso, M., Ferraris, G., Gilli, G., Gilli, P., Zanotti, G., Catti, M., 2011. In: Giacovazzo, C. (Ed.), Fundamentals of Crystallography, third ed. IUCr Texts on Crystallography, IUCr/Oxford University Press.
Glazer, A.M., 1987. The Structure of Crystals. Hilger, Bristol.
Hammond, C., 2009. The Basics of Crystallography and Diffraction. IUCr/Oxford University Press.
Hargittai, I., Hargittai, M., 1994. Symmetry: A Unifying Concept. Shelter, California.
James, R.W., 1965. The Optical Principles of the Diffraction of X-rays. Bell, London.
Janot, C., 1992. Quasicrystal: A Primer. Oxford Science Publications.
Joshua, S.J., 1991. Symmetry Principles and Magnetic Symmetry in Solid State Physics. Hilger, New York.
Julian, M.M., 2008. Foundations of Crystallography. CRC, Boca Raton.
Ladd, M. 2013. Symmetry of Crystals and Molecules, O.U.P.
McKie, D., McKie, C., 1986. Essentials of Crystallography. Blackwells, Oxford.
Megaw, H.D., 1973. Crystal Structures: A Working Approach. Saunders, Philadelphia.
Nye, J.F., 1957. Physical Properties of Crystals. Clarendon, Oxford.
Radaelli, P.G., 2011. Symmetry in Crystallography. IUCr/Oxford University Press.
Schwarzenbach, D., 1993. Crystallography. Wiley, Chichester.
Senechal, M., 1990. Crystalline Symmetries: An Informal Mathematical Introduction. Adam Hilger, Bristol.
van Smaalen, S., 2007. Incommensurate Crystallography. In: IUCr Texts on Crystallography. IUCr/Oxford University Press.
Wadhawan, V.K., 2011. Latent, Manifest, and Broken Symmetry: A Bottom-up Approach to Symmetry, with Implications for Complex Networks. CreateSpace, Charleston, SC, USA. ISBN 1463766718.
de Wolff, P.M., et al., 1989. Definition of symmetry elements in space groups and point groups. Acta Cryst. A45, 494.
Zeitschrift für Kristallographie. Akademische Verlagsgesellschaft, Frankfurt.

GROUP THEORY

Bouckaert, L.P., Smoluchowski, R., Wigner, E., 1936. Theory of Brillouin zones and symmetry properties of wave functions in crystals. Phys. Rev. 50, 58.
Bradley, C.J., Cracknell, A.P., 1972. (republished 2010) See under Tables.
Burns, G., 1977. Introduction to Group Theory with Applications. Academic Press, New York.
Koster, G.F., 1957. Space groups and their representations. Solid State Phys. 5, 173.
Lax, M., 1974. Symmetry Principles in Solid State and Molecular Physics. Wiley, New York.
Lyubarskii, G. Ya, 1960. The Application of Group Theory in Physics. Pergamon, Oxford.
Streitwolf, H.W., 1967. Group Theory in Solid State Physics. MacDonald, London.
Weinreich, G., 1965. Solids: Elementary Theory for Advanced Students. Wiley, New York.
Wondratschek, H., Neubüser, J., 1967. Determination of the symmetry elements of a space group from the general positions listed in International Tables for X-Ray Crystallography, vol. I. Acta Cryst. 23, 349.

INTERNET

International Tables for Crystallography Resources: http://it.iucr.org/resources/.

Bilbao crystallographic server: a complete set of symmetry programs. http://www.cryst.ehu.es/.

Aroyo, M.I., Perez-Mato, J.M., Orobengoa, D., Tasci, E., de la Flor, G., Kirov, A., 2011. Crystallography online: Bilbao crystallographic server. Bulg. Chem. Commun. 43 (2), 183–197.

Aroyo, M.I., Perez-Mato, J.M., Capillas, C., Kroumova, E., Ivantchev, S., Madariaga, G., Kirov, A., Wondratschek, H., 2006. Bilbao crystallographic server I: databases and crystallographic computing programs, Z. Krist. 221 (1), 15–27.

Aroyo, M.I., Kirov, A., Capillas, C., Perez-Mato, J.M., Wondratschek, H., 2006. Bilbao crystallographic server II: representations of crystallographic point groups and space groups. Acta Cryst. A62, 115–128.

E-crystallography course: http://escher.epfl.ch/eCrystallography/.

Interactive structure factor tutorial: http://www.ysbl.york.ac.uk/~cowtan/sfapplet/sfintro.html.

ISODISTORT: a tool for exploring structural distortions of crystal structures. http://stokes.byu.edu/isodistort.html.

A Hypertext book of crystallographic space group diagrams: http://img.chem.ucl.ac.uk/sgp/mainmenu.htm.

SOLUTIONS

CHAPTER 1

2. $$\begin{bmatrix} 1 & 0 & 0 \\ 0 & -1 & 0 \\ 0 & 0 & -1 \end{bmatrix} \begin{bmatrix} 0 & -1 & 0 \\ 1 & 0 & 0 \\ 0 & 0 & 1 \end{bmatrix} = \begin{bmatrix} 0 & -1 & 0 \\ -1 & 0 & 0 \\ 0 & 0 & -1 \end{bmatrix} = \{2[1\bar{1}0]\}$$

$$\begin{bmatrix} 0 & -1 & 0 \\ 1 & 0 & 0 \\ 0 & 0 & 1 \end{bmatrix} \begin{bmatrix} 1 & 0 & 0 \\ 0 & -1 & 0 \\ 0 & 0 & -1 \end{bmatrix} = \begin{bmatrix} 0 & 1 & 0 \\ 1 & 0 & 0 \\ 0 & 0 & -1 \end{bmatrix} = \{2[110]\}$$

3.

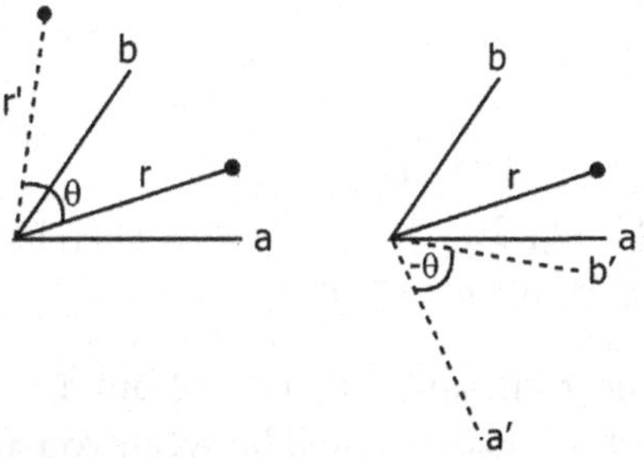

Note that in the first diagram, the position vector is rotated through an anticlockwise angle θ, whereas in the second diagram, the axes are rotated clockwise through the angle θ. Therefore, we can write that

$$R_{active} = R^{-1}_{passive}$$

Then

$$R_{active}S_{active} = T_{active}$$

$$R^{-1}_{passive}S^{-1}_{passive} = T^{-1}_{passive}$$

$$\therefore R^{-1}_{passive}S^{-1}_{passive}S_{passive}R_{passive} = T^{-1}_{passive}S_{passive}R_{passive}$$

$$\therefore I = T^{-1}_{passive}S_{passive}R_{passive}$$

$$\therefore S_{passive}R_{passive} = T_{passive}$$

4. **(a)** Since $n/2$ must be an odd integer, then n must be even. Write $S_n^{n/2} = C_n^{n/2}\sigma_h^{n/2}$. Then $C_n^{n/2} = C_2$ always. Combined with odd number of horizontal reflections, resulting in a change of hand gives a center of inversion.
   (b) If $n/2$ even then so is n. This time there are an even number of horizontal reflections, thus restoring the hand. This results in a C_2 operation.

6.

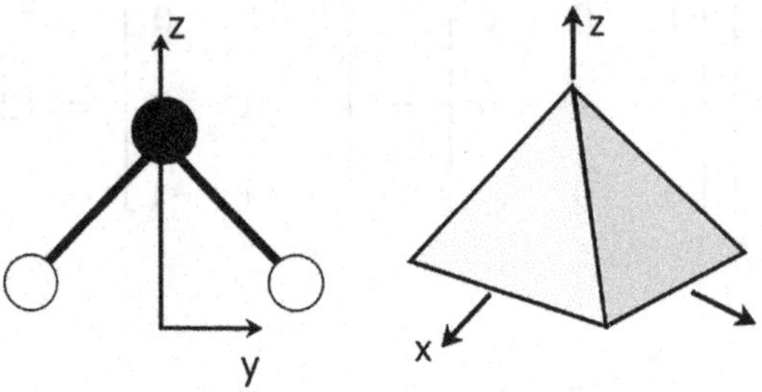

H_2O: $\{1, 2_z, m_y, m_x\}$ or $\{E, C_2(z), \sigma_v, \sigma_h\}$
Egyptian pyramid: $\{1, 4_z, 2_z, 4_z^3, m_x, m_y, m_{xy}, m_{-xy}\}$ or $\{E, C_4(z), C_2(z), C_4^3(z), \sigma_v, \sigma_v', \sigma_d, \sigma_d'\}$

7. To change left and right and top and bottom at the same time would require a center of inversion located between you and the mirror and at half your height, changing each point (x, y, z) in your body to $(\bar{x}, \bar{y}, \bar{z})$ on the other side of the mirror. However, a mirror is a reflection *plane*, so that every point in your body (x, y, z) is reflected to $(x, y, \bar{z})$ for a mirror in the xy-plane.
8. The arrow on the front of the disk is in fact at right angles to the arrow on the back. The two rotations (a) and (b) are about axes that bisect the angle between the two arrows.

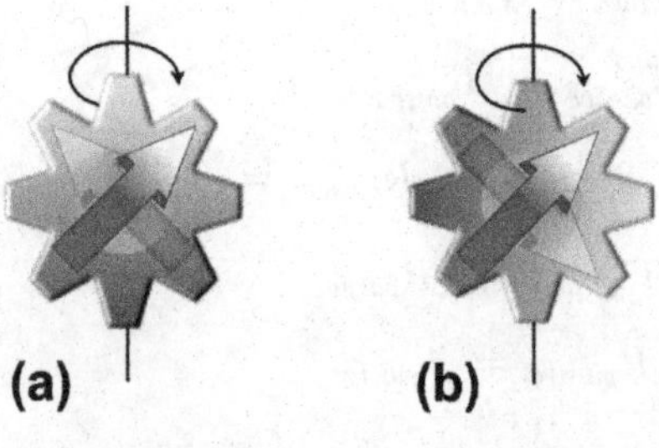

CHAPTER 2

1. **(a)** Let the axes **a**, **b** and **c** project onto orthogonal Cartesian axes as $\mathbf{a} = (a_1, a_2, a_3)$ etc. The volume is given by $V = |\det D|$ where

$$D = \begin{bmatrix} a_1 & a_2 & a_3 \\ b_1 & b_2 & b_3 \\ c_1 & c_2 & c_3 \end{bmatrix}$$

Therefore

$$V^2 = \det\left(DD^T\right) = \det\begin{bmatrix} \mathbf{a}\cdot\mathbf{a} & \mathbf{a}\cdot\mathbf{b} & \mathbf{a}\cdot\mathbf{c} \\ \mathbf{b}\cdot\mathbf{a} & \mathbf{b}\cdot\mathbf{b} & \mathbf{b}\cdot\mathbf{c} \\ \mathbf{c}\cdot\mathbf{a} & \mathbf{c}\cdot\mathbf{b} & \mathbf{c}\cdot\mathbf{c} \end{bmatrix}$$

$$= \begin{vmatrix} a^2 & ab\cos\gamma & ac\cos\beta \\ ab\cos\gamma & b^2 & bc\cos\alpha \\ ac\cos\beta & bc\cos\alpha & c^2 \end{vmatrix}$$

(b) A primitive unit cell contains one lattice point. Since a unit cell must repeat to fill all space, the primitive unit cells must be equal in volume.

(c) Use vectors:

$$\mathbf{r}_1 = x_1\mathbf{a} + y_1\mathbf{b} + z_1\mathbf{c}$$

$$\mathbf{r}_2 = x_2\mathbf{a} + y_2\mathbf{b} + z_2\mathbf{c}$$

$$\begin{aligned} \mathbf{r} &= \mathbf{r}_1 - \mathbf{r}_2 \\ &= (x_1 - x_2)\mathbf{a} + (y_1 - y_2)\mathbf{b} + (z_1 - z_2)\mathbf{c} \\ &= \Delta x\mathbf{a} + \Delta y\mathbf{b} + \Delta z\mathbf{c} \end{aligned}$$

$$\begin{aligned} r^2 &= \mathbf{r}\cdot\mathbf{r} \\ &= (\Delta x\mathbf{a} + \Delta y\mathbf{b} + \Delta z\mathbf{c})\cdot(\Delta x\mathbf{a} + \Delta y\mathbf{b} + \Delta z\mathbf{c}) \\ &= (\Delta x)^2(\mathbf{a}\cdot\mathbf{a}) + (\Delta y)^2(\mathbf{b}\cdot\mathbf{b}) + (\Delta z)^2(\mathbf{c}\cdot\mathbf{c}) \\ &\quad + 2\Delta x\Delta y(\mathbf{a}\cdot\mathbf{b}) + 2\Delta y\Delta z(\mathbf{b}\cdot\mathbf{c}) + 2\Delta z\Delta x(\mathbf{c}\cdot\mathbf{a}) \\ &= a^2(\Delta x)^2 + b^2(\Delta y)^2 + c^2(\Delta z)^2 \\ &\quad + 2\Delta x\Delta y ab\cos\gamma + 2\Delta y\Delta z bc\cos\alpha + 2\Delta z\Delta x ca\cos\beta \end{aligned}$$

2. The $\overline{4}$ operation along [001] is given by

$$\begin{bmatrix} 0 & 1 & 0 \\ -1 & 0 & 0 \\ 0 & 0 & -1 \end{bmatrix}$$

Therefore

$$R\mathbf{r} = y\mathbf{a} - x\mathbf{b} - z\mathbf{c}$$

And so we see that y and x interchange with respect to **a** and **b**, thus forcing the lengths of **a** and **b** to be equal. The changes in signs mean that the axes are perpendicular to one another.
For a mirror $m[001]$ *containing* a 2-fold axis 2[100]

$$\{m[001]\}\{2[100]\} = \begin{bmatrix} 1 & 0 & 0 \\ 0 & 1 & 0 \\ 0 & 0 & -1 \end{bmatrix} \begin{bmatrix} 1 & 0 & 0 \\ 0 & -1 & 0 \\ 0 & 0 & -1 \end{bmatrix}$$

$$= \begin{bmatrix} 1 & 0 & 0 \\ 0 & -1 & 0 \\ 0 & 0 & 1 \end{bmatrix} = m[010]$$

Thus the system is orthorhombic with $\alpha = \beta = \gamma = 90°$.

3. If the 2-fold axis is along [001], then successive operations by 3-fold symmetry about [111] give

$$\begin{bmatrix} 0 & 0 & 1 \\ 1 & 0 & 0 \\ 0 & 1 & 0 \end{bmatrix} \begin{bmatrix} 0 \\ 0 \\ 1 \end{bmatrix} \rightarrow \begin{bmatrix} 1 \\ 0 \\ 0 \end{bmatrix} \rightarrow \begin{bmatrix} 0 \\ 1 \\ 0 \end{bmatrix} \rightarrow \begin{bmatrix} 0 \\ 0 \\ 1 \end{bmatrix}$$

which means there are 2-fold axes along [100], [010] and [001]. These 2-fold operations can now act on the direction [111] of the 3-fold axis:

$$\begin{bmatrix} 1 & 0 & 0 \\ 0 & -1 & 0 \\ 0 & 0 & -1 \end{bmatrix} \begin{bmatrix} 1 \\ 1 \\ 1 \end{bmatrix} \rightarrow \begin{bmatrix} 1 \\ -1 \\ -1 \end{bmatrix}$$

$$\begin{bmatrix} -1 & 0 & 0 \\ 0 & 1 & 0 \\ 0 & 0 & -1 \end{bmatrix} \begin{bmatrix} 1 \\ 1 \\ 1 \end{bmatrix} \rightarrow \begin{bmatrix} -1 \\ 1 \\ -1 \end{bmatrix}$$

$$\begin{bmatrix} -1 & 0 & 0 \\ 0 & -1 & 0 \\ 0 & 0 & 1 \end{bmatrix} \begin{bmatrix} 1 \\ 1 \\ 1 \end{bmatrix} \rightarrow \begin{bmatrix} -1 \\ -1 \\ 1 \end{bmatrix}$$

Thus, we have four 3-fold axes along [111], $[1\bar{1}\bar{1}]$, $[\bar{1}1\bar{1}]$ and $[\bar{1}\bar{1}1]$.

5. The operation 3[111] will rotate the 4[001] axis to 4[100] and 4[010]. Each 4-fold rotation consists of 4, 4^2 and 4^3, making a total of nine 4-fold rotations. There are eight 3-fold operations corresponding to the eight <111> directions. This plus the identity operation gives $1 + 8 + 9 = 18$ operations. In addition, 3[111] acting on 4[001] gives 2[101] (use matrices). Repeating this gives six 2<110> operations to make a total of $6 + 18 = 24$. The inversion operation then doubles all of these to give 48 in total.

CHAPTER 3

2.

	No. of nearest neighbours	Distance	No. of second nearest neighhbors	Distance
cP	6	a	12	$\sqrt{2}$a
cI	8	$\frac{\sqrt{3}}{2}$a	6	a
cF	12	$\frac{a}{\sqrt{2}}$	6	a

3. cP: $V_{\text{sphere}} = \frac{4}{3}\pi\left(\frac{a}{2}\right)^3 \simeq 0.52a^3$

cI: $V_{\text{sphere}} = \frac{4}{3}\pi\left(\frac{\sqrt{3}a}{4}\right)^3 \times 2 \simeq 0.68a^3$ Factor of 2 comes from two spheres per unit cell.

cF: $V_{\text{sphere}} = \frac{4}{3}\pi\left(\frac{a}{2\sqrt{2}}\right)^3 \times 4 \simeq 0.74a^3$ Factor of 4 comes from four spheres per unit cell.

4. To convert axes $\begin{bmatrix} a_{rev} \\ b_{rev} \\ c_{rev} \end{bmatrix} = \begin{bmatrix} 1/3 & -1/3 & 1/3 \\ 1/3 & 2/3 & 1/3 \\ -2/3 & -1/3 & 1/3 \end{bmatrix} \begin{bmatrix} a_{hex} \\ b_{hex} \\ c_{hex} \end{bmatrix}$

To convert coordinates, use inverse transpose

$$\begin{bmatrix} x_{rev} \\ y_{rev} \\ z_{rev} \end{bmatrix} = \begin{bmatrix} 1 & -1 & 1 \\ 0 & 1 & 1 \\ -1 & 0 & 1 \end{bmatrix} \begin{bmatrix} x_{hex} \\ y_{hex} \\ z \end{bmatrix}$$

Therefore $(\bar{y}, x-y, z)_{hex} \to (\bar{x}+z, x-y+z, y+z)_{rev}$

5. Shortest vector in Wigner–Seitz cell of bcc is half the distance from (0, 0, 0) to $(1,1,1) = a\sqrt{3}/2$. Volume is then $\frac{4}{3}\pi\left(\frac{\sqrt{3}}{2}\right)^3 a^3 \simeq 2.72a^3$.
For fcc, Wigner–Seitz cell need half the distance from (0, 0, 0) to $(1,1,0) = a\sqrt{2}/2$. Therefore, ratio of distances cubed is $\left(\frac{\sqrt{3}}{\sqrt{2}}\right)^3 = \left(\frac{3}{2}\right)^{3/2}$.

CHAPTER 4

1. Draw stereograms or multiply operations to find the conventional group symbols: $mmm(D_{2h})$, $6/mmm(D_{6h})$, $m\bar{3}m$ (O_h), all of which are centrosymmetric.

3. $\bar{4}2m(D_{2d})$: (x,y,z) $(\bar{x},\bar{y},z)$ $(y,\bar{x},\bar{z})$ $(\bar{y},x,\bar{z})$
$(\bar{x},y,\bar{z})$ $(x,\bar{y},\bar{z})$ $(\bar{y},\bar{x},z)$ (y,x,z)

$$
\begin{array}{lllll}
622(D_6): & (x,\ y,\ z) & (\bar{y}, x-y, z) & (y-x, \bar{x}, z) & (\bar{x}, \bar{y}, z) \\
 & (y, y-x, z) & (x-y, x, z) & (y, x, \bar{z}) & (x-y, \bar{y}, \bar{z}) \\
 & (\bar{x}, y-x, \bar{z}) & (\bar{y}, \bar{x}, \bar{z}) & (y-x, y, \bar{z}) & (x, x-y, \bar{z})
\end{array}
$$

4. (a) $mm2(C_{2v})$ (b) $mmm(D_{2h})$ (c) $m\bar{3}m$ (O_h) (d) $\bar{4}3m$ (T_d) (e) 2-blades $222(D_2)$, 3-blades $32(D_3)$ (f) $mm2(C_{2v})$ (g)

A, B, C, D, E, M, T, U, V, W, Y	F, G, J, K, P, Q, R	H, I, N, X,	O	S, Z
$mm2(C_{2v})$	$m(C_{1h})$	$mmm(D_{2h})$	$\infty/mm(D_{\infty h})$	$2/m(C_{2h})$

5. Stress along [001]: $4/mmm(D_{4h})$. Stress along [111]: $\bar{3}m$ (D_{3d}). Electric field changes these to $4mm(C_{4v})$ and $3m(C_{3v})$, respectively.

CHAPTER 5

5.

6. **(a)** *c*-glide at x, ¼, z: $\{m[010]|0, ½, ½\}(x, y, z) = (x, ½-y, ½+z)$. 2_1 operation at 0, y, ¼: $\{2[010]|0, ½, ½\}(x, ½-y, ½+z) = (\bar{x}, \bar{y}, \bar{z})$, i.e. a $\bar{1}$ operation at the origin
In reverse order: same result
(b) *n*-glide at x, ¼ z: $\{m[010]|½, ½, ½\}(x, y, z) = (½+x, ½-y, ½+z)$. 4_2 screw about 0, ½, z: $\{4[001]|½, ½, ½\}(½+x, ½-y, ½+z) = (y, x, z)$, i.e. an $m[110]$ operation through the origin.

On reverse order: $\{4[001]|½, ½, ½\}(x, y, z) = (½ - y, ½ + x, ½ + z)$. Then $\{m[010]|½, ½, ½\}(½ - y, ½ + x, ½ + z) = (\bar{y}, \bar{x}, z)$, i.e. $m[1\bar{1}0]$ through the origin.

8. n-glide perpendicular to **a** gives coordinate $(½ - x, ½ + y, ½ + z)$. n-glide parallel to [110] gives coordinate $(½ + y, ½ + x, ½ + z)$.

10. $\{2[001]|\ ½, ½, ½\}$: 2_1 screw axis at ¼, ¼, 0.
$\{m[010]|\ ½, ½, 0\}$: a-glide at x, ¼, z.

$$R\begin{bmatrix} y-x \\ y \\ -z \end{bmatrix} + R\begin{bmatrix} 0 \\ 0 \\ 2/3 \end{bmatrix} + \boldsymbol{\tau} = \begin{bmatrix} y-x \\ -x \\ z \end{bmatrix} + \begin{bmatrix} 0 \\ 0 \\ 1/3 \end{bmatrix}$$

The matrix R is $\begin{bmatrix} 1 & 0 & 0 \\ 1 & -1 & 0 \\ 0 & 0 & -1 \end{bmatrix}$ and $\boldsymbol{\tau} = 0$. R is {2[210]} through the origin in the hexagonal system.

CHAPTER 6

2.

$$\begin{bmatrix} 1 & 0 & 0 \\ 0 & 1 & 0 \\ 0 & 0 & -1 \end{bmatrix}\begin{bmatrix} 1 & 0 & 0 \\ 0 & 0 & -1 \\ 0 & -1 & 0 \end{bmatrix} + \begin{bmatrix} 1/4 \\ 1/4 \\ 1/4 \end{bmatrix}$$

$$= \begin{bmatrix} 1 & 0 & 0 \\ 0 & 0 & -1 \\ 0 & 1 & 0 \end{bmatrix} + \begin{bmatrix} 1/4 \\ 0 \\ 0 \end{bmatrix} + \begin{bmatrix} 0 \\ 1/4 \\ 1/4 \end{bmatrix}$$

which is equivalent to a $4_1^3[100]$ screw rotation at x, $^1/_8$, $^1/_8$.

3. Point groups $\bar{4}m2$, $\bar{4}m2$, $3m$, $\bar{3}m$ and 23. Sites of highest symmetry: $\bar{4}m2$ 2.22 $3m$. 32 23.

4.

	b-glide	*a*-glide	*n*-glide
Origin on 2-fold	$\{m[100]\|½, ½, 0\}$	$\{m[010]\|½, ½, 0\}$	$\{m[001]\|½, ½, 0\}$
Origin on $\bar{1}$	$\{m[100]\|0, ½, 0\}$	$\{m[010]\|½, 0, 0\}$	$\{m[001]\|½, ½, 0\}$

5.

(1)	$\{1\|\mathbf{0}\}$	(2)	$\{2[001]\|\mathbf{0}\}$	(3)	$\{2[010]\|\mathbf{0}\}$	(4)	$\{2[100]\|\mathbf{0}\}$
(5)	$\{\bar{1}\|0\}$	(6)	$\{m[001]\|\mathbf{0}\}$	(7)	$\{m[010]\|\mathbf{0}\}$	(8)	$\{m[100]\|\mathbf{0}\}$

6.

$P3m1$: IIa	$R31$: I	$R3c$: IIb	$R3m$: IIc

7. The following is from the SUBGROUPGRAPH module of the Bilbao Crystallographic Server.

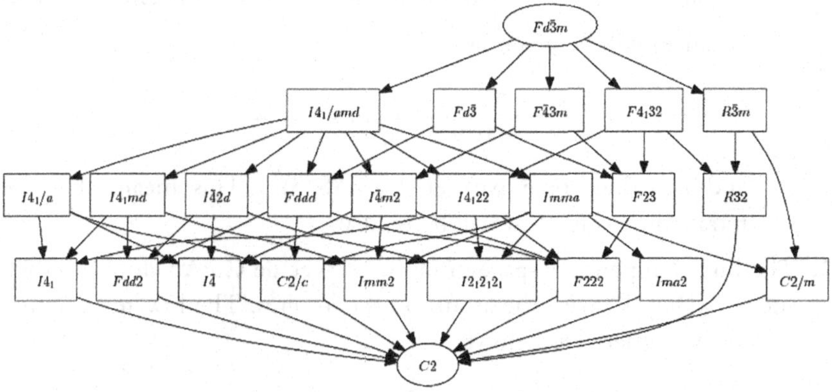

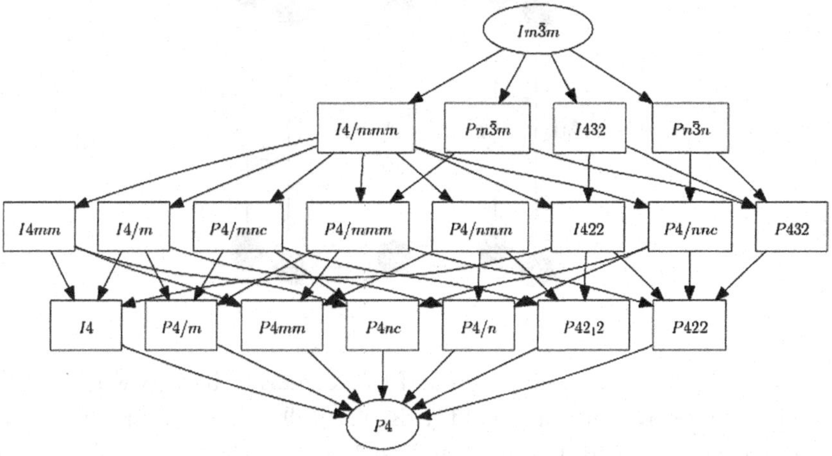

CHAPTER 7

1. **(a)** From the formula given in problem 1(c) in Chapter 2, for the hexagonal system

$$d^2 = a^2\left[(\Delta x)^2+(\Delta y)^2-\Delta x\Delta y\right] + c^2(\Delta z)^2$$

Substituting the coordinates of the 2c sites

$$d^2 = (2r)^2 = a^2 = \frac{a^2}{3}+\frac{c^2}{4}$$

This solves to give c/a = 1.633.

(c) For Zn–S, two S positions will exist in the unit cell at $\left({}^1\!/_3, {}^2\!/_3, {}^3\!/_8\right)$ and at $\left({}^2\!/_3, {}^1\!/_3, -{}^1\!/_8\right)$ with the Zn atom in between. The two Zn–S distances will be given by

$$d^2 = \frac{9}{64}c^2 \text{ and } d^2 = \frac{a^2}{3}+\frac{c^2}{64}$$

and equating these we get $c/a = (8/3)^{1/2}$. This means that the tetrahedral so formed will be regular.

2. The high-temperature space group with average Cu:Au atoms on each site is $Fm\overline{3}m(O_h^5)$ with the atoms on the 4*a* sites. The low-temperature structure is ordered as below:

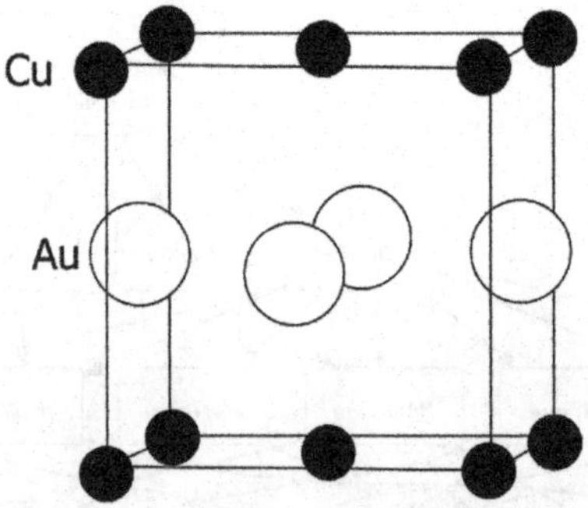

Note that the four 3-fold axes and face centering have now gone and that there remains a 4-fold axis vertically to give space group $P4/mmm(D_{4h}^1)$ with Cu on 1*a* and on 1*c*, Au on 2*e* sites.

3.

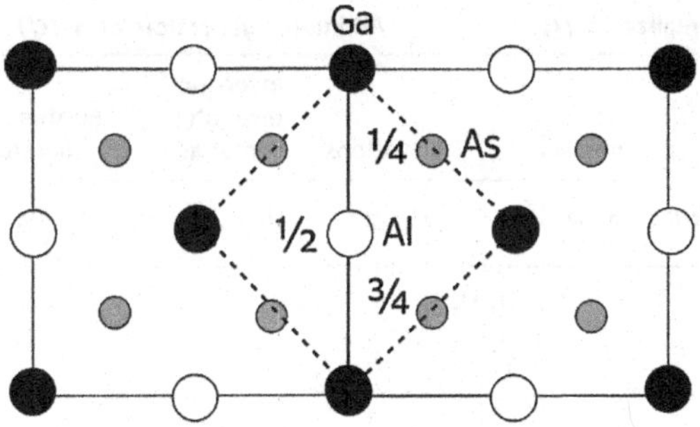

The dashed lines show the minimum unit cell, space group is $P\bar{4}m2(D_{2d}^5)$ with Ga at 1*a*: (0, 0, 0), Al at 1*c*: (½,½,½) and As at 2*g*: (0, ½, *z*) with $z = ¼$.

4. Compression changes space group $Pm\bar{3}m(O_h^7)$ to $P4/mmm(D_{4h}^1)$. In the second case, the space group becomes $P4mm(C_{4v}^1)$. This can be ferroelectric as it is a polar space group.

5.

Pb: 8*c* (¼,¼,¼) Sc: 4*a* (0, 0, 0) Ta: 4*b* (½,½,½) O: 24*e* (*x*, 0, 0) $x = ¼$

6.

Species	No orientation states	Ferroelectric	Ferroelastic
$m\bar{3}m$F222	12	-	Partial
4F1	4	Full	Full
4*mm*F2*mm*	2	-	Full
$\bar{6}$F3	2	Partial	-
4/*mmm*F4/*m*	2	-	-
6/*m*F*m*	6	Full	Partial

8.

Space Group $\mathcal{G}$	Normalizer $\mathcal{N}_{\mathcal{E}}(\mathcal{G})$		Additional generators of $\mathcal{N}_{\mathcal{E}}(\mathcal{G})$			
Symbol	**Symbol**	**Basis vectors**	**Translations**	**Inversion through center at**	**Further generators**	**Index of $\mathcal{G}$ in $\mathcal{N}_{\mathcal{E}}(\mathcal{G})$**
P321	P6/*mmm*	**a**, **b**, ½**c**	0, 0, ½	0, 0, 0	-x, −y, z	2 ·2·2
K 1*a*			0, 0, 0			0, 0, ½
Al 1*b*			0, 0, ½			0, 0, 0
S 2*d*			⅓ , ⅔ , 0.222			⅓ , ⅔ , 0.722
K 1*a*			0, 0, 0			0, 0, ½
Al 1*b*			0, 0, ½			0, 0, 0
S 2*d*			⅔ , ⅓ , 0.778			⅔ , ⅓ , 0.278
K 1*a*			0, 0, 0			0, 0, ½
Al 1*b*			0, 0, ½			0, 0, 0
S 2*d*			⅔ , ⅓ , 0.222			⅔ , ⅓ , 0.722
K 1*a*			0, 0, 0			0, 0, ½
Al 1*b*			0, 0, ½			0, 0, 0
S 2*d*			⅓ , ⅔ , 0.778			⅓ , ⅔ , 0.278

8 different settings.

CHAPTER 8

1. Black and white plane groups (see Litvin's Tables):

$p_{2a}211$ (2.4.7) $p2'mg'$ (7.4.35) $p2'm'g$ (7.4.34)

Magnetic plane groups:

$p_{2a}211$ (2.4.7) $p2'm'g$ (7.4.34) $p2'mg'$ (7.4.35)

2. Look for subgroup of $P4mm(C_{4v}^1)$.
Translations generated by: {1|1,0,0} {1|0,1,0} and {1|0,0,1}′ and the operations: {1|0,0,0} {*m*[100]|0,0,0} {*m*[010]|0,0,0}′ {2[001]|0,0,0}′
This is $P_{2c}mm'2'$ (25.10.164) with Ce at 2*a* sites, $mm'2'$ site symmetry: 0, 0, 0 [*ν*00]

3.

6mm	1	2 × 6	2 × 3	2	3 × m	3 × m	
A_1	1	1	1	1	1	1	6mm
A_2	1	1	1	1	−1	−1	$6m'm'$
B_1	1	−1	1	−1	1	−1	$6'mm'$
B_2	1	−1	1	−1	−1	1	$6'm'm$
E_1	2	1	−1	−2	0	0	
E_2	2	−1	−1	2	0	0	

For space group $P2/c$ at $\mathbf{k} = 0$ we get $P2/c$, $P2'/c'$, $P2/c'$, $P2'/c$. Reference to Bilbao Crystallographic Server program REPRES shows that 1-dimensional representations exist for Brillouin zone points at (½, 0, 0), (0, ½, 0) and (½, ½, 0). This gives $P_{2a}2/c$, $P_{2b}2/c$, P_C2/c and $P_{2b}2'/c$. All other combinations are equivalent to these, e.g. $P_{2b}2/c'$ is equivalent to $P_{2b}2/c$ with an origin shift of (0, ¼, 0). You can check the results with program MGENPOS in the Bilbao Crystallographic Server.

4. See the Figure below. Use axes as indicated. Then we find the following Table, consistent with $F_Cmm'm'$ (69.10.614) in Litvin's Tables. The BNS symbol is C_Amca (equivalent to B_Abcm with change of axes)

Fe6:	Fe8:	Fe4:	Fe3:
$\frac{3}{4},\frac{3}{4},\frac{3}{4}[0, v, 0]$	$\frac{3}{4},\frac{1}{4},\frac{1}{4}[0, -v, 0]$	$\frac{1}{4},\frac{3}{4},\frac{1}{4}[0, -v, 0]$	$\frac{1}{4},\frac{3}{4}[0, v, 0]$

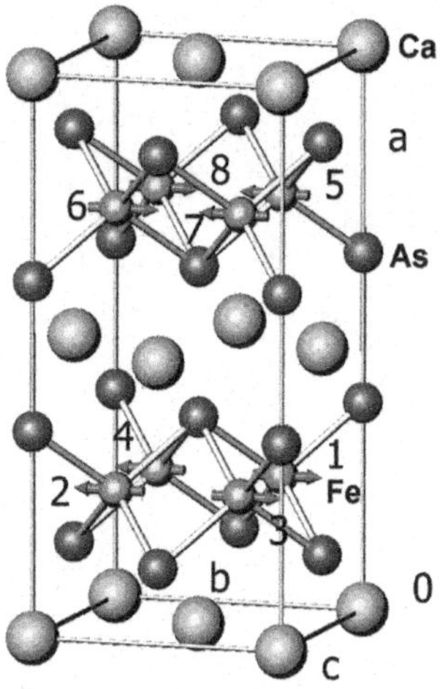

FORMULA INDEX

INDEX

G

H

I

K

L

Printed in the United States
By Bookmasters